Bio-Based Amines

This book offers a detailed exploration of bio-based amines in polymer production, focusing on their sustainable extraction and utility in creating environmentally friendly materials. It covers their chemical properties, synthesis from natural sources, and diverse applications in products such as polyurethanes, adhesives, polyamides, and epoxies. It also addresses processing challenges, life cycle assessments, and market trends, providing a comprehensive overview of the environmental impacts and potential advancements in bio-based polymer technology.

Key Features:

- Promotes a shift toward more sustainable practices within the chemical and materials science industries
- Highlights innovative approaches to synthesizing and processing
- Covers practical applications of bio-based amines in polymer products
- Addresses current market trends and future directions in bio-based polymer technology

This text serves as an essential resource for materials and chemical researchers and industry professionals aiming to advance green chemistry practices and educate readers about the potential of bio-based amines as alternatives to conventional petrochemical-derived amines.

Ram K. Gupta is Professor of Chemistry and Associate Vice President for Research and Innovation at Pittsburg State University. Previously, he worked as an assistant research professor at Missouri State University and as a senior research scientist at North Carolina A&T State University.

Rutu Patel is a research scientist at the National Institute for Materials Advancement at Pittsburg State University, specializing in the synthesis of bio-based polymers for a sustainable future. She earned her Ph.D. from Sankalchand Patel University, India.

Mayankkumar L. Chaudhary is a research scientist at the National Institute for Materials Advancement at Pittsburg State University. He earned his Ph.D. from Sankalchand Patel University, India.

Bio-Based Amines

Principles, Polymers, and Applications

Ram K. Gupta, Rutu Patel,
and Mayankkumar L. Chaudhary

CRC Press
Taylor & Francis Group
Boca Raton London New York

CRC Press is an imprint of the
Taylor & Francis Group, an **informa** business

Designed cover image: © 2026 Shutterstock

First edition published 2027
by CRC Press
2385 NW Executive Center Drive, Suite 320, Boca Raton FL 33431

and by CRC Press
4 Park Square, Milton Park, Abingdon, Oxon, OX14 4RN

CRC Press is an imprint of Taylor & Francis Group, LLC

ISBN: 978-1-041-22196-8 (hbk)
ISBN: 978-1-041-22197-5 (pbk)
ISBN: 978-1-003-73112-2 (ebk)

DOI: 10.1201/9781003731122

Typeset in Times
by codeMantra

Contents

Preface

The growing urgency to address environmental challenges and reduce dependence on fossil-based resources has accelerated the search for sustainable alternatives in materials science. Among these, bio-based amines have emerged as a promising class of compounds, offering significant potential for the development of greener polymer systems. *Bio-Based Amines: Principles, Polymers, and Applications* presents a comprehensive exploration of these materials, emphasizing their role in advancing sustainable polymer chemistry.

This book provides an in-depth overview of bio-based amines, beginning with their chemical fundamentals and methods of extraction from renewable resources. It further examines their integration into polymer production, highlighting applications in polyurethanes, adhesives, polyamides, and epoxy systems. In addition to covering synthesis and performance, the text addresses key challenges associated with processing, scalability, and material optimization. Consideration is also given to life cycle assessment, environmental impact, and evolving market trends, offering readers a balanced perspective on both opportunities and limitations.

The primary aim of this book is to encourage a transition toward more sustainable practices within the chemical and materials industries. By demonstrating the viability and advantages of bio-based amines as alternatives to conventional petrochemical-derived counterparts, this book seeks to inform and inspire researchers, engineers, and industry professionals. It showcases innovative strategies for the design, synthesis, and application of these materials, contributing to the broader advancement of green chemistry.

Ultimately, this work aspires to support the development of environmentally responsible polymer technologies by providing a solid scientific foundation and practical insights. We hope it will serve as a valuable resource for academics, researchers, and practitioners dedicated to shaping a more sustainable future.

Dr. Rutu Patel
Dr. Mayankkumar Chaudhary
Prof. (Dr.) Ram K. Gupta

1 Introduction to Bio-Based Amines

1.1 INTRODUCTION

Amines represent one of the most important and versatile classes of nitrogen-containing organic compounds in chemistry. Structurally, they are characterized by the presence of one or more amino groups in which nitrogen is bonded to carbon or hydrogen atoms. Depending on the number of carbon substituents attached to the nitrogen atom, amines are classified as primary, secondary, or tertiary, while more complex structures, such as diamines and polyamines, contain multiple amino functionalities within a single molecule. The presence of a lone pair of electrons on the nitrogen atom gives amines their distinctive chemical behavior, including basicity, nucleophilicity, and the ability to form hydrogen bonds or coordinate with metals. These features make amines exceptionally reactive and adaptable, allowing them to participate in a wide variety of chemical transformations and intermolecular interactions [1]. Because of this versatility, amines occupy a central position in both fundamental chemistry and large-scale industrial applications. They are widely used as intermediates in organic synthesis, enabling the formation of amides, imines, ureas, carbamates, and a broad range of nitrogen-containing compounds (Figure 1.1). In catalysis, amines function as organocatalysts or ligands that influence reaction pathways and selectivity. In industrial chemistry, they are indispensable in the production of pharmaceuticals, agrochemicals, dyes, surfactants, corrosion inhibitors, and specialty chemicals. Beyond small-molecule chemistry, amines play a particularly critical role in materials science, especially in the design and fabrication of polymeric materials [2,3].

Industrial amines have conventionally been manufactured using petrochemical pathways based on fossil-derived feedstocks such as crude oil and natural gas. In these routes, hydrocarbons are transformed into amines through energy-demanding processes, including nitration, hydrogenation, reductive amination, and reactions with ammonia. While these technologies are mature and enable the large-scale production of amines with high purity, they depend heavily on nonrenewable resources and often require harsh operating conditions, hazardous reagents, and significant energy input. Consequently, the production of fossil-based amines is associated with considerable environmental impacts, particularly high energy consumption and greenhouse gas emissions, which conflict with current global sustainability objectives. A typical example is the industrial synthesis of ethylenediamine via the reaction of NH_3 with 1,2-dichloroethane (DCE) [4], where DCE is obtained through the chlorination of ethylene [5]. In a similar manner, ethanolamines are produced by reacting ethylene oxide (EO) with excess NH_3 [6]. Thus, the production of both diamines and amino alcohols relies on light olefins, such

DOI: 10.1201/9781003731122-1

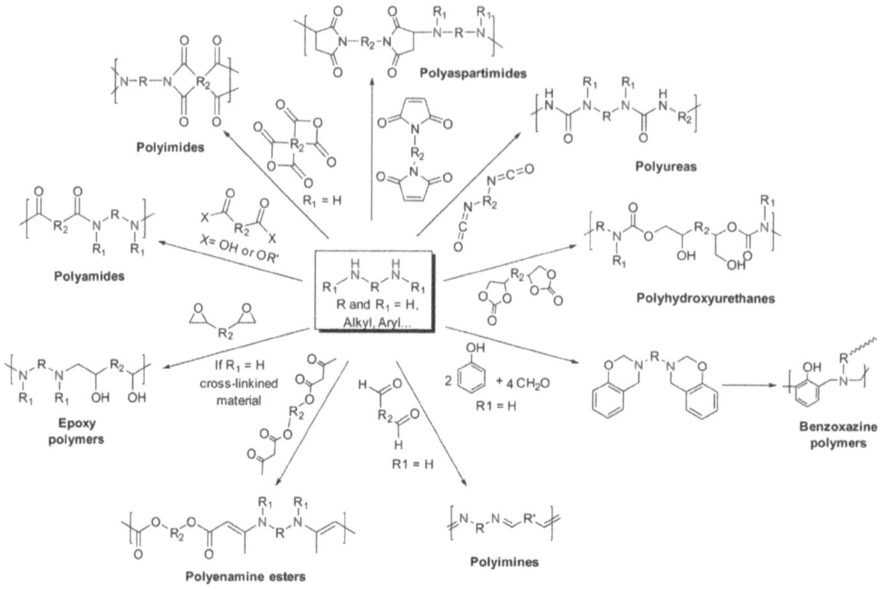

FIGURE 1.1 Summary of amine polymerization reactions. (Adapted with permission [1]. Copyright 2016 American Chemical Society.)

as ethylene, as the primary carbon source [7]. These olefins are generated through steam or catalytic cracking processes, which are among the most energy-intensive operations in the chemical industry and contribute to the release of millions of tons of CO_2 each year [8,9]. In addition to their high energy demand, these synthetic routes present further disadvantages. Pathways involving alkyl halides such as DCE raise corrosion concerns due to the formation of HCl as a by-product and lead to substantial waste salt generation, with neutralization using NaOH producing large quantities of NaCl. The use of EO in ethanolamine production introduces additional safety risks, as EO is highly toxic and explosive, making its handling particularly challenging. These limitations underscore the urgent need for more sustainable and safer approaches to amine synthesis. Ideally, such alternative catalytic routes should adhere to the 12 principles of green chemistry, emphasizing high atom efficiency, reduced waste formation, and the use of safe and renewable feedstocks [10]. In this context, biomass is increasingly recognized as a promising alternative carbon source for the chemical industry [11], providing safer and renewable precursors for amine production while reducing reliance on fossil resources and lowering the environmental footprint of these processes.

Bio-based amines have emerged as an important alternative to conventional fossil-based amines, driven by the growing need to develop more sustainable chemical and material systems. Bio-based amines are defined as amines that are derived partially or entirely from renewable biological resources rather than petrochemical feedstocks. These renewable resources include sugars and carbohydrates obtained from biomass, naturally occurring amino acids, lignin-derived aromatic compounds, and plant-based oils and fatty acids [12–17]. In many cases, bio-based amines retain the same functional groups and reactivity as their fossil-based counterparts, allowing them to be used as direct replacements, often referred to as "drop-in" alternatives. In other cases, bio-based feedstocks enable new molecular structures that are not easily accessible through traditional petrochemical pathways, opening opportunities for novel material properties and functions.

From a materials perspective, bio-based amines are especially attractive because they can be seamlessly integrated into existing polymer technologies while enhancing sustainability profiles. In polyamides, for example, bio-based diamines derived from amino acids or fermentation pathways enable the production of partially or fully bio-based nylons with mechanical and thermal properties comparable to conventional materials [18]. In epoxy systems, bio-based amine curing agents derived from lignin-based aromatics or plant oils provide alternatives to petroleum-based hardeners, often offering improved flexibility, lower toxicity, or enhanced adhesion [19,20]. Similarly, in polyurethanes and polyureas, bio-based amines serve as chain extenders or crosslinkers that influence phase separation, hydrogen bonding, and overall network performance [21–23].

This chapter provides a comprehensive foundation for understanding bio-based amines by integrating fundamental chemistry, production pathways, materials relevance, and sustainability considerations. It begins by outlining the essential chemical features of amines and their central role in industrial chemistry and materials science, followed by a critical discussion of conventional fossil-based amine production routes and their associated environmental and safety limitations. This chapter then introduces bio-based amines as sustainable alternatives, highlighting renewable biomass feedstocks and the key biochemical and chemical pathways used for their synthesis. Emphasis is placed on the relevance of bio-based amines in polymer science, where they function as monomers, curing agents, and chain extenders in polyamides, polyurethanes, epoxies, and advanced functional materials. Finally, this chapter addresses sustainability challenges related to feedstock variability, scalability, cost, and life cycle assessment, underscoring the importance of responsible and quantitative sustainability claims.

1.2 RENEWABLE FEEDSTOCKS FOR BIO-BASED AMINES

Biomass is a broad term that refers to the complete spectrum of organic macromolecules generated by living organisms, encompassing an exceptionally diverse range of chemical structures and functionalities. Unlike fossil resources, which are formed over geological timescales, biomass is considered a renewable carbon source because it is replenished through natural biological cycles within a relatively short period. Owing to this renewable nature, biomass has attracted significant attention as an alternative feedstock for the sustainable production of fuels, chemicals, and advanced materials. In general, biomass is regarded as a particularly attractive resource when it is abundantly available, has low or negligible intrinsic economic value, and can be efficiently converted into value-added products through chemical or biological transformations [24–26]. These characteristics make biomass highly relevant in the context of developing sustainable routes for the synthesis of bio-based amines.

Among the various forms of biomass, lignocellulose stands out as one of the most promising feedstocks that fulfills all these criteria. Lignocellulose constitutes the primary structural component of plant cell walls and is widely recognized as the most abundant source of organic carbon on Earth, accounting for the largest fraction of global biomass. Importantly, lignocellulosic biomass is readily available from numerous underutilized and low-value nonedible waste streams, including forestry residues, wood waste from the construction and paper industries, and agricultural by-products such as corn stover, wheat and rice straw, and sugarcane bagasse. The use of these resources is particularly advantageous because it avoids direct competition with food production and minimizes concerns related to land-use change, thereby supporting sustainable biorefinery concepts. Structurally, lignocellulose is a complex composite material composed primarily of three biopolymers: cellulose, hemicellulose, and lignin. Cellulose is a linear homopolysaccharide built from glucose units linked through β-1,4-glycosidic bonds, with a degree of polymerization that can reach up to approximately 10,000 monomer units. This highly ordered and crystalline structure provides mechanical strength to plant cell walls and represents a vast reservoir of carbohydrate-derived carbon. Hemicellulose, in contrast, is a branched heteropolysaccharide with a significantly lower degree of polymerization, typically ranging from 50 to 300 units. It is composed of a mixture of pentose sugars such as xylose and arabinose, as well as hexose sugars including glucose, mannose, and galactose. The third major component, lignin, is an irregular and highly crosslinked aromatic polymer formed through radical polymerization of three primary monolignol precursors: p-coumaryl alcohol, coniferyl alcohol, and sinapyl alcohol [27,28]. Together, these three components form a structurally robust matrix that provides both aliphatic and aromatic carbon frameworks suitable for chemical upgrading.

In the context of bio-based amine production, the carbohydrate fraction of lignocellulosic biomass, namely cellulose and hemicellulose, is of particular importance. Carbohydrates are inherently safe to handle, abundantly available, and possess highly functionalized aliphatic carbon backbones that can be selectively modified through chemical or biotechnological pathways. Their rich oxygen functionality and structural tunability make them attractive precursors for the synthesis of nitrogen-containing compounds, including amines. Through controlled transformation strategies, carbohydrate-derived intermediates can be converted into a wide variety of amine structures suitable for polymer synthesis and other applications.

Beyond lignocellulose, biomass also includes other important classes of renewable feedstocks such as proteins, plant oils and fats, and algae. Although these resources represent a smaller fraction of total biomass compared to lignocellulose, they offer distinct chemical advantages and significant potential for future integration within biorefinery frameworks. Plant oils, including those derived from coconut, palm, sunflower, and other oilseed crops, consist primarily of triglycerides. These triglycerides are composed of a glycerol backbone esterified with three fatty acid chains. The alkyl chain lengths of these fatty acids typically range from C8 to C20, with C16, C18, and C20 chains being the most prevalent [1,3,29]. The long aliphatic chains and inherent hydrophobicity of fatty acids make plant oils particularly suitable to produce fatty amines and related nitrogen-containing derivatives used in surfactants, coatings, and polymeric materials.

Proteins represent another highly valuable biomass-derived platform to produce bio-based amines. Proteins are formed through the polymerization of amino acids and inherently contain a high concentration of organically bound nitrogen within their molecular structure. This intrinsic nitrogen content makes protein-rich biomass an especially attractive substrate for amine synthesis. Several industrial and agricultural side streams are rich in proteins and therefore present promising opportunities for valorization. Notable examples include slaughterhouse waste and dried distiller's grains with solubles, which are generated as by-products of bioethanol production [30,31]. The utilization of these nitrogen-rich waste streams aligns well with circular economy principles by transforming low-value residues into high-value chemical building blocks.

Another renewable resource that is often overlooked in discussions of biomass platforms is chitin. Chitin is the second most abundant biopolymer on Earth, surpassed only by cellulose, and represents a sustainable and inexpensive source of nitrogen. Its chemical structure is based on repeating units of N-acetyl-D-glucosamine linked through β-1,4-glycosidic bonds. Chitin is widely distributed in nature and can be found in the exoskeletons of arthropods such as insects and crustaceans, in cephalopods including squids and octopuses, and in the cell walls of certain fungi [32–34]. Due to its high nitrogen content and polysaccharide backbone, chitin is of particular interest as a feedstock to produce bio-based amines and related nitrogen-containing compounds.

When comparing different biomass-derived feedstock routes for bio-based amine production, several key factors must be considered, including availability, chemical composition, nitrogen content, and ease of conversion. Lignocellulosic carbohydrates offer unparalleled abundance and structural diversity, making them ideal platforms for large-scale applications. Protein-based feedstocks provide direct access to organic nitrogen, reducing the need for external nitrogen sources. Plant oils supply long-chain aliphatic structures that are difficult to obtain from carbohydrates, while chitin offers a unique combination of carbohydrate and nitrogen functionalities within a single biopolymer. Each feedstock class, therefore, presents distinct advantages and limitations, and their effective integration within biorefinery concepts will play a crucial role in the future development of sustainable routes to bio-based amines. Figure 1.2 highlights the major renewable biomass resources used for the synthesis of bio-based amines, including natural oils, carbohydrates, proteins, lignin, and chitin. These feedstocks provide diverse carbon and nitrogen frameworks that can be converted into a wide range of aliphatic and aromatic amines through chemical or biotechnological pathways.

1.3 GENERAL PRODUCTION PATHWAYS

The production of bio-based amines generally follows two complementary strategies: biotechnological or fermentation-based pathways and chemical or catalytic conversion routes, each offering distinct benefits and limitations depending on the target molecule and intended application. Biotechnological routes make use of microorganisms or isolated enzymes to convert renewable substrates such as sugars, amino acids, or other biomass-derived intermediates into amines. These processes operate

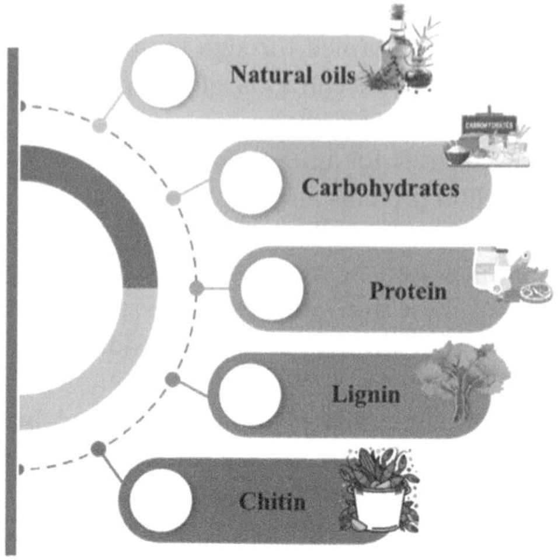

FIGURE 1.2 Renewable resources for bio-based amines.

under relatively mild conditions, typically near ambient temperature and pressure, and rely on highly selective enzymatic reactions. As a result, fermentation-based pathways often deliver products with high selectivity and purity while minimizing the formation of undesired by-products. This intrinsic selectivity reduces the need for extensive purification and aligns well with green chemistry principles, particularly those related to energy efficiency and safer processing. Biotechnological routes are especially effective for producing small, well-defined amines, such as diamines and polyamines derived from amino acids, where the nitrogen functionality is already present in the biological precursor [35]. Peters-Wendisch et al. engineered *Corynebacterium glutamicum* to enable direct fermentative production of L-serine from glucose, a process previously limited by strong intracellular degradation and metabolic flux constraints. By overexpressing the serine biosynthesis genes (*serA*, *serB*, *serC*), removing feedback inhibition, and suppressing competing pathways through deletion of *sdaA* and reduction of *glyA*, they achieved high serine accumulation and productivity. This work demonstrates that effective L-serine production requires simultaneous enhancement of biosynthesis and restriction of serine consumption, highlighting the central metabolic role of serine. Importantly, L-serine is a key precursor for downstream chemicals such as 2-aminoethanol (ethanolamine), underscoring the relevance of this strategy for producing bio-based amines from renewable glucose feedstocks [36]. The study by Rontein et al. reports that plants can synthesize ethanolamine (2-aminoethanol) directly from free L-serine, rather than exclusively through membrane lipid turnover. The authors identified and characterized a pyridoxal-5′-phosphate–dependent serine decarboxylase, demonstrating its ability to catalyze the direct decarboxylation of serine to ethanolamine using biochemical assays and genetic complementation. This work establishes a previously unrecognized biosynthetic route for ethanolamine in plants and highlights the role of serine as a key renewable precursor for bio-based amines [37]. Moreover, *Pseudomonas putida* S12 was metabolically modified to enable the production of monoethanolamine (MEA) from glucose by directing metabolic flux through L-serine, which is subsequently converted to MEA via decarboxylation catalyzed by L-serine decarboxylase [38].

Despite these advantages, fermentation-based production of bio-based amines faces several challenges that limit its universal applicability. The range of amine structures accessible through biological systems is constrained by metabolic compatibility, making it difficult to produce complex, bulky, or highly hydrophobic amines. In addition, fermentation processes often suffer from relatively low product titers and slower reaction rates compared to chemical synthesis, which can negatively impact overall productivity and economics. The separation and purification of amines from aqueous fermentation broths can also be energy intensive and technically challenging, particularly for volatile or water-soluble products. Furthermore, maintaining sterile conditions and controlling biological systems on an industrial scale require careful process management and add operational complexity.

Chemical and catalytic conversion routes provide an alternative and often complementary approach to bio-based amine synthesis. These pathways involve the transformation of biomass-derived platform molecules into amines using heterogeneous or homogeneous catalysts [39,40]. Typical starting materials include alcohols, aldehydes, ketones, carboxylic acids, and nitriles obtained from renewable feedstocks such as lignocellulosic carbohydrates, plant oils, and lignin-derived aromatics. Through well-established reactions such as reductive amination, hydrogen borrowing (Figure 1.3a) [41], amination of alcohols (Figure 1.3b) [41], and nitrile hydrogenation (Figure 1.3c) [42], chemical routes enable the synthesis of a wide variety of aliphatic and aromatic amines with controlled molecular structures. One of the key advantages of catalytic pathways is their compatibility with existing chemical infrastructure, which allows for continuous processing, high throughput, and straightforward scale-up to industrial production levels. Moreover, chemical routes offer greater flexibility in molecular design, enabling access to amines that are difficult or impossible to obtain through biological pathways, including aromatic and long-chain amines important for polymer and materials applications. The study by Murugesan et al. reports a Ni-based homogeneous catalyst named as Ni-triphos complex for reductive amination of carbonyl compound and hydrogenation to make different kind of amines [43]. In another study, Yasukawa et al. developed heterogeneous iridium–chiral phosphoric acid catalyst systems for the efficient asymmetric hydrogenation of imines and direct reductive amination of ketones under hydrogen. Using these catalysts, a wide range of chiral aromatic and aliphatic amines, including pharmaceutical intermediates, were produced with high yields and excellent enantioselectivities. Importantly, the catalysts enabled stable continuous-flow synthesis of chiral amines for over 2 days, operating effectively at lower pressures than comparable batch processes [44].

However, chemical and catalytic approaches are not without limitations. Many of these reactions require elevated temperatures, high pressures, or the use of hydrogen and other reagents, leading to higher energy consumption and potentially increased environmental impact if not carefully managed. Catalyst stability, selectivity, and recyclability are also critical challenges, as catalyst deactivation or poor selectivity can lead to by-product formation and reduced process efficiency. From a

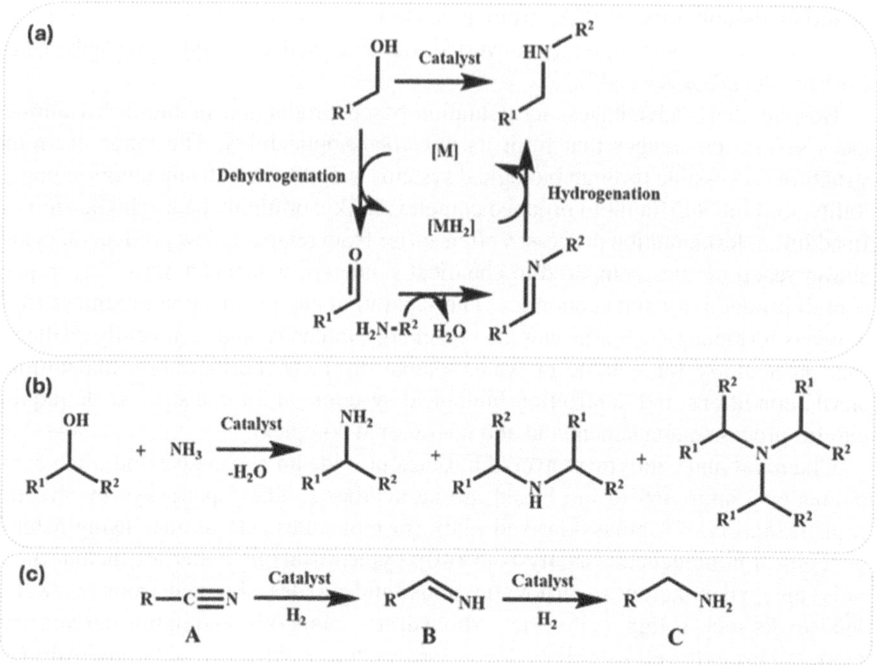

FIGURE 1.3 (a) Borrowing hydrogen in the amination of alcohols. (b) Amination of alcohol with ammonia. (Adapted with permission [41]. Copyright 2011 Wiley.) (c) Hydrogenation of nitriles A to intermediate imine B to primary amines C. (Adapted with permission [42]. Copyright 2010 American Chemical Society.)

sustainability standpoint, the overall environmental performance of chemical routes depends strongly on catalyst design, solvent choice, and the origin of energy and reducing agents. Consequently, significant research efforts are focused on developing robust, selective, and recyclable catalytic systems that operate under milder conditions and use greener reagents.

In practice, the most effective strategies for producing bio-based amines increasingly combine elements of both biotechnological and chemical approaches. Hybrid bio-chemical pathways may involve the fermentation of biomass-derived sugars into a renewable intermediate, followed by catalytic upgrading into the desired amine. Such integrated routes capitalize on the selectivity of biological processes and the scalability and versatility of chemical catalysis. By balancing efficiency, flexibility, and sustainability, these combined strategies are expected to play a central role in enabling the large-scale production of bio-based amines for advanced materials and polymer applications.

1.4 ROLE OF BIO-BASED AMINES IN POLYMER SCIENCE

Bio-based amines play a central and increasingly influential role in polymer science by acting as renewable, multifunctional building blocks that enable the development of sustainable polymeric materials while maintaining, and in many cases enhancing, material performance. Their structural diversity, which ranges from simple aliphatic and aromatic amines to multifunctional diamines and polyamines derived from biomass, allows them to be tailored for a wide spectrum of polymerization reactions. As monomers, curing agents, and chain extenders, bio-based amines participate in step-growth and network-forming reactions with carboxylic acids, epoxides, isocyanates, anhydrides, and cyclic carbonates, offering precise control over polymer architecture, molecular weight distribution, and crosslink density. The literature consistently demonstrates that amines derived from renewable feedstocks can be successfully integrated into established polymer platforms such as polyamides, polyurethanes, polyureas, and epoxy resins, either as direct replacements for fossil-based amines or as structurally novel components that impart additional functionality [45].

In polyamide systems, bio-based diamines originating from amino acids, carbohydrates, or fermentation pathways have been employed to produce partially or fully bio-based nylons with mechanical and thermal properties comparable to those of conventional petroleum-derived analogues. These studies highlight the ability of bio-based amines to maintain crystallinity, tensile strength, and thermal resistance while increasing renewable carbon content. Similarly, in polyurethane and polyurea materials, bio-based amines have been widely used as chain extenders and crosslinkers, where their molecular structure influences phase separation, hydrogen bonding interactions, and hard-segment formation. The reviewed literature shows that careful selection of amine structure enables fine tuning of elasticity, toughness, and thermal stability, making bio-based amines particularly attractive for composites, coatings, elastomers, foams, and adhesive applications (Figure 1.4) [18].

In epoxy thermosets, bio-based amines have emerged as key curing agents capable of replacing conventional petroleum-derived hardeners. Amine curing agents derived from lignin-based aromatics, fatty amines, or amino acid platforms have

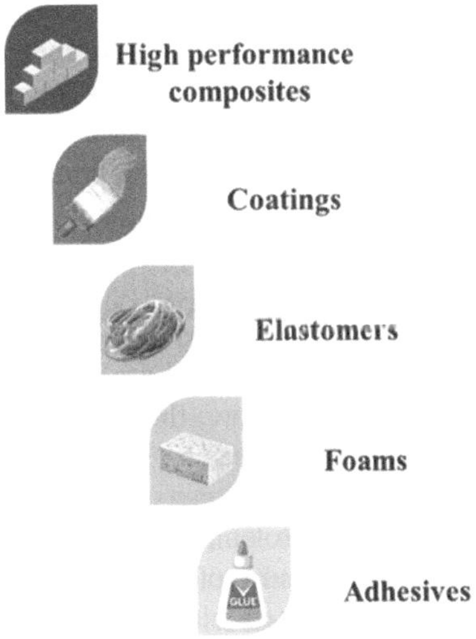

FIGURE 1.4 Applications of polymeric materials made from bio-based amines.

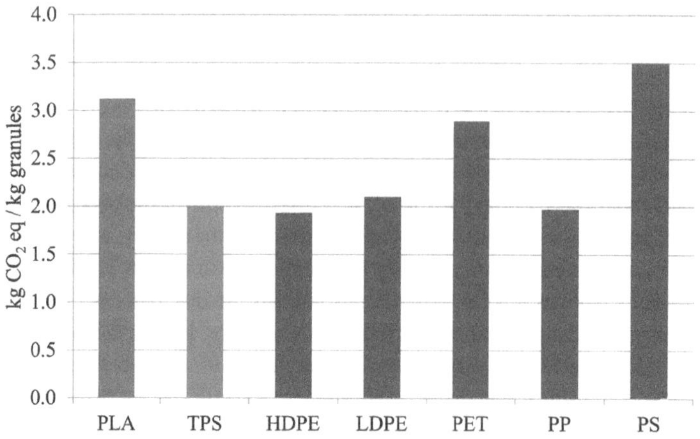

FIGURE 1.5 Comparison of greenhouse gas emissions (kg CO_2-eq per kg of polymer) for selected bio-based and conventional polymers, illustrating how inclusion of end-of-life treatment can increase overall environmental impact compared to cradle-to-gate assessments. HDPE = high-density poly ethylene, LDPE = low-density polyethylene, PET = polyethylene terephthalate, PP = polypropylene, PS = polystyrene. (Adapted with permission [50]. Copyright 2013 Elsevier.)

avoid competition with food production, but seasonal availability and geographical constraints can complicate supply logistics and increase transportation emissions. Likewise, certain biomass streams may require intensive pretreatment or harsh chemical processing to unlock usable intermediates, which can offset environmental gains. Without robust strategies to manage feedstock quality and preprocessing, the sustainability profile of bio-based supply chains may suffer from increased energy use and waste generation.

Scalability and economic viability also pose significant barriers to widespread adoption of bio-based amines. Renewable feedstocks are not inherently cost-competitive with well-established petroleum sources, especially when the latter benefit from decades of optimization and economies of scale. Bio-based production routes frequently require specialized fermentation facilities, tailored catalysts, or integrated biorefinery systems, all of which demand substantial capital investment and continued optimization to achieve cost parity. The development of advanced catalysts and engineered microbes that enhance yield and conversion efficiency can help, but these innovations remain in the research or pilot phase for many target amines. Until large-scale, cost-effective production is demonstrated, many bio-based chemicals including amines destined for polymer applications will remain niche or premium products.

Moreover, responsible sustainability claims must contend with life cycle implications beyond emissions, such as land use, water consumption, nutrient runoff, and biodiversity impacts associated with cultivating feedstock sources at scale. For example, expanding acreage for biomass production could inadvertently displace native ecosystems or deepen competition with food crops if not carefully managed. These socio-environmental risks must be integrated into sustainability assessments and weighed against potential benefits. This integration often requires multi-criteria evaluation frameworks that go beyond greenhouse gas metrics to capture broader ecological and social dimensions of sustainability.

In response to these challenges, researchers stress the importance of transparent, rigorous, and comprehensive sustainability evaluations that combine LCAs with techno-economic analysis and resource footprint assessments. Such evaluations help elucidate where renewable feedstocks and bio-based amines can deliver genuine environmental benefits and where further technological refinement is necessary. Only through this integrated approach can responsible claims be made about the sustainability of bio-based chemicals and materials and can policy, industry, and research communities align toward more impactful implementation. As the field advances, emphasis on end-of-life solutions, recycling infrastructure, and circular economy integration will be prerequisites for translating renewable feedstocks into truly sustainable material systems.

1.6 CONCLUSION

Amines are a chemically versatile class of nitrogen-containing compounds that play a central role in industrial chemistry and polymer science due to their reactivity, basicity, and ability to form diverse molecular architectures. Conventional industrial amines are predominantly produced from fossil-derived feedstocks through energy-intensive petrochemical routes that rely on light olefins and hazardous intermediates, leading to high greenhouse gas emissions, safety concerns, and waste generation. These limitations have driven increasing interest in bio-based amines as sustainable alternatives derived from renewable resources. Biomass provides a wide range of platforms for bio-based amine synthesis, including lignocellulosic carbohydrates, plant oils, proteins, and nitrogen-rich biopolymers such as chitin. Lignocellulose, composed of cellulose, hemicellulose, and lignin, is the most abundant renewable carbon source and offers both aliphatic and aromatic structures suitable for chemical upgrading. Proteins and amino acids provide intrinsic organic nitrogen, while plant oils supply long-chain aliphatic structures, and chitin offers a combined carbohydrate–nitrogen framework. These feedstocks can be converted into amines through biotechnological routes, chemical catalytic routes, or integrated bio-chemical strategies. Biotechnological pathways use microorganisms or enzymes to transform renewable substrates into amines with high selectivity under mild conditions, though they are often limited by product scope, titers, and downstream separation challenges. Chemical and catalytic routes enable broader structural diversity and industrial scalability through reactions such as reductive amination, hydrogen borrowing, amination of alcohols, and nitrile hydrogenation, but may require elevated temperatures, pressures, or hydrogen. Hybrid approaches combine the advantages of both strategies and are increasingly adopted. Bio-based amines are widely used in polymer science as monomers, curing agents, and chain extenders in polyamides, polyurethanes, polyureas, and epoxy resins. Their incorporation enables control over polymer architecture, crosslink density, and intermolecular interactions, resulting in materials with competitive mechanical, thermal, and functional properties. Bio-based amines also facilitate the development of advanced polymer systems, including recyclable, reprocessable, and self-healing materials through dynamic covalent chemistry. The sustainability benefits of bio-based amines depend strongly on feedstock selection, production pathways, scalability, and end-of-life management. Life cycle assessments show that renewable feedstocks can reduce fossil resource use and greenhouse gas emissions, but these advantages may diminish when end-of-life treatments, processing inefficiencies, or feedstock variability are considered. Consequently, quantitative life cycle assessment and responsible sustainability claims are essential to accurately evaluate the environmental impact of bio-based amines and their derived polymeric materials.

REFERENCES

1. V. Froidevaux, C. Negrell, S. Caillol, J.P. Pascault, B. Boutevin, Biobased amines: From synthesis to polymers; Present and future, *Chem. Rev.* 116 (2016) 14181–14224.
2. S.A. Lawrence, *Amines: Synthesis, Properties and Applications*, Cambridge University Press, Cambridge, 2004.
3. M. Pelckmans, T. Renders, S. Van de Vyver, B.F. Sels, Bio-based amines through sustainable heterogeneous catalysis, *Green Chem.* 19 (2017) 5303–5331.
4. R. Rossbacher, H.H. K. Eller, E. Henkes, *Ullmann's Encyclopedia of Industrial Chemistry*, Wiley VCH Verlag GmbH Co., Weinheim, 2012: pp. 647–698.
5. K. K. Beutel., T.M.M. Rossberg, W. Lendle, G. Pfleiderer, A. Tögel, E.-L. Dreher, E. Langer, H. Rassaerts, P. Kleinschmidt, H. Strack, R. Cook, U. Beck, K.-A. Lipper, T. R. Torkelson, E. Löser, Chlorinated hydrocarbons, in: *Ullmann's Encyclopedia of Industrial Chemistry*, Wiley VCH Verlag GmbH Co., Weinheim, 2006: pp. 1–186.
6. R. Rossbacher, H.H. M. Frauenkron, J.-P. Melder, G. Ruider, *Ullmann's Encyclopedia of Industrial Chemistry*, Wiley-VCH Verlag GmbH Co., Weinheim, 2012: pp. 405–431.
7. K. Weissermel, H.-J. Arpe, *Industrial Organic Chemistry*, Wiley-VCH Verlag GmbH, Weinheim, 2007: pp. 143–190.
8. K. Weissermel, H.-J. Arpe, *Industrial Organic Chemistry*, Wiley-VCH Verlag GmbH, Weinheim, 2007: pp. 59–89.
9. T. Ren, M.K. Patel, K. Blok, Steam cracking and methane to olefins: Energy use, CO_2 emissions and production costs, *Energy*. 33 (2008) 817–833.
10. P.T. Anastas, J.C. Warner, *Green Chemistry: Theory and Practice*, Oxford University Press, Oxford, 2000.
11. M. Dusselier, M. Mascal, B.F. Sels, Top chemical opportunities from carbohydrate biomass: A chemist's view of the biorefinery, in: K. M. Nicholas (Ed.), *Selective Catalysis for Renewable Feedstock Chemistry*, Springer International Publishing, Cham, 2014: pp. 1–40.
12. B. Vermeeren, S. Van Praet, W. Arts, T. Narmon, Y. Zhang, C. Zhou, H.P. Steenackers, B.F. Sels, From sugars to aliphatic amines: as sweet as it sounds? Production and applications of bio-based aliphatic amines, *Chem. Soc. Rev.* 53 (2024) 11804–11849.
13. N.K. Gupta, P. Reif, P. Palenicek, M. Rose, Toward renewable amines: recent advances in the catalytic amination of biomass-derived oxygenates, *ACS Catal.* 12 (2022) 10400–10440.
14. J. Zhang, J. Yang, X. Li, H. Liu, X. Yao, C. Xia, Z. Huang, Recent advances in the efficient synthesis of useful amines from biomass-based furan compounds and their derivatives over heterogeneous catalysts, *Catalysts*. 13 (2023) 528.
15. B. Ganem, New chemistry of naturally occurring polyamines, *Am. Chem. Soc.* 1082 (1982) 290–298.
16. A.A. Castillo-Garcia, K. Barta, Strategies for obtaining value-added aromatic amines via lignin-first biorefinery approaches, *Curr. Opin. Green Sustain. Chem.* 53 (2025) 101011.

17. D. Betancourt-Jimenez, J.P. Youngblood, C.J. Martinez, Synthesis and characterization of fatty acid amides from commercial vegetable oils and primary alkyl amines for phase change material applications, *ACS Sustain. Chem. Eng.* 8 (2020) 13683–13691.
18. J.A. Lee, J.H. Ahn, I. Kim, S. Li, S.Y. Lee, Synthesis, characterization, and application of fully biobased and biodegradable nylon-4,4 and -5,4, *ACS Sustain. Chem. Eng.* 8 (2020) 5604–5614.
19. J. Li, Z. Weng, Q. Cao, Y. Qi, B. Lu, S. Zhang, J. Wang, X. Jian, Synthesis of an aromatic amine derived from biomass and its use as a feedstock for versatile epoxy thermoset, *Chem. Eng. J.* 433 (2022) 134512.
20. C. Pappa, E. Feghali, K. Vanbroekhoven, K.S. Triantafyllidis, Recent advances in epoxy resins and composites derived from lignin and related bio-oils, *Curr. Opin. Green Sustain. Chem.* 38 (2022) 100687.
21. A.-S. Mora, R. Tayouo, B. Boutevin, G. David, S. Caillol, Synthesis of pluri-functional amine hardeners from bio-based aromatic aldehydes for epoxy amine thermosets, *Molecules.* 24 (2019) 3285.
22. H. Blattmann, R. Mülhaupt, Multifunctional β-amino alcohols as bio-based amine curing agents for the isocyanate- and phosgene-free synthesis of 100% bio-based polyhydroxyurethane thermosets, *Green Chem.* 18 (2016) 2406–2415.
23. Z. Lan, R. Daga, R. Whitehouse, S. McCarthy, D. Schmidt, Structure–properties relations in flexible polyurethane foams containing a novel bio-based crosslinker, *Polymer (Guildf).* 55 (2014) 2635–2644.
24. A. Corma, S. Iborra, A. Velty, Chemical routes for the transformation of biomass into chemicals, *Chem. Rev.* 107 (2007) 2411–2502.
25. M.J. Biddy, C. Scarlata, C. Kinchin, *Chemicals from Biomass: A Market Assessment of Bioproducts with Near-term Potential*, National Renewable Energy Lab.(NREL), Golden, CO, 2016.
26. G. Hayes, M. Laurel, D. MacKinnon, T. Zhao, H.A. Houck, C.R. Becer, Polymers without petrochemicals: Sustainable routes to conventional monomers, *Chem. Rev.* 123 (2023) 2609–2734.
27. M. Aresta, A. Dibenedetto, F. Dumeignil, *Biorefinery: From Biomass to Chemicals and Fuels: Towards Circular Economy*, Walter de Gruyter GmbH & Co KG, 2021.
28. Y. Liao, S.-F. Koelewijn, G. Van den Bossche, J. Van Aelst, S. Van den Bosch, T. Renders, K. Navare, T. Nicolaï, K. Van Aelst, M. Maesen, H. Matsushima, J.M. Thevelein, K. Van Acker, B. Lagrain, D. Verboekend, B.F. Sels, A sustainable wood biorefinery for low–carbon footprint chemicals production, *Science.* 367 (2020) 1385–1390.
29. J.A. Melero, J. Iglesias, A. Garcia, Biomass as renewable feedstock in standard refinery units. Feasibility, opportunities and challenges, *Energy Environ. Sci.* 5 (2012) 7393–7420.
30. C.O. Tuck, E. Pérez, I.T. Horváth, R.A. Sheldon, M. Poliakoff, Valorization of biomass: Deriving more value from waste, *Science.* 337 (2012) 695–699.
31. E. Scott, F. Peter, J. Sanders, Biomass in the manufacture of industrial products. The use of proteins and amino acids, *Appl. Microbiol. Biotechnol.* 75 (2007) 751–762.
32. M. Rinaudo, Chitin and chitosan: Properties and applications, *Prog. Polym. Sci.* 31 (2006) 603–632.
33. M.N.V. Ravi Kumar, A review of chitin and chitosan applications, *React. Funct. Polym.* 46 (2000) 1–27.
34. N. Yan, X. Chen, Sustainability: Don't waste seafood waste, *Nature.* 524 (2015) 155–157.
35. A.J.J. Straathof, Transformation of biomass into commodity chemicals using enzymes or cells, *Chem. Rev.* 114 (2014) 1871–1908.
36. P.-W. Petra, S. Michael, E. Helga, K. Nicole, S. Hermann, E. Lothar, Metabolic engineering of corynebacterium glutamicum for l-serine production, *Appl. Environ. Microbiol.* 71 (2005) 7139–7144.

37. D. Rontein, I. Nishida, G. Tashiro, K. Yoshioka, W.-I. Wu, D.R. Voelker, G. Basset, A.D. Hanson, Plants synthesize ethanolamine by direct decarboxylation of serine using a pyridoxal phosphate enzyme *, *J. Biol. Chem.* 276 (2001) 35523–35529.
38. D. Rontein, D. Rhodes, A.D. Hanson, Evidence from engineering that decarboxylation of free serine is the major source of ethanolamine moieties in plants, *Plant Cell Physiol.* 44 (2003) 1185–1191.
39. C. de Bellefon, P. Fouilloux, Homogeneous and heterogeneous hydrogenation of nitriles in a liquid phase: Chemical, mechanistic, and catalytic aspects, *Catal. Rev.* 36 (1994) 459–506.
40. M. José Climent, A. Corma, S. Iborra, Homogeneous and heterogeneous catalysts for multicomponent reactions, *RSC Adv.* 2 (2012) 16–58.
41. S. Bähn, S. Imm, L. Neubert, M. Zhang, H. Neumann, M. Beller, The catalytic amination of alcohols, *ChemCatChem.* 3 (2011) 1853–1864.
42. R. Reguillo, M. Grellier, N. Vautravers, L. Vendier, S. Sabo-Etienne, Ruthenium-catalyzed hydrogenation of nitriles: Insights into the mechanism, *J. Am. Chem. Soc.* 132 (2010) 7854–7855.
43. K. Murugesan, Z. Wei, V.G. Chandrashekhar, H. Jiao, M. Beller, R. V Jagadeesh, General and selective synthesis of primary amines using Ni-based homogeneous catalysts, *Chem. Sci.* 11 (2020) 4332–4339.
44. T. Yasukawa, R. Masuda, S. Kobayashi, Development of heterogeneous catalyst systems for the continuous synthesis of chiral amines via asymmetric hydrogenation, *Nat. Catal.* 2 (2019) 1088–1092.
45. Y. Xie, S. Gao, D. Zhang, C. Wang, F. Chu, Bio-based polymeric materials synthesized from renewable resources: A mini-review, *Resour. Chem. Mater.* 2 (2023) 223–230.
46. H. Nabipour, X. Wang, L. Song, Y. Hu, A high performance fully bio-based epoxy thermoset from a syringaldehyde-derived epoxy monomer cured by furan-derived amine, *Green Chem.* 23 (2021) 501–510.
47. H. Fulcrand, L. Rouméas, G. Billerach, C. Aouf, E. Dubreucq, Advances in bio-based thermosetting polymers, *Recent Adv. Polyphen. Res.* 6 (2019) 285–334.
48. G.I.C. Righetti, F. Faedi, A. Famulari, Embracing sustainability: The World of bio-based polymers in a mini review, *Polymers (Basel).* 16 (2024) 950.
49. A. Arif, N. Chanchaona, C.H. Lau, Comparing the environmental impacts of using bio-renewable and fossil-derived solvent in polymer membrane fabrications, *Adv. Membr.* 3 (2023) 100079.
50. T.A. Hottle, M.M. Bilec, A.E. Landis, Sustainability assessments of bio-based polymers, *Polym. Degrad. Stab.* 98 (2013) 1898–1907.

2 Natural Sources and Synthesis of Bio-Based Amines

2.1 INTRODUCTION

The increasing demand for sustainable chemicals and materials has placed renewable feedstocks at the center of modern chemical manufacturing and polymer science. Growing environmental concerns, resource depletion, and climate change have accelerated efforts to replace fossil-derived chemicals with renewable and environmentally benign alternatives. Among nitrogen-containing compounds, amines occupy a particularly important position due to their extensive use as intermediates in pharmaceuticals, agrochemicals, surfactants, dyes, and, most importantly, polymeric materials such as polyamides, polyurethanes, epoxy resins, polyureas, and curing agents for thermosetting systems. Their high reactivity, structural diversity, and ability to participate in hydrogen bonding and crosslinking reactions make amines indispensable building blocks for tailoring polymer structure and performance [1].

Conventional industrial production of amines relies almost exclusively on petrochemical feedstocks and energy-intensive catalytic processes. Typical synthetic routes include the reductive amination of aldehydes and ketones, hydrogenation of nitriles, and alkylation of ammonia, often conducted under high pressures of hydrogen and elevated temperatures in the presence of metal catalysts [2]. These methods frequently suffer from limited selectivity toward primary amines, poor atom economy, and the generation of significant amounts of inorganic salts and solvent waste. Moreover, their dependence on nonrenewable resources and carbon-intensive processing contributes to greenhouse gas emissions and long-term sustainability concerns. In this context, the development of bio-based amines derived from renewable resources has emerged as a central objective in green chemistry and sustainable materials research [1].

Bio-based amines are generally obtained either from naturally occurring nitrogen-containing compounds or through the catalytic and biotechnological transformation of biomass-derived carbon platforms. Renewable feedstocks such as plant biomass, agricultural and forestry residues, marine waste, and protein-rich by-products provide abundant and diverse starting materials [3–9]. Naturally occurring amines, amino acids, alkaloids, and polyamines can sometimes be isolated directly, while chitin and chitosan derived from shellfish waste represent important amine-rich biopolymers for further functionalization [10,11]. More commonly, bio-based amines are synthesized indirectly by converting biomass-derived intermediates such as aldehydes, ketones, alcohols, nitriles, and furan derivatives into amine products. Advances in biomass fractionation and catalytic upgrading have significantly expanded the portfolio of renewable carbon platforms available for amination chemistry [1].

DOI: 10.1201/9781003731122-2

In parallel, major progress has been achieved in catalytic strategies for amine synthesis that are consistent with green chemistry principles. Reductive amination of bio-derived carbonyl compounds remains one of the most widely applied routes, providing direct access to primary, secondary, and tertiary amines using ammonia or amine donors [12]. Hydrogenation of nitriles continues to play a key role in industrial amine production, although its application in bio-based systems requires careful control of selectivity and catalyst deactivation [13]. Borrowing-hydrogen or hydrogen-autotransfer amination of alcohols has emerged as a particularly attractive strategy for renewable feedstocks obtained from fermentation or biomass upgrading [14,15]. In this approach, the alcohol substrate is temporarily dehydrogenated to a carbonyl intermediate, which reacts with ammonia or an amine to form an imine that is subsequently reduced in situ, ideally generating only water as a by-product. Such routes offer high atom efficiency and reduced waste formation, making them well suited for sustainable processing.

Biotechnological approaches provide complementary pathways for the selective production of bio-based amines. Microbial fermentation and whole-cell bioconversion have enabled the synthesis of important diamines such as putrescine and cadaverine from renewable sugars and amino acids through engineered metabolic pathways [16–18]. These diamines are particularly valuable as monomers for fully bio-based polyamides and polyurethanes, offering renewable alternatives to conventional petroleum-derived diamines. Enzyme-catalyzed amination reactions further provide opportunities for mild reaction conditions and high chemo- and stereoselectivity, although challenges related to enzyme stability, cofactor regeneration, and downstream separation remain significant [19,20].

Despite substantial progress, several challenges continue to limit the large-scale implementation of bio-based amine production. Biomass-derived feedstocks often exhibit variable composition and contain impurities that can deactivate catalysts or inhibit microbial growth. Achieving high selectivity toward primary amines, particularly in ammonia-based systems, remains difficult due to competing over-alkylation and condensation reactions. In fermentation-based routes, downstream processing and purification dominate both cost and environmental impact because dilute aqueous streams must be concentrated and neutralized. Furthermore, the economic competitiveness of bio-based amines relative to established petrochemical processes depends strongly on feedstock availability, process integration, catalyst lifetime, and market acceptance.

Within this broader context, bio-based amines occupy a strategic position at the interface of green chemistry, catalysis, biotechnology, and sustainable polymer science. Their successful development not only reduces dependence on fossil carbon but also enables the design of novel polymer architectures with improved environmental performance. Renewable amines have already been incorporated into partially and fully bio-based polyamides, polyurethanes, epoxy curing agents, and non-isocyanate polyurethane systems, demonstrating properties comparable to or exceeding those of conventional materials. Consequently, bio-based amines represent key enabling molecules in the transition from green chemistry concepts to commercially viable sustainable polymer technologies.

This chapter presents an overview of the natural sources, extraction methods, and synthetic pathways for bio-based amines. Emphasis is placed on renewable feedstocks, catalytic and biotechnological routes, and the sustainability considerations governing process design and scale-up. By integrating recent advances in biomass conversion, catalysis, and fermentation technology, this chapter highlights both the opportunities and remaining challenges in establishing bio-based amines as essential building blocks for next-generation sustainable polymers.

2.2 NATURAL SOURCES OF BIO-BASED AMINES

2.2.1 Plant-Derived Feedstocks

Plant biomass represents the most abundant and versatile renewable source to produce bio-based amines. Major plant-derived feedstocks include carbohydrates, lignocellulosic components, vegetable oils, and plant proteins. Carbohydrates such as glucose, fructose, and sucrose obtained from sugarcane, sugar beet, and corn are readily converted into platform chemicals including furfural, 5-hydroxymethylfurfural (HMF), levulinic acid, and various polyols, which serve as important intermediates for reductive amination and borrowing-hydrogen amination reactions. Lignocellulosic biomass from wood and crop residues provides cellulose and hemicellulose for sugar-based intermediates as well as lignin-derived aromatic aldehydes and phenols that can be transformed into aromatic amines. In 1955, Hass highlighted the potential of sugars as inexpensive and abundant raw materials for industrial synthesis [21]. He introduced the concept of "sucrochemistry," referring to organic chemistry based on sucrose and its derivatives, as a promising strategy to reduce dependence on fossil resources. This concept has since evolved into the modern biorefinery approach. The present section focuses on the conversion of nonedible carbohydrate feedstocks, particularly cellulose, hemicellulose, and their derivatives. Carbohydrate processing yields two main classes of platform intermediates, namely alcohols and carbonyl compounds. Direct amination of alcohols is currently a well-established route to produce short-chain alkylamines, including mono-, di-, and trialkylamines. Methylamines, for example, are key intermediates in the chemical industry [22]. Biomass processing also provides a wide range of renewable diols through hydrogenation of (hemi)cellulose. Ethylene glycol (EG), which can be catalytically produced from cellulose [23,24], represents a versatile substrate for amination reactions [25]. Baiker and co-workers reported the gas-phase amination of EG with dimethylamine (DMA) in a continuous fixed-bed reactor using an Al_2O_3-supported Cu catalyst under atmospheric pressure [26]. Hydrogen borrowing amination has likewise been applied to higher aliphatic diols such as propane, butane, and hexane diols.

For substrates containing carbonyl groups, amination preferentially occurs at the more reactive C=O functionality. Glycolaldehyde, the smallest sugar containing both a carbonyl and a hydroxyl group, can be obtained either by retro-aldol cleavage of glucose or by partial oxidation of EG [27]. Recently, Zhang and co-workers demonstrated a two-step synthesis of monoethanolamine (MEOA) from cellulose. In the first step, cellulose (1 wt.%) was converted into glycolaldehyde (21%) via retro-aldol cleavage using H_2WO_4 in water at 290°C. The resulting glycolaldehyde solution was then subjected to reductive amination with NH_3 over a bifunctional Ru/ZrO_2 catalyst at 75°C and 30 bar H_2, affording MEOA in 52% yield and an overall cellulose-to-MEOA yield of approximately 10%. The same Ru/ZrO_2 catalyst was also effective for converting other bio-derived carbonyl compounds into primary amines, including the transformation of 5-methylfurfural into 5-methylfurfurylamine (61%), cyclohexanone into cyclohexylamine (99%), and glyceraldehyde into 1-amino-2,3-propanediol (82%) [28].

Lignin represents an abundant renewable source of aromatic compounds and offers unique opportunities for the synthesis of aromatic amines and nitrogen-containing derivatives. Depolymerization of lignin yields phenolic aldehydes, ketones, and acids such as vanillin and related compounds, which can be further functionalized through reductive amination or related transformations. In addition, fatty acids derived from natural oils can be coupled to aromatic ring systems obtained from lignin to produce synthetic capsaicinoids with applications in the food and pharmaceutical industries [29].

An important class of lignin- and fat-derived amine products discussed in this section is quaternary ammonium salts, which constitute industrially relevant higher aliphatic amine derivatives. It was estimated that in 2004 approximately 80,000 metric tons of natural fatty acids were consumed annually to produce quaternary ammonium salts. Within this class, esterquats are widely used biodegradable cationic surfactants. Their synthesis involves esterification of long-chain fatty acids (mainly C16 and C18) with tertiary bio-based aminoalcohols, followed by quaternization with alkylating agents to yield cationic esterquats [30,31]. Owing to their excellent surface-active properties, esterquats are extensively applied as fabric softeners [32]. The combination of lignin-derived aromatics, natural fats, and bio-based aminoalcohols thus provides an efficient route toward fully renewable cationic surfactants and specialty amine derivatives.

Vegetable oils from soybean, castor, palm, rapeseed, and sunflower are widely used industrial feedstocks that yield fatty acids, fatty alcohols, and nitriles, which are subsequently hydrogenated or aminated to produce long-chain primary amines, diamines, and amine-functional polyols. Natural oils and fats represent important renewable feedstocks to produce higher aliphatic amines and quaternary ammonium derivatives. Long-chain fatty acids (C16–C18) are first esterified with tertiary bio-based aminoalcohols such as triethanolamine, followed by quaternization with alkylating agents to form quaternary ammonium ester salts [30]. In addition, fatty acids and fatty acid methyl esters can be hydrogenated to fatty alcohols, which are subsequently aminated with DMA to produce tertiary alkyldimethylamines [33]. These tertiary amines are further converted into quaternary ammonium salts or amine oxides used in detergents and personal care formulations [30]. Compared with conventional halide-based routes that generate large amounts of salt waste, hydrogen borrowing amination offers a more sustainable alternative by enabling one-pot alcohol amination with water as the only by-product [33].

Plant proteins obtained from oilseed meals and cereal by-products can be hydrolyzed into amino acids, which serve as direct nitrogen-containing precursors for diamines and polyamines. Most amino acids are currently produced industrially by microbial fermentation of carbohydrates using nitrogen sources, with glutamic acid and lysine being the most prominent examples produced on a large scale using *Corynebacterium glutamicum* [34,35]. In addition to fermentation, large quantities of residual proteins are generated as side streams from biofuel and food industries, including dried distiller's grains with solubles, sugar beet vinasse, soybean meal, and poultry feather meal [7,36]. After hydrolysis of these protein-rich residues, mixtures of amino acids are obtained. The separation of individual amino acids represents a major technical and economic challenge, and current methods mainly rely on electrodialysis based on pH-dependent charge differences, followed by enzymatic or reactive extraction for further purification [37]. Alternatively, amination reactions can be carried out directly on protein hydrolysates, and the resulting amine products can be separated downstream more efficiently than their zwitterionic amino acid precursors [38].

Among individual amino acids, glutamic acid receives particular attention because it is the most abundant amino acid in plant biomass and is readily available from fermentation. The conversion of amino acids into amines typically proceeds through defunctionalization reactions such as decarboxylation, enabling the production of primary amines with high atom efficiency while retaining the biogenic nitrogen content. Owing to their abundance, renewability, and structural diversity, plant-derived feedstocks constitute the primary raw materials for the sustainable synthesis of bio-based amines.

2.2.2 Agricultural By-Products and Residues

Agricultural by-products and residues are promising low-cost, widely available feedstocks for the sustainable production of bio-based amines. These materials are central to the circular bioeconomy because they add value to waste streams while reducing competition with food resources and minimizing disposal impacts. Common examples of agricultural residues include crop leftovers such as corn stover, wheat straw, rice husk, sugarcane bagasse, and oilseed cakes, as well as processing wastes from fruit and vegetable production [39–41]. These residues are rich in lignocellulosic components and, in some cases, proteins, making them suitable sources of fermentable sugars, organic acids, amino acids, and nitrogen-containing intermediates for downstream amine synthesis.

The lignocellulosic fraction of these agro-residues must first be pretreated and hydrolyzed to liberate fermentable sugars (glucose, xylose) and other intermediates that can serve as carbon sources for microbial or catalytic conversion. Hydrolysis followed by fermentation can yield alcohol and carbonyl compounds, which can be aminated catalytically to produce alkylamines, amino alcohols, and other amine products. Although much of the literature focuses on biofuel production from residues such as corn stover and sugarcane bagasse, recent research points toward integrated biorefinery schemes where fermentation products can be further converted into value-added chemicals including amines. Valorization of lignocellulosic waste for high-value chemicals is the subject of ongoing work and demonstrates the potential of such residues as chemical feedstock sources [42].

Protein-rich residues such as soybean meal, rapeseed cake, and other oilseed cakes additionally provide nitrogen sources that can be converted into amino acids and subsequently transformed into diamines (for example putrescine and cadaverine) through enzymatic or microbial routes. These diamines are important as bio-based

quinine, caffeine, and nicotine. These compounds not only demonstrate the chemical versatility of biologically incorporated nitrogen but also provide valuable templates for the synthesis of functional amine derivatives and nitrogen-containing heterocycles with tailored properties [50].

In addition to polyamines and alkaloids, amino acids, peptides, and nitrogen-containing heterocycles are ubiquitous in living organisms and constitute an important reservoir of renewable nitrogen. These compounds can be isolated from biomass or selectively modified through chemical and enzymatic transformations to obtain functional amines, amino alcohols, and diamines for specialty applications [49]. Although the natural concentrations of such amines are generally low and therefore unsuitable for direct use as bulk industrial feedstocks, they are highly valuable as high-purity specialty intermediates, chiral building blocks, and reference compounds.

Overall, naturally occurring amines provide important insight into sustainable nitrogen incorporation and biological synthesis strategies, and they serve as key inspiration for the development of bio-based routes to nitrogen-containing chemicals. While their limited availability restricts their large-scale industrial exploitation, their structural diversity and intrinsic functionality continue to play a crucial role in high-value chemical production and in advancing the understanding of renewable nitrogen chemistry.

2.3 EXTRACTION, CONVERSION, AND CHEMICAL SYNTHESIS OF BIO-BASED AMINES

2.3.1 Extraction and Isolation from Renewable Resources

Protein hydrolysis represents a key enabling step in the valorization of renewable nitrogen-containing biomass to produce bio-based amines. As reviewed by Wouters et al., enzymatic hydrolysis efficiently converts plant proteins into mixtures of peptides and free amino acids through cleavage of peptide bonds by specific proteases under mild aqueous conditions. This process increases solubility, liberates ionizable amino groups, and allows control over molecular weight distribution through enzyme selection and reaction conditions. Compared with chemical hydrolysis, enzymatic routes better preserve amino acid integrity and offer higher selectivity, making them particularly suitable for downstream chemical or biotechnological transformations. In the context of this chapter, protein hydrolysates constitute an important renewable nitrogen pool, providing amino acids that can be directly converted into primary amines through defunctionalization reactions such as decarboxylation or further transformed into diamines and polyamines via catalytic or microbial routes. However, as highlighted in the review, separation of individual amino acids from complex hydrolysates remains a major challenge, often dominating process cost. This limitation has motivated emerging strategies in which amino acid conversion to amines is performed directly on crude hydrolysates, followed by easier downstream separation of the amine products. Thus, protein hydrolysis serves not only as a nitrogen release step but also as a critical bridge between renewable biomass and sustainable amine synthesis pathways. Figure 2.2 shows the principal effects of hydrolysis on protein structure [51].

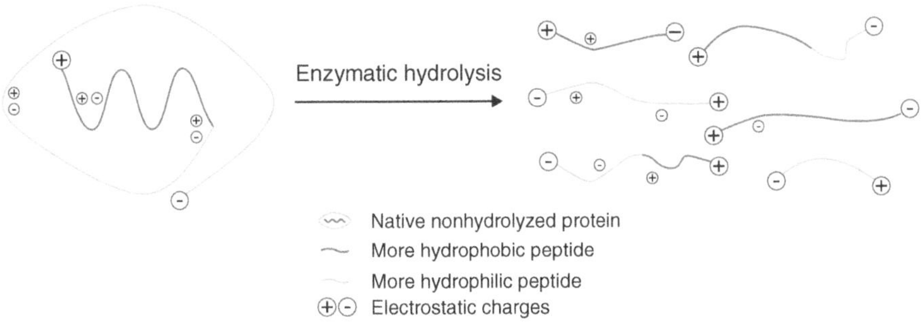

FIGURE 2.2 Schematic overview of enzymatic hydrolysis of protein. (Adapted with permission [51]. Copyright 2016 Institute of Food Technologists.)

Efficient separation and recovery of amines from aqueous streams represent a major bottleneck in the sustainable production of bio-based amines, particularly when these compounds are generated via biotechnological or biomass-derived routes. As demonstrated by Doeker et al., conventional physical liquid–liquid extraction is often ineffective under biotechnologically relevant conditions because amines are predominantly present in their protonated, hydrophilic form at neutral pH, resulting in very low partitioning into organic solvents. To overcome this limitation, the authors highlight reactive extraction as a powerful alternative, in which fatty acids such as oleic acid act as liquid cation exchangers and form neutral complexes with protonated amines, enabling high extraction efficiencies even at low product titers and moderate pH values. Using an oleic acid/1-octanol system, extraction yields of up to 99% and efficient back-extraction of the amines were achieved, demonstrating full reversibility and recyclability of the organic phase. This approach is particularly relevant for bio-based amine production, as it allows in situ product removal, mitigates product inhibition and toxicity in fermentation systems, and simplifies downstream processing [52].

Crustacean shell waste, such as shrimp and prawn shells, represents an important renewable and nitrogen-rich biomass resource, primarily composed of chitin, calcium carbonate, proteins, lipids, and pigments. As demonstrated by Acosta et al., chitin can be efficiently isolated from a variety of crustacean residues through sequential demineralization, deproteination, and depigmentation steps, typically yielding 14%–25% chitin on a dry-weight basis depending on the shell source and processing conditions [45]. The isolated chitin exhibits a well-defined polymeric structure with a high content of acetamide functionalities, making it a valuable precursor for nitrogen-containing materials. Further chemical treatment of chitin by alkaline deacetylation leads to the formation of chitosan, in which acetamide groups are converted into free primary amine groups. Mohammed et al. showed that controlled alkaline deacetylation not only increases the degree of deacetylation but also induces partial depolymerization, allowing the production of chitosan with tunable molecular weight and amine density [44]. The ability to tailor these structural parameters is particularly important for downstream processing, as chitosan can be directly employed as an amine-rich biopolymer or further converted into low-molecular-weight nitrogen-containing platform molecules. Within the framework of this chapter, the extraction and transformation of chitin into chitosan illustrate how marine by-products can be efficiently valorized into renewable amine-containing intermediates, bridging biomass extraction with sustainable chemical pathways for bio-based amine synthesis.

The isolation and purification of amino acids from renewable protein-derived streams is a critical step in the production of bio-based amines and related nitrogen-containing chemicals. As demonstrated in the work on electrodialysis by Kattan Readi et al., amino acids are particularly well suited for separation using electro-membrane processes because of their zwitterionic nature and pH-dependent charge behavior. By carefully adjusting the pH, amino acids can be selectively transported through ion-exchange membranes under an applied electric field, enabling their fractionation from complex aqueous mixtures. The study shows that basic amino acids such as lysine and arginine can be isolated efficiently, and that separation can be further enhanced by enzymatic modification, for example, through decarboxylation of lysine into 1,5-pentanediamine, which alters charge behavior and facilitates selective recovery. Control of pH during electrodialysis was identified as a key parameter for achieving high recovery, high current efficiency, and low energy consumption. This combined strategy of enzymatic conversion and electrodialysis illustrates an effective downstream processing approach for biorefineries, allowing renewable amino acids to be isolated as pure streams or converted into diamines and other amines suitable for further chemical transformation. In the context of this chapter, electrodialysis thus represents a powerful and scalable purification technique that bridges protein hydrolysis and sustainable chemical pathways for bio-based amine synthesis. Figure 2.3 presents a schematic overview of how different amino acids are separated based on their charge characteristics [53].

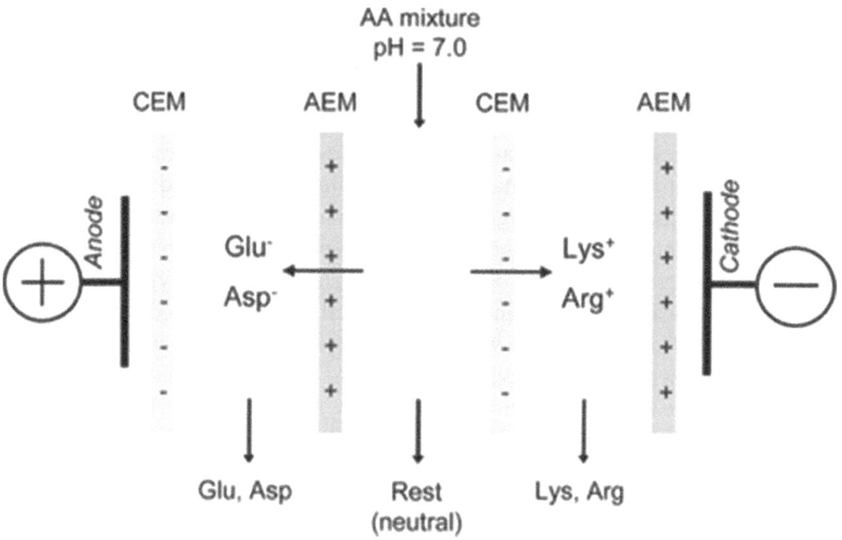

FIGURE 2.3 Separation of amino acids by electrodialysis. ([Glu-glutamic acid; Asp-aspartic acid; Lys-lysine; Arg-arginine; CEM-cation exchange membrane; AEM-anion exchange membrane; AA-amino acid]. Adapted with permission [53]. Copyright 2012 American Chemical Society.)

2.3.2 Chemical Pathways and Catalytic Strategies for the Synthesis of Bio-Based Amines

The synthesis of bio-based amines from renewable feedstocks relies on a combination of well-established chemical pathways and advanced catalytic systems designed to maximize selectivity, efficiency, and sustainability. Among these routes, reductive amination of bio-based carbonyl compounds, hydrogenation of nitriles, borrowing-hydrogen amination of alcohols, and amination of biomass-derived platform molecules represent the most important strategies for converting renewable carbon substrates into nitrogen-containing products.

Reductive amination of bio-based carbonyls is one of the most widely applied and versatile methods for amine synthesis. In this approach, aldehydes or ketones derived from biomass, such as furfural, HMF, glycolaldehyde, or glyceraldehyde, react with ammonia or amines to form imine intermediates that are subsequently reduced to the corresponding amines. This route is particularly attractive because it proceeds under relatively mild conditions and allows high selectivity toward primary amines when excess ammonia and suitable catalysts are employed. As a result, reductive amination plays a central role in the conversion of sugar-derived carbonyl compounds into amino alcohols, diamines, and other polymer-relevant intermediates [54]. Vermeeren et al. investigated the catalytic reductive amination of biomass-derived α-hydroxy carbonyl compounds as a sustainable route to multifunctional aliphatic amines. The study addresses the challenge of product selectivity arising from competing reaction pathways that can yield alkanolamines, isomeric alkanolamines, or polyamines from a single renewable substrate (Figure 2.4). Through detailed mechanistic analysis, the authors identified a key α-amino ketone intermediate and demonstrated that selectivity can be effectively controlled by tuning catalyst choice, reaction temperature,

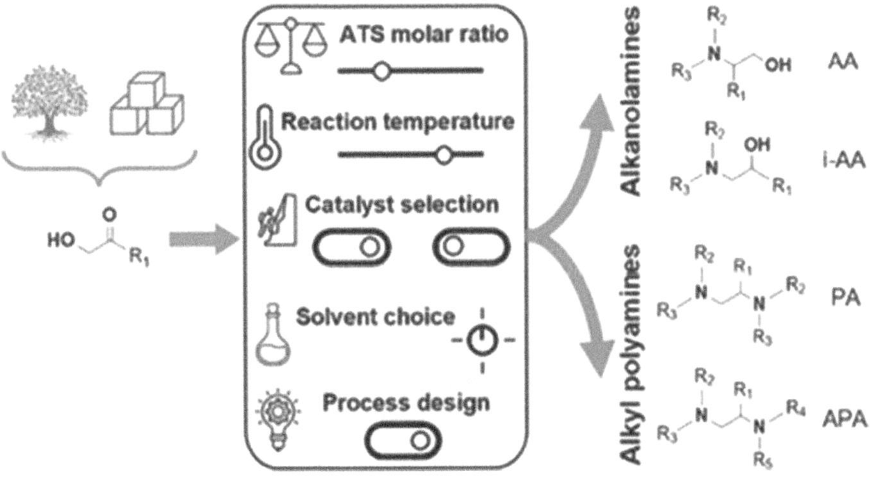

FIGURE 2.4 Selective synthesis of alkanolamines and alkyl polyamines from bio-based α-hydroxy carbonyl compounds by tuning catalytic reductive amination conditions. (Adapted with permission [55]. Copyright 2024 American Chemical Society.)

solvent, amine-to-substrate ratio, and process design. By combining these parameters, highly selective and universally applicable strategies were developed for the targeted synthesis of specific amine products from bio-based oxygenates [55].

Hydrogenation of nitriles represents another industrially important pathway to amines and is commonly used to produce primary amines. Nitriles can be obtained from biomass-derived intermediates or introduced through functionalization of renewable substrates, followed by catalytic hydrogenation over metal catalysts. Although this route is robust and well established, careful catalyst and process design are required to suppress side reactions and ensure selective formation of primary amines, particularly when multifunctional bio-based substrates are involved. The schematic of the hydrogenation of nitriles and possible side reactions is given in Figure 2.5 [56]. Chatterjee et al. investigated the hydrogenation of nitriles to amines using supercritical carbon dioxide ($scCO_2$) as a green reaction medium, demonstrating that solvent effects can be exploited to control amine selectivity without additives. Using benzonitrile as a model substrate and Pd-based heterogeneous catalysts, the study showed that both conversion and selectivity toward primary or secondary amines can be finely tuned by adjusting CO_2 pressure, hydrogen pressure, temperature, and catalyst support. Pd/MCM-41 enabled high selectivity toward benzylamine, while higher CO_2 pressures favored dibenzylamine formation due to changes in phase behavior and intermediate solubility. The work highlights $scCO_2$ as an effective and recyclable green medium for nitrile hydrogenation, offering a cleaner alternative to conventional solvent- and additive-intensive processes [57].

Borrowing-hydrogen amination of alcohols, also referred to as hydrogen-autotransfer amination, has emerged as a highly atom-efficient and green alternative for amine synthesis from renewable alcohols. In this process, the alcohol is temporarily dehydrogenated to a carbonyl intermediate, which reacts with ammonia or an amine to form an imine that is subsequently hydrogenated using the hydrogen released in the first step. Because no external hydrogen source is required and water is the only stoichiometric by-product, borrowing-hydrogen amination is particularly

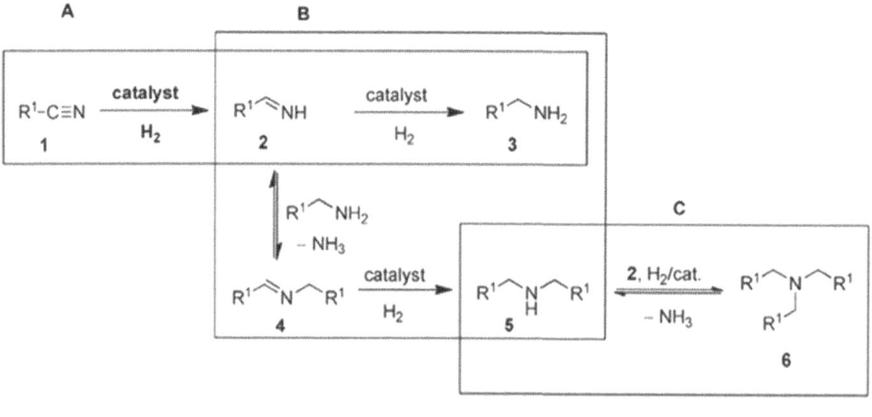

FIGURE 2.5 Hydrogenation of nitriles and possible side reactions. (Adapted with permission [56]. Copyright 2015 Wiley.)

attractive for bio-based alcohols such as ethanol, polyols, and sugar-derived diols. This strategy enables direct access to amino alcohols and diamines from renewable feedstocks [14,15]. Hameury et al. provided a comprehensive review of the borrowing-hydrogen (hydrogen-autotransfer) amination of bio-based alcohols as a sustainable route to amines, emphasizing its relevance for biomass valorization due to the high abundance of alcohol functionalities in renewable substrates. The authors summarized advances in both homogeneous and heterogeneous catalytic systems and highlighted the transformation of diverse bio-derived alcohols, including fatty alcohols, terpenoids, furan derivatives, and carbohydrate-based polyols. Particular attention was given to structure–reactivity relationships, functional group tolerance, and selectivity challenges arising from multifunctional natural substrates. Overall, the review demonstrates that borrowing-hydrogen amination enables atom-efficient amine synthesis with water as the only by-product, while also identifying remaining limitations related to catalyst robustness and selectivity when processing complex biomass-derived molecules [58].

In addition to these individual pathways, **amination of biomass-derived platform molecules** integrates multiple functional transformations within a biorefinery context. Platform chemicals obtained from carbohydrates, lignin, oils, and proteins can be selectively aminated through combinations of oxidation, reduction, and C–N bond-forming reactions, allowing the synthesis of structurally diverse amines tailored for polymer, surfactant, and fine chemical applications.

The efficiency and sustainability of these chemical pathways are strongly governed by the **choice of catalytic systems.** Both homogeneous and heterogeneous catalysts have been extensively explored, with heterogeneous metal catalysts often preferred for industrial applications due to their ease of separation, recyclability, and compatibility with continuous processing. Recent advances in catalyst design have focused on improving activity, selectivity, and resistance to deactivation when processing complex biomass-derived feeds, as well as reducing reliance on precious metals in favor of earth-abundant alternatives such as Ni, Fe, Co, and Cu [15,59].

Yan, Feringa, and Barta reported a homogeneous iron-catalyzed borrowing-hydrogen strategy for the direct alkylation of amines with alcohols (Figure 2.6), addressing a key sustainability challenge in C–N bond formation by replacing precious metals with earth-abundant iron. The study demonstrates that a well-defined iron cyclopentadienone complex can efficiently mediate alcohol dehydrogenation, imine formation, and imine hydrogenation within a single catalytic cycle, producing secondary and tertiary amines with water as the sole stoichiometric by-product. A broad substrate scope was established, including aromatic and aliphatic amines as well as primary alcohols and diols, enabling selective mono-alkylation and the formation of five-, six-, and seven-membered nitrogen heterocycles. Importantly, the methodology was shown to be compatible with biomass-derived alcohols and employed cyclopentyl methyl ether as a green solvent, further enhancing its sustainability profile. The work highlights the potential of iron-based catalysis for atom-economic amine synthesis and provides a significant advance toward the valorization of renewable alcohols in environmentally responsible chemical transformations [59].

By combining efficient chemical pathways with advanced catalytic systems and green chemistry principles, the synthesis of bio-based amines can be aligned with

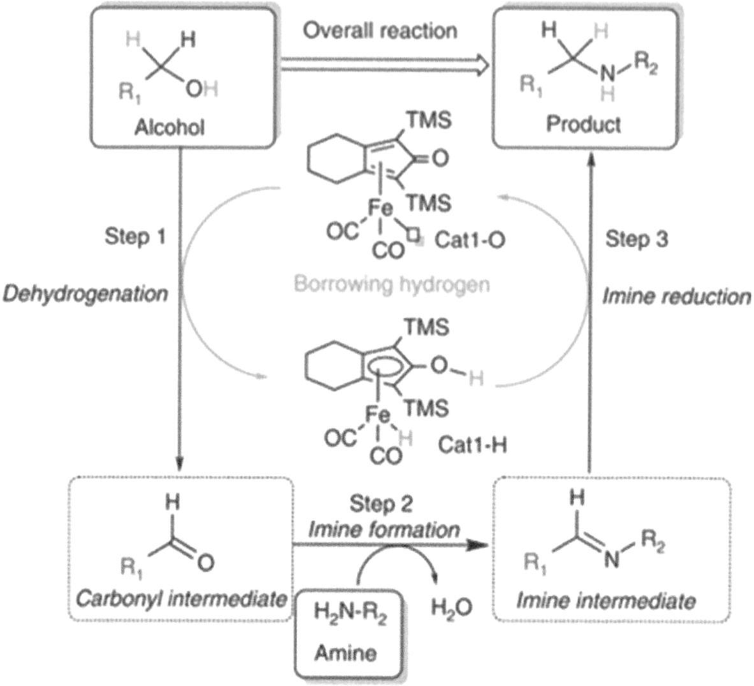

FIGURE 2.6 The overall transformation is the direct coupling of an alcohol (R_1–OH) with an amine (R_2–NH_2) to form the product amine (R_1–NH–R_2). (Adapted with permission [59]. Copyright 2014 Springer Nature.)

the goals of sustainable chemical manufacturing. These integrated approaches form a crucial bridge between renewable feedstocks and high-value nitrogen-containing products, enabling the development of next-generation bio-based polymers and materials.

2.4 BIOTECHNOLOGICAL ROUTES

2.4.1 Microbial Production of Diamines and Polyamines

Microbial production of amines has emerged as a promising sustainable alternative to conventional petrochemical synthesis, driven by advances in metabolic engineering and industrial biotechnology. Microorganisms can convert renewable carbon and nitrogen sources into a variety of amines through naturally occurring or engineered metabolic pathways. These biological routes typically operate under mild conditions in aqueous environments and offer high selectivity, reducing the need for harsh reagents and energy-intensive processing. By leveraging enzymatic specificity and cellular regulation, microbial systems enable precise control over amine structure and functionality. Consequently, microbial amine production plays an increasingly important role in the development of bio-based chemicals within integrated biorefinery frameworks.

Kind et al. demonstrated the microbial production of the bio-based diamine cadaverine (1,5-diaminopentane) using a systems-engineered strain of *Corynebacterium glutamicum*, establishing a sustainable biological route from sugars to polyamide building blocks. Through extensive metabolic pathway optimization and elimination of by-product formation, the engineered strain achieved high titers, yields, and productivities under fed-batch conditions. The study highlights the feasibility of integrating microbial diamine production with downstream purification and polymerization, underscoring the industrial relevance of fermentation-based amine synthesis [60]. Moreover, In an earlier study, Kind et al. reported the systematic metabolic engineering of *Corynebacterium glutamicum* for the microbial production of diaminopentane from lysine via enzymatic decarboxylation, providing a foundational framework for biological diamine synthesis. By redesigning central carbon metabolism, enhancing precursor supply, and optimizing lysine decarboxylase expression, the authors significantly improved diamine yields while minimizing competing pathways. This work represents a key milestone in the development of microbial platforms for renewable amine production [61].

2.4.2 Enzymatic Amination Processes

Enzymatic amination processes provide a selective and sustainable approach to produce diamines from renewable substrates. These routes rely on highly specific enzymes, such as aminotransferases, dehydrogenases, decarboxylases, and amine oxidoreductases, to introduce or transform nitrogen functionalities under mild aqueous conditions. By exploiting enzyme specificity and stereocontrol, enzymatic amination enables precise formation of C–N bonds while minimizing side reactions and waste generation. Such processes can be implemented as standalone biocatalytic reactions or integrated into multi-step enzymatic cascades, allowing efficient conversion of bio-based intermediates into diamines. As a result, enzymatic amination represents a promising strategy within biotechnological platforms for producing renewable diamines with high selectivity and reduced environmental impact.

Wu et al. reported an enzymatic approach for the highly enantioselective synthesis of α- and β-amino acids via phenylalanine aminomutase (PAM)-catalyzed ammonia addition to cinnamic acid derivatives. The study demonstrated that PAM, a 5-methylene-3,5-dihydroimidazol-4-one-dependent enzyme, can catalyze direct C–N bond formation with excellent regio- and stereoselectivity under mild conditions. This work expanded the scope of biocatalytic amination beyond conventional ammonia lyases and highlighted the potential of aminomutases for producing chiral nitrogen-containing building blocks [62]. In addition, Liu et al. developed an engineered cytochrome P450-based biocatalytic platform for the enantioselective primary amination of 1-aryl-2-alkyl alkynes via nitrene C(sp^3)–H insertion. Through directed evolution, a P411 enzyme variant was obtained that enabled highly selective formation of propargyl amines with high turnover numbers and enantiomeric excess (Figure 2.7). This work illustrates how enzyme engineering can unlock new-to-nature amination reactions and provides a powerful example of enzymatic strategies for selective C–N bond formation [63].

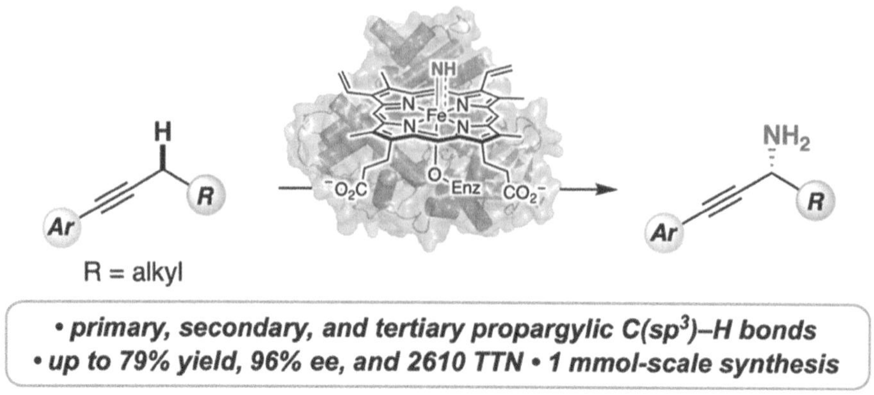

FIGURE 2.7 Primary amination of 1-aryl-2-alkyl alkynes. (Adapted with permission [63]. Copyright 2021 American Chemical Society.)

2.5 SCALABILITY AND PROCESS DEVELOPMENT

The transition of bio-based amine synthesis from laboratory research to industrial implementation requires careful attention to scalability and process development. Process design and reactor selection are central to this transition and must be tailored to the specific amination route, feedstock characteristics, and catalytic system. Batch reactors are commonly employed during early development due to their operational flexibility and ease of parameter optimization, whereas continuous reactor configurations such as fixed-bed, slurry, or flow reactors are favored at larger scales for improved productivity, reproducibility, and process control. Reactor design must also account for efficient heat and mass transfer, particularly in reactions involving ammonia and hydrogen, where pressure, temperature, and safety considerations are critical. In parallel, downstream processing and purification play a decisive role in overall process economics and sustainability. Bio-based amines are frequently produced in aqueous or multicomponent reaction media, making separation from unreacted substrates, catalysts, solvents, and by-products challenging. Conventional techniques such as distillation, extraction, and crystallization are often combined with advanced separation methods including membrane filtration or electrodialysis to achieve the required product purity. These steps can dominate energy consumption and operating costs, especially when dealing with dilute product streams typical of biotechnological processes. Technical challenges during scale-up further complicate process development and include catalyst deactivation, fouling, mass-transfer limitations, and variability in biomass-derived feedstocks. In addition, changes in reaction selectivity and by-product formation may occur upon scale-up, necessitating further optimization of reaction conditions and separation strategies. Successful industrial deployment therefore requires an integrated approach that couples reaction engineering with separation technology, process intensification, and sustainability assessment. Addressing these challenges holistically is essential to ensure that bio-based amine production processes are economically viable, operationally robust, and environmentally responsible at commercial scale.

2.6 CONCLUSION

Bio-based amines can be produced from a wide range of renewable resources through integrated chemical and biotechnological approaches. Plant biomass, agricultural residues, marine by-products, and protein-rich wastes serve as important sources of carbon and nitrogen, enabling access to amines either directly or through suitable intermediates. Effective extraction and isolation strategies, including protein hydrolysis, chitin and chitosan recovery, acid–base extraction, and advanced separation techniques, allow nitrogen-containing compounds to be obtained from complex biomass matrices. Multiple chemical pathways enable the transformation of these renewable intermediates into amines. Reductive amination of biomass-derived carbonyl compounds offers a versatile route to primary and multifunctional amines under relatively mild conditions, while hydrogenation of nitriles remains a robust method for producing primary amines with industrial relevance. Borrowing-hydrogen amination of alcohols provides an atom-efficient alternative that avoids external reducing agents and is particularly suitable for renewable alcohols and polyols. These pathways are supported by both homogeneous and heterogeneous catalytic systems, with increasing emphasis on earth-abundant metals, catalyst stability, and compatibility with complex bio-based feeds. Biotechnological routes complement chemical methods by enabling highly selective amine and diamine production under mild aqueous conditions. Microbial fermentation and enzymatic amination processes allow efficient conversion of renewable substrates into amines through well-defined metabolic and enzymatic pathways. Together with chemical catalysis, these approaches expand the range of accessible bio-based amines and offer flexibility in process design. At larger scales, process development considerations such as reactor selection, downstream purification, and feedstock variability become critical. Separation and purification often dominate energy and cost requirements, while scale-up introduces challenges related to mass transfer, catalyst lifetime, and process integration. Addressing these factors through integrated reaction and separation design is essential for viable implementation. Overall, the combination of renewable feedstocks, established and emerging synthesis routes, catalytic innovation, and biotechnological strategies enables a diverse and adaptable platform for producing bio-based amines suitable for polymers, surfactants, and fine chemical applications.

REFERENCES

1. V. Froidevaux, C. Negrell, S. Caillol, J.P. Pascault, B. Boutevin, Biobased amines: From synthesis to polymers; Present and future, *Chem. Rev.* 116 (2016) 14181–14224.
2. R.V. Jagadeesh, K. Murugesan, A.S. Alshammari, H. Neumann, M.-M. Pohl, J. Radnik, M. Beller, MOF-derived cobalt nanoparticles catalyze a general synthesis of amines, *Science (80-.).* 358 (2017) 326–332.
3. B. Vermeeren, S. Van Praet, W. Arts, T. Narmon, Y. Zhang, C. Zhou, H.P. Steenackers, B.F. Sels, From sugars to aliphatic amines: As sweet as it sounds? Production and applications of bio-based aliphatic amines, *Chem. Soc. Rev.* 53 (2024) 11804–11849.
4. H. Wu, H. Li, Z. Fang, Hydrothermal amination of biomass to nitrogenous chemicals, *Green Chem.* 23 (2021) 6675–6697.
5. G. Mahadevan, S. Valiyaveettil, Engineered polyaramides for extraction of bioamines from water, *Sci. Rep.* 15 (2025) 36577.

46. H. Huang, S. Liu, X. Guo, H. Jiang, Y. Cai, Z. Tan, G. Zhou, X. Cai, M. Zhuang, S. Xie, Sustainable ammonia and amines from chitin, *Bioresour. Technol.* 414 (2024) 131582.
47. K.W. Omari, L. Dodot, F.M. Kerton, A simple one-pot dehydration process to convert N-acetyl-D-glucosamine into a nitrogen-containing compound, 3-acetamido-5-acetylfuran, *ChemSusChem.* 5 (2012) 1767–1772.
48. R.A. Franich, S.J. Goodin, A.L. Wilkins, Acetamidofurans, acetamidopyrones, and acetamidoacetaldehyde from pyrolysis of chitin and n-acetylglucosamine, *J. Anal. Appl. Pyrolysis.* 7 (1984) 91–100.
49. N.H. Ishak, N.M. Sarbon, A review of protein hydrolysates and bioactive peptides deriving from wastes generated by fish processing, *Food Bioprocess Technol.* 11 (2018) 2–16.
50. N. Kerru, L. Gummidi, S. Maddila, K.K. Gangu, S.B. Jonnalagadda, A review on recent advances in nitrogen-containing molecules and their biological applications, *Molecules.* 25 (2020) 1909.
51. A.G.B. Wouters, I. Rombouts, E. Fierens, K. Brijs, J.A. Delcour, Relevance of the functional properties of enzymatic plant protein hydrolysates in food systems, *Compr. Rev. Food Sci. Food Saf.* 15 (2016) 786–800.
52. M. Doeker, V. Hüttche, A. Jupke, Reactive extraction for the recovery of primary amines from aqueous streams, *Sep. Purif. Technol.* 277 (2021) 118229.
53. O.M.K. Readi, M. Gironès, W. Wiratha, K. Nijmeijer, On the isolation of single basic amino acids with electrodialysis for the production of biobased chemicals, *Ind. Eng. Chem. Res.* 52 (2013) 1069–1078.
54. D. Max Roundhill, Transition metal and enzyme catalyzed reactions involving reactions with ammonia and amines, *Chem. Rev.* 92 (1992) 27.
55. B. Vermeeren, S. Van Praet, A. De Ridder, B.F. Sels, Selective multifunctional aliphatic amine synthesis: Unraveling universal control strategies in catalytic reductive amination of biobased α-hydroxy carbonyl oxygenates, *ACS Sustain. Chem. Eng.* 12 (2024) 10209–10220.
56. D.B. Bagal, B.M. Bhanage, Recent advances in transition metal-catalyzed hydrogenation of nitriles, *Adv. Synth. Catal.* 357 (2015) 883–900.
57. M. Chatterjee, H. Kawanami, M. Sato, T. Ishizaka, T. Yokoyama, T. Suzuki, Hydrogenation of nitrile in supercritical carbon dioxide: A tunable approach to amine selectivity, *Green Chem.* 12 (2010) 87–93.
58. S. Hameury, H. Bensalem, K. De Oliveira Vigier, Sustainable amination of bio-based alcohols by hydrogen borrowing catalysis, *Catalysts.* 12 (2022) 1306.
59. T. Yan, B.L. Feringa, K. Barta, Iron catalysed direct alkylation of amines with alcohols, *Nat. Commun.* 5 (2014) 5602.
60. S. Kind, S. Neubauer, J. Becker, M. Yamamoto, M. Völkert, G. von Abendroth, O. Zelder, C. Wittmann, From zero to hero – Production of bio-based nylon from renewable resources using engineered Corynebacterium glutamicum, *Metab. Eng.* 25 (2014) 113–123.
61. S. Kind, W.K. Jeong, H. Schröder, C. Wittmann, Systems-wide metabolic pathway engineering in Corynebacterium glutamicum for bio-based production of diaminopentane, *Metab. Eng.* 12 (2010) 341–351.
62. B. Wu, W. Szymanski, P. Wietzes, S. de Wildeman, G.J. Poelarends, B.L. Feringa, D.B. Janssen, Enzymatic synthesis of enantiopure α- and β-amino acids by phenylalanine aminomutase-catalysed amination of cinnamic acid derivatives, *ChemBioChem.* 10 (2009) 338–344.
63. Z. Liu, Z.-Y. Qin, L. Zhu, S. V Athavale, A. Sengupta, Z.-J. Jia, M. Garcia-Borràs, K.N. Houk, F.H. Arnold, An enzymatic platform for primary amination of 1-aryl-2-alkyl alkynes, *J. Am. Chem. Soc.* 144 (2022) 80–85.

3 Chemical Properties and Reactivity of Bio-Based Amines

3.1 INTRODUCTION

Nitrogen-containing compounds occupy a central position in all areas of chemistry. Among them, amines and amide derivatives, including amino acids and nucleotides, are the most fundamental molecular motifs selected by nature for the construction and sustenance of life. Beyond biological systems, amines are indispensable in modern society due to their widespread use in bulk and fine chemical industries. They serve as core structural units in agrochemicals, dyes, pharmaceuticals, natural products, and polymeric materials. A comprehensive treatment of their synthesis, properties, and applications underscores their importance across both academic research and industrial manufacturing [1]. Despite their broad utility, most commercially available amines are produced from depleting fossil resources. These materials are commonly manufactured through petrochemical routes such as nitrile hydrogenation, reductive amination, or ammonia alkylation, which often rely on harsh reaction conditions, multistep processes, and significant energy input. Such routes are frequently associated with high waste generation and environmental burden, raising serious concerns regarding their long-term sustainability [2,3].

As global demand for amines continues to increase, there is a growing imperative to replace these unsustainable practices with cleaner, more efficient, and environmentally responsible alternatives based on renewable resources. In response to these challenges, recent years have witnessed a rapid expansion in the development of bio-based amines, which are derived wholly or partially from biomass feedstocks such as amino acids, carbohydrates, fatty acids, lignin, terpenes, and other naturally occurring molecules. Figure 3.1 shows the overview of synthesis of green amines. These renewable amines are increasingly viewed not only as substitutes for petroleum-derived analogues but also as structurally distinctive compounds capable of offering new chemical functionalities. Extensive studies have demonstrated the feasibility of producing a wide range of nitrogen-containing chemicals from renewable carbon sources using catalytic and green chemistry approaches [4,5]. This transition represents a critical step toward reducing dependence on fossil resources while maintaining access to chemically versatile amine functionalities. A defining feature of bio-based amines is their structural complexity, which often reflects the molecular architecture of their biological precursors. In contrast to simple aliphatic amines, renewable amines frequently incorporate aromatic rings, heterocycles, stereogenic centers, and additional functional groups such as hydroxyls or ethers. These structural attributes exert a profound influence on their chemical properties and reactivity,

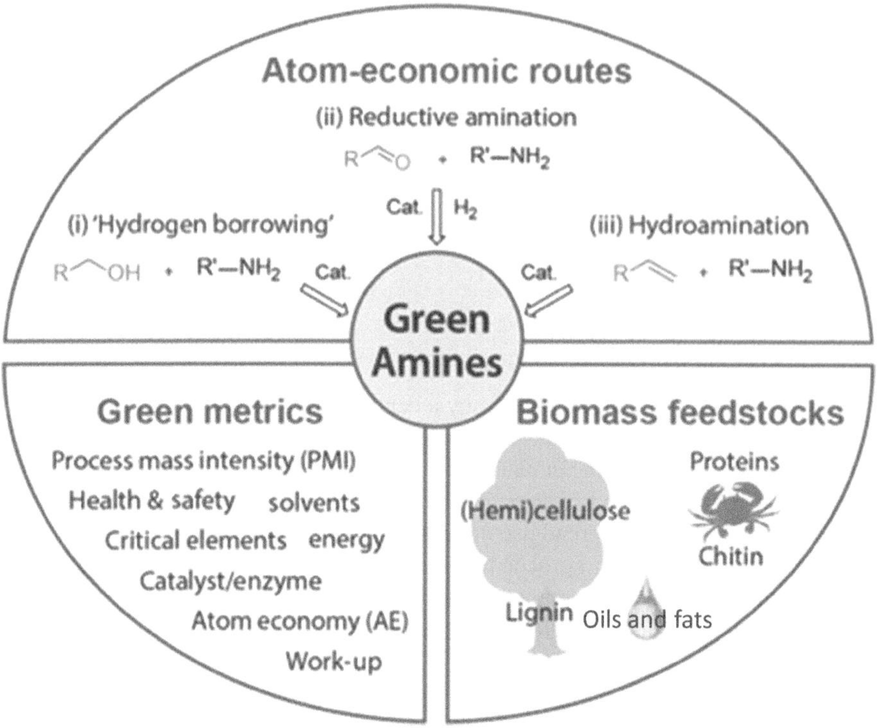

FIGURE 3.1 Synthesis of green amines. (Adapted with permission [4]. Copyright 2025 Royal Society of Chemistry. Licensed under CC BY 4.0. https://creativecommons.org/licenses/by/4.0/.)

affecting parameters such as basicity, nucleophilicity, hydrogen-bonding capability, solubility, and thermal stability. As a result, the behavior of bio-based amines cannot be reliably predicted by analogy to conventional petrochemical amines, necessitating a dedicated examination of their fundamental chemical characteristics.

From a classical standpoint, bio-based amines can still be categorized as primary, secondary, or tertiary amines. However, the electronic and steric environments surrounding the nitrogen atom are often markedly different. Electron-donating or electron-withdrawing substituents originating from biomass-derived frameworks can significantly alter electron density at the amine center, thereby modulating acid–base equilibria and nucleophilic strength. Comparative studies have shown that renewable amines may exhibit either enhanced or diminished reactivity depending on their specific molecular context, highlighting the inadequacy of simplified structure–property assumptions [6]. These findings underscore the need to treat bio-based amines as a chemically diverse and tunable family rather than as direct replacements for traditional amines. Among the intrinsic properties of amines, basicity plays a central role in governing their reactivity, salt formation, and interaction with other chemical species. In bio-based amines, basicity is influenced not only by substitution at the nitrogen atom but also by intramolecular hydrogen bonding, steric hindrance, and resonance interactions arising from adjacent functional groups. These factors can substantially affect proton affinity and buffering behavior, with important implications for reaction kinetics, catalyst compatibility, and long-term material stability.

Foundational analyses of structure–basicity relationships in amines provide critical insights that are particularly valuable when extended to multifunctional, renewable systems [7]. Closely linked to basicity is nucleophilicity, which determines the reactivity of amines toward electrophilic species such as epoxides, isocyanates, cyclic carbonates, aldehydes, and activated esters. This property is especially important in polymer chemistry, where amines function as curing agents, chain extenders, and crosslinkers. Bio-based amines may exhibit altered nucleophilic behavior due to steric congestion or electronic delocalization, leading to modified reaction rates or selectivity. Understanding these effects is essential for optimizing reaction conditions, particularly under mild, solvent-free, or low-energy processes that align with sustainable manufacturing principles. Hydrogen bonding and intermolecular interactions further distinguish bio-based amines from their conventional counterparts. The frequent presence of hydroxyl, ether, or aromatic functionalities enables the formation of extensive intra- and intermolecular hydrogen-bond networks. These interactions influence boiling points, viscosity, miscibility, and molecular organization in both solution and solid states. Importantly, hydrogen bonding can either enhance or suppress chemical reactivity by stabilizing reactive conformations or shielding the amine functionality, thereby affecting both processing behavior and final material performance [6].

Solubility and miscibility are equally critical considerations for the industrial deployment of renewable amines. Biomass-derived structures often introduce increased polarity, which can enhance water solubility and favor aqueous or low-volatile organic compounds (VOCs) formulations. However, excessive polarity may reduce compatibility with hydrophobic matrices, requiring careful molecular design or formulation strategies. The balance between hydrophilicity and hydrophobicity, governed by functional group distribution and molecular architecture, plays a decisive role in applications such as coatings, adhesives, and polymer blends. Thermal and oxidative stability represent additional factors that influence the practical applicability of bio-based amines. Certain renewable structures may contain thermally labile or oxidation-sensitive moieties that limit their use under demanding conditions. Nevertheless, recent studies have shown that appropriately designed bio-based amines, particularly those incorporating aromatic or heterocyclic units, can achieve thermal stability comparable to or even exceeding that of fossil-derived analogues [8].

These findings challenge early assumptions regarding the performance limitations of renewable amines. The importance of these chemical properties becomes especially evident in the synthesis of polymers and functional materials. Bio-based amines are increasingly employed in polyurethanes, non-isocyanate polyurethanes, epoxy resins, polyamides, and dynamic covalent networks. In such systems, small variations in amine structure, basicity, or hydrogen-bonding capacity can profoundly influence polymer architecture, mechanical strength, thermal behavior, and durability. Consequently, a fundamental understanding of amine chemistry is essential for the rational design of high-performance, sustainable polymeric materials. Beyond intrinsic molecular characteristics, the behavior of bio-based amines under different reaction conditions is of practical significance. Reaction temperature, pH, and the presence of catalysts can markedly influence reactivity and selectivity. Renewable

3.2.1 Primary, Secondary and Tertiary Amines

As with conventional amines, bio-based amines can be categorized into primary, secondary, and tertiary classes according to the number of carbon substituents bonded to the nitrogen atom (Figure 3.3). This classification remains chemically relevant, as it dictates the number of available N–H bonds, steric environment, and intrinsic nucleophilicity. However, in renewable systems, this classification must be interpreted in conjunction with the broader molecular framework. Primary bio-based amines are commonly obtained from amino acids, fatty acid derivatives, and carbohydrate-derived intermediates through decarboxylation, reductive amination, or catalytic hydrogenation routes. These amines possess one organic substituent and one or two N–H bonds, enabling strong hydrogen-bond donation and high nucleophilicity. In multifunctional bio-based structures, primary amines often coexist with polar groups such as hydroxyls or ethers, which can participate in intramolecular hydrogen bonding. Such interactions may partially shield the nitrogen lone pair, reducing its effective reactivity despite the formal classification as a primary amine [17,18]. Secondary bio-based amines typically arise from further functionalization of primary renewable amines or from reductive amination of bio-derived aldehydes and ketones. These amines often exhibit enhanced basicity relative to primary analogues due to increased electron donation from alkyl substituents. However, steric hindrance and conformational constraints imposed by cyclic or aromatic biomass-derived frameworks can significantly influence accessibility to the nitrogen center. In many cases, secondary bio-based amines display reactivity profiles that deviate from classical trends, particularly in condensed phases or polymerizing systems. Tertiary bio-based amines are frequently encountered in lignin-derived aromatics, terpene-based scaffolds, and alkaloid-inspired structures. The absence of N–H bonds eliminates hydrogen-bond donation but preserves basicity and coordination ability. In renewable architectures, tertiary amines often engage in intramolecular electronic interactions with aromatic or heterocyclic moieties, leading to delocalization of the nitrogen lone pair. This delocalization can suppress nucleophilicity while maintaining catalytic or proton-scavenging functionality. Such behavior is particularly relevant when tertiary bio-based amines are employed as catalysts or accelerators in polymer synthesis [5]. Overall, while the primary–secondary–tertiary classification remains a useful starting point, the functional performance of bio-based amines is ultimately dictated by the broader molecular context rather than nitrogen substitution alone [19,20].

3.2.2 Unique Functional Groups and Derivatives

A distinguishing characteristic of bio-based amines is the frequent integration of nitrogen functionality with additional polar, aromatic, or heterocyclic groups within

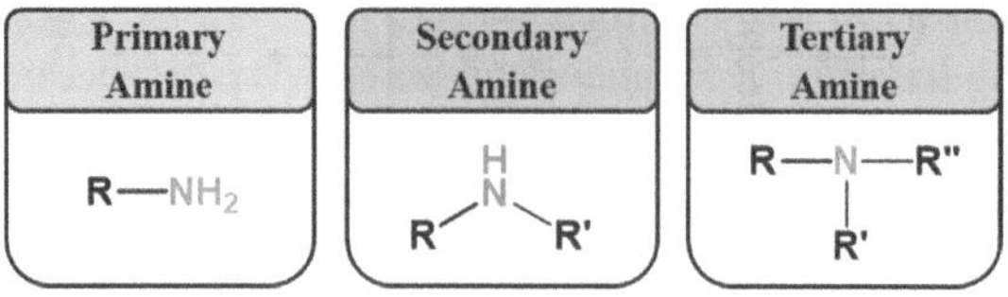

FIGURE 3.3 Classification of amines.

the same molecule. This multifunctionality arises directly from biomass precursors and represents a fundamental departure from the structural simplicity of many petrochemical amines. Common motifs include amino alcohols, aromatic amines derived from lignin fragments, furan-containing amines from carbohydrate platforms, and polyfunctional amines bearing ether or ester linkages. The presence of hydroxyl groups adjacent to amine functionalities is particularly impactful. Such arrangements promote strong intra- and intermolecular hydrogen bonding, which can stabilize specific conformations and reduce the effective availability of the nitrogen lone pair. As a result, these amines may exhibit lower apparent basicity or reduced nucleophilicity compared to structurally simpler analogues. Detailed investigations have demonstrated that hydrogen-bond networks in bio-based amines play a decisive role in controlling viscosity, miscibility, and reaction kinetics, especially in condensed phases and polymer formulations [6]. Aromatic and heterocyclic substituents further complicate the electronic environment of bio-based amines. Conjugation between the nitrogen lone pair and adjacent aromatic systems can delocalize electron density, altering both acid–base behavior and electrophilic reactivity. In lignin-derived or furan-based amines, such delocalization often leads to reduced nucleophilicity but enhanced thermal stability and compatibility with aromatic polymer matrices. These effects are particularly advantageous in epoxy, polyurethane, and polyamide systems, where aromatic content contributes to mechanical strength and thermal resistance. Beyond their intrinsic structures, bio-based amines also serve as versatile precursors for a wide range of nitrogen-containing derivatives, including amides, ureas, carbamates, imines, and quaternary ammonium salts. The multifunctional nature of many renewable amines enables simultaneous participation in covalent bonding and non-covalent interactions, promoting network formation and interfacial adhesion. Such behavior is rarely achievable with mono-functional petrochemical amines and represents a key advantage of bio-based systems. In summary, the chemical structure of bio-based amines is defined not only by nitrogen substitution but by the synergistic influence of biomass-derived functional groups. This structural complexity governs their physicochemical properties and underpins their unique reactivity patterns. A rigorous understanding of these structure–function relationships is essential for rationalizing the behavior of bio-based amines in chemical synthesis and forms the foundation for the subsequent discussion of their basicity, nucleophilicity, intermolecular interactions, and performance in polymeric materials [21,22]. Table 3.1 highlights how specific structural features in bio-based amines govern their electronic environment and steric profile, thereby modulating reactivity. Electron-donating groups, conjugation, hydrogen bonding, and backbone rigidity collectively influence basicity, nucleophilicity, selectivity, and stability. Understanding these structure–property relationships enables rational molecular design for targeted applications in polymer synthesis and catalysis.

3.3 BASICITY OF BIO-BASED AMINES

Basicity is one of the most fundamental physicochemical properties of amines and governs their behavior in acid–base equilibria, salt formation, catalysis, coordination chemistry, and nucleophilic reactions. In practical applications, amine basicity directly

TABLE 3.1
Structure–Property Relationships in Bio-Based Amines [5–7,9]

Functional Group/ Structural Feature	Dominant Electronic or Steric Effect	Impact on Amine Reactivity
Aliphatic alkyl substituents	+I electron donation	Increases basicity and nucleophilicity relative to unsubstituted amines
Hydroxyl group (–OH) adjacent to amine	Intramolecular hydrogen bonding	Reduces effective nucleophilicity; may lower apparent basicity due to lone-pair shielding
Ether linkage (–O–)	Weak electron donation, conformational flexibility	Moderate modulation of basicity; enhances solubility and processability
Aromatic ring (lignin-derived)	π-conjugation and resonance delocalization	Decreases nucleophilicity; improves thermal stability and aromatic compatibility
Heterocyclic framework (furan, pyridine-like)	Lone-pair delocalization and rigidity	Alters acid–base behavior; can enhance selectivity in catalytic systems
Cyclic or rigid backbone	Steric hindrance	Reduces reaction rate with bulky electrophiles; improves dimensional stability
Multiple functional groups (multifunctional amines)	Cooperative electronic and steric effects	Enables network formation, strong intermolecular interactions, and enhanced adhesion
Absence of N–H bonds (tertiary amines)	No hydrogen-bond donation	Reduced nucleophilicity but retained basicity; useful as catalysts or proton scavengers

affects reaction kinetics, selectivity, catalyst efficiency, and compatibility with other formulation components. For bio-based amines, basicity acquires additional significance because renewable feedstocks frequently introduce multifunctional, conformationally constrained, or electronically complex molecular environments around the nitrogen center. Consequently, classical basicity trends derived from simple petrochemical amines cannot always be applied directly to bio-based analogues, necessitating a structure-specific and physically grounded analysis. From a thermodynamic standpoint, amine basicity reflects the ability of the nitrogen lone pair to accept a proton and is commonly quantified by the pK_a of the corresponding ammonium ion in a given medium. Proton exchange processes in amines are rapid and occur under thermodynamic control, such that kinetic effects are generally negligible in acid–base equilibria. These governing principles are well established in physical organic chemistry and provide the conceptual framework for interpreting basicity trends in both conventional and bio-based amines [23,24].

Electronic effects represent the primary intrinsic factor controlling amine basicity. Alkyl substituents increase basicity through inductive electron donation, which stabilizes the positively charged ammonium species formed upon protonation by increasing the availability of the nitrogen lone pair. This effect is well documented for simple aliphatic amines, where increased alkyl substitution generally correlates with enhanced proton affinity. In bio-based amines, alkyl substitution patterns are often inherited from carbohydrate, fatty acid, or terpene backbones, resulting in non-uniform electron donation across the molecule. While extended aliphatic segments may locally enhance basicity, their influence can be attenuated or counterbalanced by neighboring heteroatoms or polar functionalities. In contrast, electron-withdrawing substituents decrease amine basicity by reducing electron density at the nitrogen center via inductive or resonance effects. Aromatic amines represent a classical example, in which conjugation between the nitrogen lone pair and the aromatic π-system leads to significant delocalization and reduced basicity compared to aliphatic amines. This behavior is particularly relevant for lignin-derived aromatic amines, where resonance stabilization is inherent to the molecular structure. In aromatic systems, basicity is further modulated by substituents on the aromatic ring; electron-donating substituents decrease lone-pair participation in mesomeric structures and consequently increase basicity, whereas electron-withdrawing substituents have the opposite effect. These relationships between substituent electronics and amine basicity have been extensively analyzed in both experimental and theoretical studies and remain directly applicable to bio-based systems when structural context is properly considered [25,26]. Hydrogen bonding plays a decisive role in modulating the basicity of many bio-based amines. Renewable feedstocks frequently introduce hydroxyl, ether, or carbonyl groups near amine functionality, enabling intramolecular hydrogen bonding. Such interactions can stabilize specific molecular conformations and partially sequester the nitrogen lone pair, thereby reducing its availability for protonation. Amino alcohols and alkanolamines derived from sugars, glycerol, or amino acid platforms provide clear examples of this behavior. Despite the presence of electron-donating alkyl substituents, these compounds often exhibit lower apparent basicity than expected due to strong O–H···N hydrogen bonding. Experimental studies have shown that hydrogen bonding can significantly alter protonation equilibria, particularly in condensed phases where intramolecular interactions compete effectively with solvent stabilization [27]. The influence of hydrogen bonding on basicity is highly medium-dependent. In polar solvents, solvation of the protonated ammonium ion can partially offset intramolecular hydrogen bonding, whereas in less polar solvents or solvent-free systems, such interactions dominate. This behavior is especially relevant in polymer synthesis and bulk processing, where bio-based amines often operate outside dilute solution conditions [28].

Steric effects provide an additional layer of control over amine basicity by influencing both proton accessibility and stabilization of the protonated species. Increased substitution at nitrogen generally enhances electron donation, but excessive steric hindrance can impede protonation and destabilize the resulting ammonium ion. In tertiary amines, this balance often leads to lower apparent basicity than predicted by electronic considerations alone. Bio-based amines frequently incorporate rigid or cyclic frameworks derived from sugars, terpenes, or aromatic biomass components. Such conformational constraints can limit the ability of the molecule to adopt geometries favorable for protonation or solvation, further reducing basicity. These steric and conformational effects are less prominent in simple petrochemical amines but are common in renewable systems, making them an important consideration in structure–basicity relationships. Measured amine basicity is inherently dependent on the surrounding medium. In aqueous systems, strong solvation of the protonated ammonium ion typically enhances apparent basicity, particularly for small and polar amines. However, many bio-based amines exhibit limited water solubility or form extensive hydrogen-bond networks with solvent molecules, leading to deviations from expected pK_a values. In organic solvents, polymer melts, or solid matrices, solvation effects are significantly reduced and intrinsic molecular factors dominate. Under such conditions, bio-based amines containing multiple polar groups may exhibit lower effective basicity than simple aliphatic amines, even if solution-phase measurements suggest comparable proton affinity. These medium-dependent variations highlight the importance of evaluating basicity under conditions relevant to the intended application rather than relying solely on aqueous pK_a data. Broadly, primary amines tend to exhibit higher basicity than secondary and tertiary analogues when steric and hydrogen-bonding effects are minimal. However, bio-based amines frequently deviate from this trend due to multifunctionality and conformational constraints. Alkanolamines derived from carbohydrates often display reduced basicity because of intramolecular hydrogen bonding, whereas long-chain aliphatic amines derived from fatty acids may approach or exceed the basicity of conventional petrochemical amines due to strong inductive effects. Aromatic bio-based amines, particularly those derived from lignin, generally exhibit lower basicity due to resonance delocalization of the nitrogen lone pair. Heterocyclic bio-based amines show intermediate behavior that depends sensitively on ring electronics and substitution pattern [29].

Amine basicity has direct consequences for reactivity and performance in chemical and materials applications. In polymer synthesis, basicity influences curing kinetics, catalyst efficiency, and network formation in systems such as polyurethanes, non-isocyanate polyurethanes, epoxies, and polyamides. Excessively basic amines may lead to uncontrolled reaction rates or side reactions, while insufficient basicity can require elevated temperatures or additional catalysts. Beyond reactivity, basicity governs salt formation, corrosion behavior, and compatibility with acidic additives or fillers in formulated systems. The ability to tune basicity through molecular design by controlling electronic effects, hydrogen bonding, and steric environment represents a powerful strategy for optimizing bio-based amines for specific chemical and industrial applications. In summary, the basicity of bio-based amines arises from a complex interplay of electronic effects, hydrogen bonding, steric constraints, and medium-dependent interactions. While the underlying principles are rooted in classical amine chemistry, the multifunctional and rigid architectures characteristic of renewable feedstock frequently lead to non-classical behavior. A rigorous understanding of these factors is therefore essential for predicting reactivity and enabling the rational design of next-generation bio-based amines for sustainable chemical processes and advanced material systems.

3.4 NUCLEOPHILICITY AND REACTIVITY

Nucleophilicity is a kinetic property that describes the ability of a chemical species to donate an electron pair to an electrophilic center during the rate-determining step of a reaction. Unlike basicity, which is defined thermodynamically through equilibrium proton transfer, nucleophilicity is governed by reaction rates and therefore depends strongly on molecular structure, steric accessibility, solvent environment, and the nature of the electrophile. As a result, nucleophilicity and basicity are related but fundamentally distinct properties, and correlations between them are often system-dependent rather than universal. This distinction is particularly important for bio-based amines, which frequently possess multifunctional architectures, hydrogen-bonding groups, and rigid or heterocyclic frameworks inherited from renewable feedstocks. Such structural features can significantly modify nucleophilic behavior, often leading to reactivity trends that deviate from classical expectations based solely on pK_a values.

A rigorous and widely accepted quantitative framework for describing nucleophilicity is provided by the Mayr nucleophilicity scale, which relates reaction rate constants to intrinsic nucleophilicity parameters and electrophile strength according to the equation:

$$\log K = s\left(N + E\right)$$

where N is the nucleophilicity parameter of the nucleophile, E is the electrophilicity parameter of the electrophile, and s is a sensitivity factor reflecting the response of the nucleophile to changes in electrophile strength [30]. This formalism emphasizes that nucleophilicity is an intrinsic kinetic property that must be evaluated within a defined reaction family and solvent system. Systematic kinetic studies on primary and secondary amines using this framework have demonstrated that nucleophilicity varies over several orders of magnitude depending on substitution pattern, hydrogen-bonding ability, and solvent effects [31]. Importantly, these studies show that amines with similar basicities can display markedly different nucleophilic reactivities, reinforcing the need to treat nucleophilicity independently from acid–base behavior.

Several experimental studies have directly examined the relationship between amine nucleophilicity and basicity and have consistently shown that simple correlations between pK_a and reaction rate constants are unreliable. Kinetic investigations of nucleophilic addition reactions reveal that basicity alone cannot predict nucleophilic reactivity, particularly when steric effects, solvation, or hydrogen bonding are significant [31]. A quantitative analysis of alkylamine reactivities demonstrated that nucleophilicity correlates with basicity only after correcting for steric hindrance and solvation effects using extended linear free-energy relationships [32]. Even in such cases, the increase in nucleophilicity with increasing basicity is substantially attenuated, indicating that proton affinity contributes only partially to overall nucleophilic strength. These findings are especially relevant for bio-based amines, where bulky substituents and polar functional groups are common.

Electronic factors influence nucleophilicity through modulation of lone-pair energy and availability, but steric effects often dominate the reactivity of amines toward electrophilic centers. Kinetic studies of nucleophilic addition to activated olefins and nitriles show that increased steric congestion near the nitrogen atom leads to pronounced reductions in reaction rates, even when electronic donation is favorable. Comparative studies of primary, secondary, and tertiary amines reveal that secondary amines are frequently more nucleophilic than primary amines of similar basicity due to reduced solvation penalties, whereas tertiary amines often appear weakly nucleophilic because steric shielding hinders effective orbital overlap in the transition state [31]. These trends are highly relevant to bio-based amines derived from sugars, terpenes, or lignin, which often feature rigid scaffolds and substituted nitrogen centers. Hydrogen bonding plays a decisive role in controlling the nucleophilicity of many amines. Amines bearing hydroxyl or ether groups can form strong intra- or intermolecular hydrogen bonds that stabilize ground states and reduce the effective nucleophilic participation of the nitrogen lone pair. Kinetic studies have shown that hydroxyl-substituted amines, such as diethanolamine and related structures, are significantly less nucleophilic than predicted from their basicity. Experimental investigations using model coordination systems have further demonstrated that nucleophilicity is strongly influenced by steric accessibility and hydrogen-bonding capability rather than proton affinity alone [33]. These findings provide a mechanistic explanation for the often-reduced reactivity of multifunctional bio-based amines in substitution and addition reactions, despite their moderate to high basicity. Nucleophilicity is inherently medium-dependent, and solvent effects can dramatically alter reactivity trends. Studies conducted in aprotic solvents such as acetonitrile show significantly higher nucleophilicity for alkyl amines compared to protic or highly hydrogen-bonding environments [31]. Conversely, in aqueous or strongly polar solvents, extensive solvation suppresses nucleophilic reactivity, particularly for small or highly polar amines. These solvent effects are especially important when considering bio-based amines in practical applications such as polymer synthesis, where reactions often occur in bulk, in low-polarity media, or under solvent-free conditions. Under such environments, intrinsic steric and electronic factors dominate nucleophilicity, leading to reactivity patterns that differ substantially from aqueous pK_a-based expectations. Detailed kinetic analyses of nucleophilic addition reactions have shown that the identity of the rate-determining step can vary with amine structure. For primary and secondary amines, nucleophilic attack on the electrophile is often rate-limiting, whereas for tertiary amines the rate-determining step may shift to proton transfer or intermediate stabilization. Such mechanistic shifts explain why tertiary amines can display unexpectedly low nucleophilic reactivity even when they are strongly basic. These mechanistic insights are critical for understanding and predicting the behavior of bio-based amines in reactions involving activated carbon–carbon multiple bonds, epoxides, isocyanates, or cyclic carbonates, where subtle changes in structure can alter the reaction pathway. The collective evidence from kinetic, thermodynamic, and mechanistic studies demonstrates that nucleophilicity in amines is a multifactorial property governed by electronic effects, steric accessibility, hydrogen bonding, and medium effects. For bio-based amines, multifunctionality and conformational constraints frequently lead to suppressed or non-classical nucleophilic behavior, even when basicity is moderate or high. Understanding these relationships is essential for the rational design of bio-based amines for use in polymer synthesis, catalysis, and functional materials. By controlling steric environment, hydrogen-bonding capability, and reaction medium, it is possible to tailor nucleophilicity independently of basicity, thereby optimizing reactivity without compromising stability or selectivity. The comparison of the nucleophilic reactivities of amines in acetonitrile with other nucleophiles is given in Figure 3.4 [31].

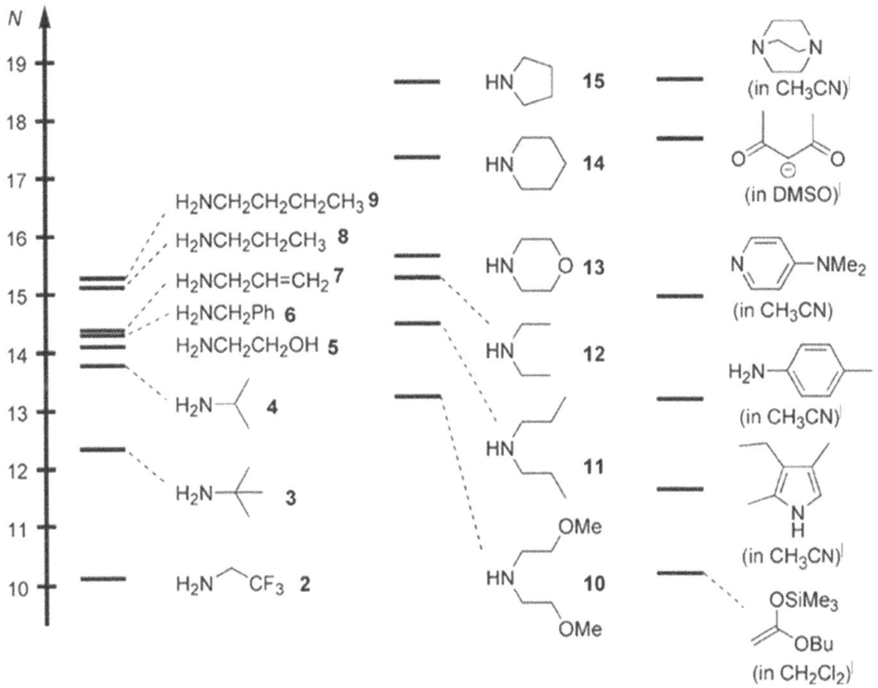

FIGURE 3.4 Comparison of the nucleophilic reactivities of amines in acetonitrile with other nucleophiles. (Adapted with permission [31]. Copyright 2009 Wiley.)

3.5 HYDROGEN BONDING AND INTERMOLECULAR INTERACTION

Hydrogen bonding and related intermolecular interactions play a central role in determining the physicochemical behavior of bio-based amines, particularly under concentrated, bulk, or solvent-free conditions relevant to industrial processing and materials applications. While hydrogen bonding has already been discussed in earlier sections in the context of basicity and nucleophilicity, its broader influence on molecular association, phase behavior, transport properties, and material performance warrants a distinct and focused discussion. Bio-based amines frequently possess multifunctional architectures in which the amine group coexists with hydroxyl, ether, carbonyl, or heterocyclic functionalities derived from renewable feedstocks such as sugars, amino acids, lignin, and glycerol. This structural complexity enables both intramolecular hydrogen bonding, which primarily affects reactivity, and intermolecular hydrogen bonding, which governs collective behavior in the condensed phase. The balance between these two modes of interaction is a defining characteristic of many bio-based amines and distinguishes them from simpler petrochemical analogues [6]. Intermolecular hydrogen bonding between amine molecules can lead to significant self-association, particularly in primary and secondary amines that act as both hydrogen-bond donors and acceptors. In multifunctional bio-based systems, this effect is amplified by additional donor–acceptor sites, resulting in extended hydrogen-bond networks. Such association increases boiling points, viscosity, and cohesive energy density, while simultaneously reducing volatility and diffusivity. These effects are frequently observed in alkanolamines and sugar-derived amines, where extensive hydrogen bonding contributes to low vapor pressure and enhanced thermal stability but can also limit mass transport and reaction rates in bulk systems [6]. Hydrogen bonding also plays a critical role in controlling solubility and miscibility. Bio-based amines with multiple polar functionalities often display enhanced affinity for water and other polar solvents due to favorable hydrogen-bond interactions, which can be advantageous for aqueous or low-VOC formulations. Conversely, strong intermolecular association can reduce compatibility with nonpolar matrices, leading to phase separation or incomplete dispersion in hydrophobic polymer systems. The resulting balance between hydrophilicity and self-association must therefore be carefully managed through molecular design or formulation strategies [5]. Beyond amine–amine interactions, hydrogen bonding between bio-based amines and surrounding species such as solvents, polymers, fillers, or substrates strongly influences macroscopic performance. In polymer systems, secondary hydrogen bonding between amine groups and carbonyls, hydroxyls, or heteroatoms in the polymer backbone can act as a reversible physical crosslink, enhancing adhesion, toughness, and cohesive strength without permanent covalent bonding. Such interactions are particularly important in coatings, adhesives, and composite materials, where bio-based amines are increasingly employed as curing agents, chain extenders, or reactive modifiers [9]. The competitive nature of hydrogen-bonding interactions must also be considered. In many bio-based amines, intramolecular hydrogen bonding stabilizes specific conformations, while intermolecular hydrogen bonding promotes aggregation. Changes in temperature, concentration, or medium polarity can shift this balance, leading to pronounced changes in viscosity, solubility, and effective reactivity. These effects are especially relevant in solvent-free or high-solids processing, where hydrogen bonding is not disrupted by dilution and therefore exerts a dominant influence on system behavior [31]. From a materials perspective, hydrogen bonding and intermolecular interactions contribute significantly to the sustainability-related advantages of bio-based amines. Reduced volatility and enhanced adhesion can improve workplace safety and durability, while reversible hydrogen-bond networks may enable self-healing, recyclability, or reprocessability in advanced polymer systems. At the same time, excessive intermolecular association can pose challenges for processing and scale-up, underscoring the need for careful structure–property optimization. In summary, hydrogen bonding and intermolecular interactions in bio-based amines extend far beyond their influence on basicity and nucleophilicity. They govern self-association, phase behavior, solubility, transport properties, and material performance, particularly in concentrated and solid-state environments. A clear understanding of these interactions is therefore essential for translating the molecular advantages of bio-based amines into reliable and scalable chemical and materials technologies.

3.6 SOLUBILITY AND MISCIBILITY

Solubility and miscibility are critical physicochemical properties that govern the processability, formulation flexibility, and end-use performance of bio-based amines. Unlike basicity or nucleophilicity, which primarily influence chemical reactivity, solubility and miscibility determine how effectively amines can be incorporated into solvents, polymer matrices, coatings, adhesives, and composite systems. For bio-based amines, these properties are strongly influenced by multifunctional molecular architectures and extensive intermolecular interactions inherited from renewable feedstocks. From a thermodynamic perspective, solubility is dictated by the balance between cohesive forces within the solute and favorable interactions between solute and solvent. Classical solubility concepts based on cohesive energy density and solubility parameters remain valuable tools for rationalizing amine behavior. The Hildebrand and Hansen solubility parameter frameworks provide a quantitative description of dispersive, polar, and hydrogen-bonding contributions to solubility, offering insight into why polar and multifunctional bio-based amines often show high affinity for protic and polar aprotic solvents but limited compatibility with nonpolar media [34].

Bio-based amines frequently exhibit enhanced polarity due to the presence of hydroxyl, ether, or heterocyclic groups, which increases their solubility in water, alcohols, and polar organic solvents. While this behavior can be advantageous for aqueous processing and low-VOC formulations, it may hinder miscibility with hydrophobic polymers such as polyolefins or nonpolar elastomers. In such cases, unfavorable enthalpic interactions and strong amine–amine association can lead to phase separation or incomplete dispersion, limiting material homogeneity. Solvent–solute interactions in amine systems are often better described using multiparameter approaches that explicitly account for hydrogen bonding and dipolarity. The Kamlet–Taft solvatochromic parameters (α, β, π^*) provide a useful framework for understanding how hydrogen-bond donating and accepting abilities of both solvent and amine influence solubility and miscibility [35]. These parameters are particularly relevant for bio-based amines, where strong hydrogen-bond acceptor and donor capabilities can dominate solvation behavior. In recent years, predictive models such as COSMO-RS have gained increasing attention for estimating solubility and phase behavior of multifunctional organic compounds. COSMO-RS combines quantum chemical calculations with statistical thermodynamics to predict activity coefficients and miscibility without extensive experimental data. Such approaches have proven effective for polar and hydrogen-bonding compounds, including amines, and offer valuable guidance for solvent selection and formulation design involving bio-based amines [36,37].

Miscibility considerations extend beyond simple solvent systems to polymer–amine interactions. In reactive polymer formulations, partial miscibility can be sufficient if diffusion and reaction occur on comparable timescales, whereas in nonreactive blends, poor miscibility may result in long-term phase instability. Understanding solubility and miscibility through molecular descriptors therefore enables informed selection and design of bio-based amines tailored for specific processing environments. In summary, solubility and miscibility of bio-based amines are governed by a complex interplay of polarity, hydrogen bonding, and cohesive energy contributions. Classical solubility theories, complemented by modern multiparameter and computational approaches, provide a robust framework for predicting and tuning these properties. Such understanding is essential for translating the chemical sustainability of bio-based amines into practical, scalable materials and formulations.

environments. Under acidic conditions, protonation of the nitrogen lone pair suppresses nucleophilicity and limits participation in substitution or addition reactions. In contrast, basic conditions favor the free amine form and enhance nucleophilic reactivity but may also promote competing base-catalyzed side reactions. For bio-based amines, pH effects are often more complex due to the presence of multiple ionizable or hydrogen-bonding groups. Amines derived from amino acids or carbohydrates may exist in zwitterionic or multi-protonated forms over specific pH ranges, leading to nonlinear reactivity profiles. Buffer composition and ionic strength further modulate reactivity by altering solvation and ion pairing. Studies on amine reactivity in buffered systems have demonstrated that pH-dependent speciation, rather than intrinsic basicity alone, governs reaction outcomes. These considerations are particularly important for bio-based amines used in aqueous synthesis, biomedical materials, or environmentally benign processing routes [43].

3.8.3 Catalyst Effect

Catalysts play a crucial role in controlling the reactivity of amines by lowering activation barriers, stabilizing transition states, or activating reaction partners. In reactions involving bio-based amines, catalytic effects can be especially pronounced because multifunctional structures may interact strongly with catalyst surfaces or coordination sites. Both homogeneous and heterogeneous catalysts have been shown to influence amine reactivity through acid–base interactions, hydrogen bonding, or metal–nitrogen coordination. Acidic catalysts, such as Brønsted or Lewis acids, can activate electrophiles toward nucleophilic attack by amines while simultaneously modulating amine basicity through reversible coordination or protonation. Conversely, basic catalysts may enhance deprotonation steps or promote amine activation in reactions involving weak electrophiles. In heterogeneous systems, catalyst surface polarity and functionalization strongly affect adsorption strength and orientation of bio-based amines, thereby influencing reaction rates and selectivity. Recent studies have highlighted that catalyst choice can mitigate limitations associated with reduced nucleophilicity or steric hindrance in bio-based amines. By tailoring catalyst acidity, coordination environment, or surface chemistry, it is possible to selectively enhance desired reaction pathways while suppressing degradation or side reactions. Such strategies are increasingly important for sustainable chemical processes where mild conditions and high selectivity are required [44,45].

3.9 RECENT RESEARCH DEVELOPMENTS AND INDUSTRIAL PERSPECTIVES

Recent advances in bio-based amines reflect a clear shift from proof-of-concept synthesis toward scalable, application-driven technologies. Comprehensive analyses highlight how renewable feedstocks such as sugars, lignin fragments, fatty acids, and amino acid derivatives are now being converted into amines through increasingly efficient catalytic routes. Modern hydrogenation, reductive amination, and borrowing-hydrogen strategies have matured significantly, enabling higher selectivity, broader substrate scope, and reduced waste generation compared with traditional petrochemical pathways. These developments underscore a transition from laboratory novelty to process-relevant chemistry aligned with green chemistry principles. From a research standpoint, a notable trend is the move toward structure-tailored bio-based amines rather than simple fossil-derived replacements. Recent studies emphasize designing amines that exploit the intrinsic functionality of biomass, such as aromaticity from lignin or rigidity from carbohydrate backbones, to deliver performance advantages. This approach allows bio-based amines to serve not only as sustainable alternatives but also as value-added building blocks with tunable reactivity, hydrogen-bonding capacity, and thermal behavior. The integration of catalysis with feedstock design has therefore become a central research focus, enabling better control over amine structure–property relationships [4]. Industrial perspectives are increasingly shaped by the successful incorporation of bio-based amines into polymer and materials markets. Recent polymer-focused studies demonstrate that renewable amines can function effectively as curing agents, chain extenders, and reactive modifiers in polyurethanes, epoxies, and polyamides, often achieving mechanical and thermal performance comparable to or exceeding that of petrochemical counterparts. Importantly, these materials benefit from reduced volatility, improved adhesion, and enhanced polarity control, which are attractive for coatings, adhesives, and composite applications. Despite these advances, industrial adoption remains influenced by practical considerations such as feedstock availability, cost competitiveness, and process compatibility with existing infrastructure. Current research increasingly addresses these challenges by emphasizing drop-in compatibility, catalyst recyclability, and continuous processing. The polymers literature highlights that bio-based amines are most readily adopted when they integrate seamlessly into established formulation windows without requiring extensive process redesign [9]. Looking forward, the convergence of catalytic innovation and materials performance positions bio-based amines as key contributors to sustainable chemical manufacturing. Ongoing research efforts aimed at improving efficiency, consistency, and lifecycle performance are expected to further accelerate industrial uptake. Together, recent developments demonstrate that bio-based amines have progressed beyond niche applications and are now emerging as viable components of mainstream industrial chemistry.

3.10 CONCLUSION

To conclude, the chemical properties and reactivity of bio-based amines from a structure–property–application perspective, highlighting both their similarities to and distinctions from conventional petrochemical amines. By integrating fundamental concepts such as basicity, nucleophilicity, hydrogen bonding, and solubility with the unique structural features imparted by renewable feedstocks, it becomes clear that bio-based amines represent a chemically diverse and highly tunable class of compounds. Multifunctionality, conformational constraints, and strong intermolecular interactions frequently lead to non-classical behavior, necessitating careful molecular-level analysis rather than direct analogy to fossil-derived systems. The discussion of stability and reaction-condition dependence further underscores the importance of understanding how temperature, pH, and catalytic environments influence reactivity and long-term performance. From an applied standpoint, the successful integration of bio-based amines into polymer systems and functional materials demonstrates their growing industrial relevance, while also revealing remaining challenges related to processing and consistency. Continued advances in catalysis, molecular design, and predictive modeling are expected to further enhance the viability of bio-based amines. Collectively, the insights presented in this chapter provide a foundation for the rational development of next-generation sustainable amine-based chemistries.

REFERENCES

1. S.A. Lawrence, *Amines: Synthesis, Properties and Applications*, Cambridge University Press, 2004.
2. F. Roschangar, R.A. Sheldon, C.H. Senanayake, Overcoming barriers to green chemistry in the pharmaceutical industry – The Green Aspiration Level™ concept, *Green Chem.* 17 (2015) 752–768.
3. P.G. Levi, J.M. Cullen, Mapping global flows of chemicals: From fossil fuel feedstocks to chemical products, *Environ. Sci. Technol.* 52 (2018) 1725–1734.
4. A.M. Afanasenko, N. Deak, J. October, R. Sole, K. Barta, 'Green' synthesis of amines from renewable resources? A detailed analysis of case studies using the CHEM21 green metrics toolkit, *Green Chem.* 27 (2025) 5947–5981.
5. M. Pelckmans, T. Renders, S. Van de Vyver, B.F. Sels, Bio-based amines through sustainable heterogeneous catalysis, *Green Chem.* 19 (2017) 5303–5331.
6. A.-S. Mora, R. Tayouo, B. Boutevin, G. David, S. Caillol, Synthesis of pluri-functional amine hardeners from bio-based aromatic aldehydes for epoxy amine thermosets, *Molecules.* 24 (2019) 3285.
7. V. Froidevaux, C. Negrell, S. Caillol, J.P. Pascault, B. Boutevin, Biobased amines: From synthesis to polymers; present and future, *Chem. Rev.* 116 (2016) 14181–14224.
8. T.N. Borhani, M. Short, Prediction of thermal and oxidative degradation of amines to improve sustainability of CO_2 absorption process, *Sustainability.* 17 (2025) 10311.
9. M. Diaz-Galbarriatu, J. Sánchez-Bodón, E. Hernáez-Laviña, J.L. Vilas-Vilela, I. Moreno-Benítez, Biobased polyamides: A journey from the biomass towards cutting edge materials, *Polymers (Basel).* 17 (2025) 2599.
10. O.D.Y.H. Lamborn, Economics of sucrose, *Ind. Eng. Chem.* 47 (1955) 1392–1397.
11. Z. Cao, J. Engelhardt, M. Dierks, M.T. Clough, G. Wang, E. Heracleous, A. Lappas, R. Rinaldi, F. Schüth, Catalysis meets nonthermal separation for the production of (alkyl) phenols and hydrocarbons from pyrolysis oil, *Angew. Chem.* 129 (2017) 2374–2379.
12. F. Ma, M.A. Hanna, Biodiesel production: A review, *Bioresour. Technol.* 70 (1999) 1–15.
13. J.M. Marchetti, V.U. Miguel, A.F. Errazu, Possible methods for biodiesel production, *Renew. Sustain. Energy Rev.* 11 (2007) 1300–1311.
14. K.W. Omari, L. Dodot, F.M. Kerton, A simple one-pot dehydration process to convert N-acetyl-D-glucosamine into a nitrogen-containing compound, 3-acetamido-5-acetylfuran, *ChemSusChem.* 5 (2012) 1767–1772.
15. Y. Teng, E.L. Scott, S.C.M. Witte-van Dijk, J.P.M. Sanders, Simultaneous and selective decarboxylation of l-serine and deamination of l-phenylalanine in an amino acid mixture—A means of separating amino acids for synthesizing biobased chemicals, *N. Biotechnol.* 33 (2016) 171–178.
16. L. Claes, R. Matthessen, I. Rombouts, I. Stassen, T. De Baerdemaeker, D. Depla, J.A. Delcour, B. Lagrain, D.E. De Vos, Bio-based nitriles from the heterogeneously catalyzed oxidative decarboxylation of amino acids, *ChemSusChem.* 8 (2015) 345–352.

17. A.F. Abdel-Magid, K.G. Carson, B.D. Harris, C.A. Maryanoff, R.D. Shah, Reductive amination of aldehydes and ketones with sodium triacetoxyborohydride. Studies on direct and indirect reductive amination procedures, *J. Org. Chem.* 61 (1996) 3849–3862.
18. A.S. Touchy, S.M.A. Hakim Siddiki, K. Kon, K. Shimizu, Heterogeneous Pt catalysts for reductive amination of levulinic acid to pyrrolidones, *ACS Catal.* 4 (2014) 3045–3050.
19. D.M. Roundhill, Transition metal and enzyme catalyzed reactions involving reactions with ammonia and amines, *Chem. Rev.* 92 (1992) 27.
20. M. Pera-Titus, F. Shi, Catalytic amination of biomass-based alcohols, *ChemSusChem.* 7 (2014) 720–722.
21. R. Lan, J.T.S. Irvine, S. Tao, Synthesis of ammonia directly from air and water at ambient temperature and pressure, *Sci. Rep.* 3 (2013) 1145.
22. J. Verduyckt, D.E. De Vos, Controlled defunctionalisation of biobased organic acids, *Chem. Commun.* 53 (2017) 5682–5693.
23. A.R. Katritzky, C.A. Ramsden, J.A. Joule, V.V Zhdankin, *Handbook of Heterocyclic Chemistry*, 3rd Ed., Elsevier, Amsterdam, 2010.
24. A.T. Balaban, D.C. Oniciu, A.R. Katritzky, Aromaticity as a cornerstone of heterocyclic chemistry, *Chem. Rev.* 104 (2004) 2777–2812.
25. P. Love, R.B. Cohen, R.W. Taft, Polar substituent effects in gas-phase Lewis acid-base equilibriums. I. Intrinsic basicity of amines, *J. Am. Chem. Soc.* I (1968) 2455–2462.
26. A.D. Headley, Quantitative analysis of solvation effects and the influence of alkyl substituents on the basicity of amines, *J. Org. Chem.* 56 (1991) 3688–3691.
27. M.H. Abraham, Hydrogen bonding. 31. Construction of a scale of solute effective or summation hydrogen-bond basicity, *J. Phys. Org. Chem.* 6 (1993) 660–684.
28. C. Reichardt, T. Welton, *Solvents and Solvent Effects in Organic Chemistry*, John Wiley & Sons, 2011.
29. I. Dzidic, Relative gas-phase basicities of some amines, anilines, and pyridines. Application of some Broensted acids as reactants in chemical ionization mass spectrometry, *J. Am. Chem. Soc.* 94 (1972) 8333–8335.
30. H. Mayr, T. Bug, M.F. Gotta, N. Hering, B. Irrgang, B. Janker, B. Kempf, R. Loos, A.R. Ofial, G. Remennikov, H. Schimmel, Reference scales for the characterization of cationic electrophiles and neutral nucleophiles, *J. Am. Chem. Soc.* 123 (2001) 9500–9512.
31. T. Kanzian, T.A. Nigst, A. Maier, S. Pichl, H. Mayr, Nucleophilic reactivities of primary and secondary amines in acetonitrile, *Eur. J. Org. Chem.* 2009 (2009) 6379–6385.
32. H.K. Hall, R.B. Bates, Correlation of alkylamine nucleophilicities with their basicities, *Tetrahedron Lett.* 53 (2012) 1830–1832.
33. V.P. Andreev, Y.P. Nizhnik, N.S. Lebedeva, New basicity/nucleophilicity scale on the basis of parameters of formation of axial n,ν-complexes derived from tetra phenylporphyrinatozinc(II) and base/nucleophile as ligand, *Russ. J. Org. Chem.* 44 (2008) 906–915.
34. C.M. Hansen, *Hansen Solubility Parameters: A User's Handbook*, CRC Press, Taylor & Francis, London, New York, 2007.
35. D. Gerlich, S. Horning, Experimental investigation of radiative association processes as related to interstellar chemistry, *Chem. Rev.* 92 (1992) 1509–1539.
36. A. Klamt, F. Eckert, W. Arlt, COSMO-RS: An alternative to simulation for calculating thermodynamic properties of liquid mixtures, *Annu. Rev.* 1 (2010) 101–122.
37. A. Klamt, The COSMO and COSMO-RS solvation models, *WIREs Comput. Mol. Sci.* 8 (2018) e1338.
38. C. Biliyok, A. Lawal, M. Wang, F. Seibert, Dynamic modelling, validation and analysis of post-combustion chemical absorption CO_2 capture plant, *Int. J. Greenh. Gas Control.* 9 (2012) 428–445.
39. T. Lichtenegger, S. Pirker, Recurrence CFD – A novel approach to simulate multiphase flows with strongly separated time scales, *Chem. Eng. Sci.* 153 (2016) 394–410.

long campaigns, reproducibility, separation efficiency, energy demand, and equipment compatibility dominate overall performance [9]. For many bio-based amines, dilute product streams and strong interactions with water and salts lead to energy-intensive downstream processing that can outweigh the benefits of renewable feedstocks. Downstream processing is particularly critical for short-chain diamines and multifunctional amines, which are highly polar, form stable salts, and exhibit overlapping physical properties. Achieving polymer-grade or pharmaceutical-grade purity often requires multistep purification involving clarification, pH-controlled separation, evaporation, and distillation. These operations introduce cumulative yield losses, high energy consumption, and increased waste generation, frequently determining whether a process progresses beyond pilot scale [10,11]. As a result, many promising bio-based amine platforms stall during scale-up despite strong upstream performance. Continuous processing and flow-based technologies have been proposed as potential solutions to improve productivity and consistency. Immobilized enzyme reactors and hybrid continuous-batch configurations can enhance control and reproducibility, but they also introduce new constraints related to fouling, enzyme deactivation, equipment cost, and integration with downstream purification [12,13]. Industrial experience suggests that fully continuous bio-based amine production remains challenging, and hybrid approaches are currently the most practical route to commercialization. Against this background, the purpose of this chapter is to provide a process-centered analysis of the scale-up challenges associated with bio-based amines, rather than a chemistry-focused review. This chapter positions fermentation-derived amines as benchmark systems, examines laboratory-to-industrial translation in reactor operation and reaction engineering, and places particular emphasis on downstream separation, energy demand, safety, and techno-economic constraints. By integrating insights from industrial case studies, biocatalytic process development, and sustainability assessments, this chapter highlights the system-level considerations required for successful industrial deployment of bio-based amines. Its placement within this book reflects the recognition that advances in synthesis must be matched by equally rigorous attention to processing and scale-up in order to realize the full potential of renewable amine technologies.

4.2 BIO-BASED AMINE PRODUCTION PATHWAYS AT SCALE

Fermentation represents the most mature and industrially validated route for the large-scale production of bio-based amines, particularly amino acids and linear aliphatic diamines. Among these, L-lysine production serves as a benchmark system, with annual global production reaching several million tons. Industrial L-lysine manufacturing relies on large-scale aerobic fermentation using engineered *Corynebacterium glutamicum* or *Escherichia coli*, followed by downstream recovery as lysine salts or hydrochloride forms. Extensive optimization of strain metabolism, oxygen transfer, and nutrient feeding strategies has enabled high titers and yields; however, these systems also highlight the scale-dependent challenges associated with heat removal, oxygen mass transfer, and long-term process stability [3,4]. Beyond amino acids, increasing attention has been directed toward the fermentative production of bio-based diamines such as cadaverine (1,5-diaminopentane), putrescine (1,4-diaminobutane), and other C3–C6 diamines, which are important monomers for polyamides and specialty polymers. These processes typically involve the decarboxylation or reductive amination of amino acid precursors through engineered microbial pathways. While laboratory-scale demonstrations report promising titers, industrial scale-up remains constrained by metabolic burden, by-product formation, and sensitivity of engineered strains to process fluctuations [5,6]. A critical bottleneck in fermentation-based amine production lies in downstream processing. Diamines are often produced in complex fermentation broths containing salts, residual sugars, and biomass-derived impurities. Separation and purification at scale require energy-intensive steps such as pH adjustment, evaporation, distillation, or crystallization, which significantly impact process economics and sustainability [9]. These downstream challenges often dominate total production cost, underscoring the importance of integrated process design rather than focusing solely on upstream productivity.

Biocatalytic amination has emerged as a powerful alternative or complementary strategy to fermentation, particularly for the synthesis of chiral and functional amines. Enzymes such as ω-transaminases, amine dehydrogenases, and related reductive aminases enable highly selective conversion of carbonyl compounds into amines under mild conditions. From a processing standpoint, these routes offer advantages such as reduced reaction severity, high selectivity, and compatibility with aqueous systems [14]. Despite these advantages, scale-up of biocatalytic amination faces intrinsic challenges. Transaminase-catalyzed reactions are thermodynamically reversible, often requiring excess amine donors or coupled co-product removal strategies to drive conversions. At industrial scale, this translates into higher material usage, additional separation steps, and increased waste generation. Moreover, enzyme stability, substrate inhibition, and sensitivity to solvent composition become increasingly critical at higher substrate loadings [7,8]. Several studies have demonstrated kilogram- to pilot-scale applications of transaminase-based processes, highlighting the feasibility of industrial implementation when reaction conditions are carefully optimized. These examples emphasize the importance of process intensification strategies such as enzyme immobilization, continuous-flow operation, and integrated cofactor recycling. Nevertheless, the economic viability of such systems remains highly dependent on enzyme lifetime, raw material costs, and downstream purification efficiency [12].

Conventional petrochemical amine production relies on well-established processes such as hydrogenation of nitriles, reductive amination of aldehydes and ketones, or ammonolysis of alcohols. These routes benefit from decades of industrial optimization, high space–time yields, and robust continuous processing infrastructure. From a purely processing perspective, petrochemical routes often outperform bio-based alternatives in terms of throughput, operational simplicity, and cost efficiency. However, petrochemical amine production is typically associated with high reaction temperatures and pressures, reliance on fossil-derived feedstocks, and the use of metal catalysts that may raise environmental and regulatory concerns. In contrast, bio-based amine pathways offer improved sustainability profiles and feedstock renewability but suffer from increased process complexity, particularly in upstream fermentation control and downstream purification [1,2]. A realistic comparison

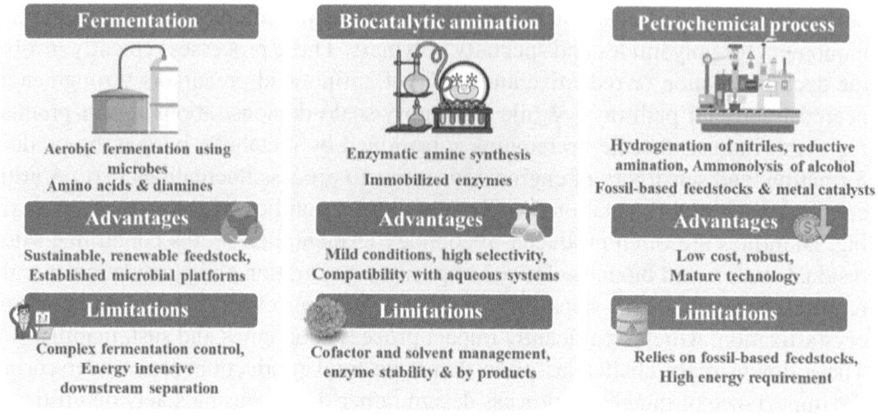

FIGURE 4.1 Industrial production routes for bio-based amines, comparing fermentation, biocatalytic amination, and conventional petrochemical processes with their key advantages, limitations, and challenges.

between bio-based and petrochemical amine routes reveals that the primary challenges for bio-based systems are not rooted in reaction chemistry alone, but in the integration of multiple processing steps at scale. Feedstock variability, lower product concentrations, and energy-intensive separations often offset the environmental advantages unless carefully addressed through process integration and intensification. As a result, hybrid strategies combining biotechnological and chemocatalytic steps are increasingly viewed as promising pathways to bridge the gap between sustainability and industrial feasibility [13].

At industrial scale, fermentation and biocatalysis provide viable routes to bio-based amines, but their success depends strongly on downstream processing, process integration, and long-term operational stability. Compared to petrochemical routes, bio-based pathways offer sustainability advantages but require further advances in separation technologies, enzyme robustness, and system-level process design to achieve broad commercial adoption. Figure 4.1 illustrates the three principal industrial routes for amine production and compares their processing characteristics, advantages, and limitations.

4.3 LABORATORY-SCALE VS INDUSTRIAL-SCALE PROCESSING

Laboratory-scale processing of bio-based amines is typically conducted in well-controlled environments using small bioreactors or batch reactors, where parameters such as temperature, pH, substrate concentration, and agitation can be rapidly adjusted. Under these conditions, reaction performance is often optimized for yield or selectivity, and short operational times minimize the impact of catalyst or enzyme deactivation. However, many of these advantages diminish as reactor volumes increase and operational complexity grows. At industrial scale, reactor operation must prioritize robustness, safety, and long-term stability rather than maximum laboratory performance. Large fermenters and chemical reactors exhibit pronounced gradients in temperature, pH, and dissolved oxygen, particularly in highly viscous or biomass-rich systems. In fermentation-based amine production, maintaining homogeneous oxygen transfer becomes increasingly difficult, directly influencing microbial productivity and metabolic stability. These limitations are well documented in industrial L-lysine production, where scale-dependent mass and heat transfer constraints significantly affect process reliability [3,4]. For biocatalytic amination, enzyme stability and tolerance to process fluctuations represent major concerns during scale-up. While laboratory studies often operate under mild and idealized conditions, industrial processes must withstand prolonged operation, variable feedstock quality, and mechanical stress. Enzyme deactivation due to shear forces, impurities, or by-product accumulation becomes more pronounced at scale, necessitating conservative operating windows that may reduce productivity compared to laboratory benchmarks [7,8].

Batch processing remains the most common mode for laboratory studies of bio-based amine synthesis due to its operational simplicity and flexibility. Batch systems allow rapid screening of reaction conditions and enzyme or strain variants; however, they are inherently limited in productivity and scalability. At industrial scale, batch processes often suffer from low space–time yields, inconsistent product quality, and inefficient resource utilization. Fed-batch operation has emerged as the dominant strategy for large-scale fermentation of amino acids and bio-based diamines. Controlled feeding of substrates mitigates substrate inhibition, reduces osmotic stress, and enables higher final product concentrations. Nevertheless, fed-batch processes introduce additional control complexity, requiring precise monitoring of nutrient supply, oxygen demand, and metabolic state over extended operating periods. Small deviations in feed strategy can lead to significant performance losses, particularly when scaling from pilot to commercial reactors [5,6]. Continuous processing offers attractive advantages, including improved productivity, steady-state operation, and reduced reactor downtime. In biocatalytic amine synthesis, continuous flow systems combined with immobilized enzymes have demonstrated enhanced reproducibility and operational efficiency at intermediate scales. However, industrial adoption remains limited due to challenges related to enzyme lifetime, fouling, cofactor management, and the need for highly reliable process control. Continuous fermentation systems face similar constraints, particularly with respect to contamination risk and long-term genetic stability of production strains [12,14].

Reproducibility represents one of the most underestimated challenges when transitioning bio-based amine processes from laboratory to industrial scale. Laboratory experiments are often conducted over short durations and under idealized conditions, masking slow degradation phenomena such as catalyst poisoning, enzyme denaturation, or microbial mutation. In contrast, industrial processes must operate continuously or semi-continuously over weeks or months, amplifying even minor instabilities. In fermentation-based systems, scale-dependent fluctuations in oxygen transfer, pH control, and nutrient gradients can trigger metabolic shifts that reduce product yield or increase by-product formation. Such effects are particularly pronounced in engineered microbial strains, where high metabolic flux toward amine production may compromise genetic and phenotypic stability over long campaigns [9]. Similarly, biocatalytic amination processes often exhibit declining performance over

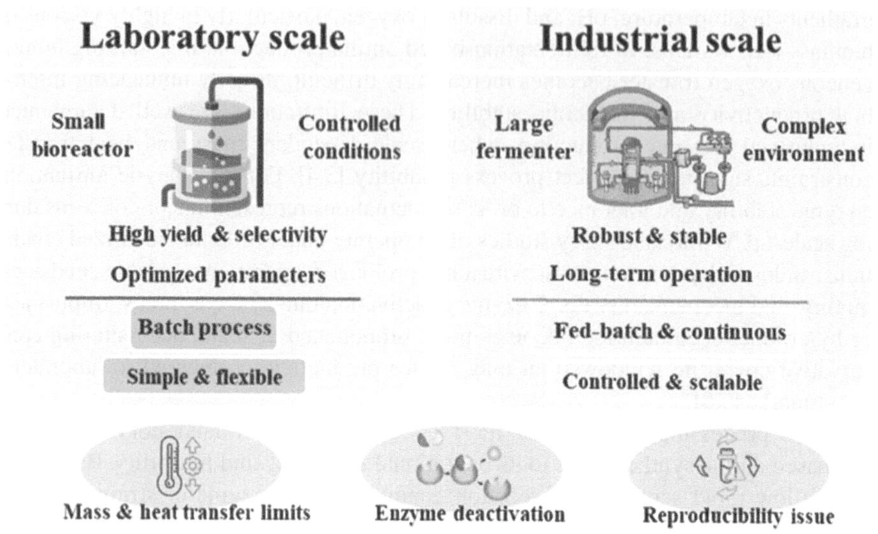

FIGURE 4.2 Comparison of laboratory and industrial processing of bio-based amines and key scale-up challenges.

extended operation due to enzyme inactivation or accumulation of inhibitory species. While laboratory-scale studies frequently report high conversions and selectivities, these metrics may not be reproducible at scale without extensive process optimization and monitoring. Successful industrial implementations therefore emphasize conservative design margins, real-time process analytics, and robust quality control frameworks to ensure consistent performance [13]. The transition from laboratory-scale to industrial-scale processing of bio-based amines involves fundamental changes in reactor operation, control philosophy, and performance metrics. While laboratory studies prioritize flexibility and optimization, industrial processes demand robustness, reproducibility, and long-term stability. Understanding these differences is essential for designing scalable bio-based amine production routes that are technically and economically viable. Figure 4.2 explains the key differences between laboratory-scale and industrial-scale processing of bio-based amines, emphasizing how operational priorities shift from optimization and flexibility to robustness, stability, and long-term performance during scale-up.

4.4 REACTION ENGINEERING AND KINETIC LIMITATIONS

Reaction equilibrium represents a fundamental limitation in both biocatalytic and chemocatalytic amination processes. In many bio-based routes, amination reactions are reversible, and achieving high conversions requires careful manipulation of reaction conditions. This challenge is particularly pronounced in transaminase-catalyzed systems, where the position of equilibrium is dictated by the relative thermodynamic stability of the amine donor and acceptor pairs. At laboratory scale, equilibrium limitations are often overcome by using excess amine donors or coupling reactions to remove co-products. However, these strategies become increasingly problematic at industrial scale due to increased material costs and downstream separation burdens [14]. In fermentation-based amine production, equilibrium constraints manifest indirectly through metabolic feedback regulation. High intracellular concentrations of amine products or pathway intermediates can inhibit key enzymes or redirect metabolic flux toward competing pathways. While metabolic engineering can partially address these issues, complete elimination of equilibrium-driven limitations is rarely achievable at scale [5,6]. Chemocatalytic amination routes derived from biomass feedstocks also encounter equilibrium challenges, particularly in reductive amination reactions involving alcohols or carbonyl compounds. These systems often require elevated temperatures, pressures, or hydrogen availability to shift equilibrium toward amine formation, which complicates reactor design and increases energy demand during scale-up [1,2].

Substrate inhibition is a recurrent issue in high-concentration amination processes and becomes more pronounced as systems transition from laboratory to industrial scale. In biocatalytic processes, enzymes such as transaminases and amine dehydrogenases often exhibit reduced activity at high substrate or product concentrations, limiting achievable space–time yields. Laboratory-scale studies frequently operate at dilute conditions to avoid inhibition, but such conditions are economically unfavorable for industrial implementation [7]. By-product accumulation further complicates scale-up by reducing selectivity and increasing purification complexity. In fermentation-based routes, side metabolites generated under oxygen-limited or nutrient-imbalanced conditions can accumulate and interfere with amine recovery. These effects are difficult to predict from small-scale experiments, where residence times are short and process conditions are tightly controlled [3]. In chemocatalytic amination, by-products such as oligomers, over-alkylated amines, or catalyst-deactivating species can accumulate over prolonged operation. At scale, even low rates of side reactions can lead to significant catalyst fouling or loss of selectivity, necessitating frequent regeneration or replacement and reducing overall process efficiency [15].

Heat and mass transfer limitations are among the most critical scale-dependent factors influencing reaction performance in bio-based amine production. Laboratory reactors benefit from high surface-area-to-volume ratios and efficient mixing, enabling rapid heat dissipation and uniform reactant distribution. As reactor volume increases, these advantages diminish, leading to temperature gradients, concentration heterogeneities, and localized hot spots. In aerobic fermentation processes, oxygen transfer often becomes the rate-limiting step, directly impacting microbial productivity and metabolic balance. Insufficient oxygen availability can trigger unwanted metabolic pathways, increasing by-product formation and reducing amine yield. Such effects are well documented in industrial amino acid fermentation systems and are a major driver of scale-up complexity [4]. Biocatalytic and chemocatalytic amination reactions are similarly affected by mass transfer constraints, particularly when operating with poorly soluble substrates or gases such as hydrogen or ammonia. Inadequate mass transfer can mask intrinsic reaction kinetics, leading to misleading laboratory-scale performance metrics that do not translate to industrial reactors [11].

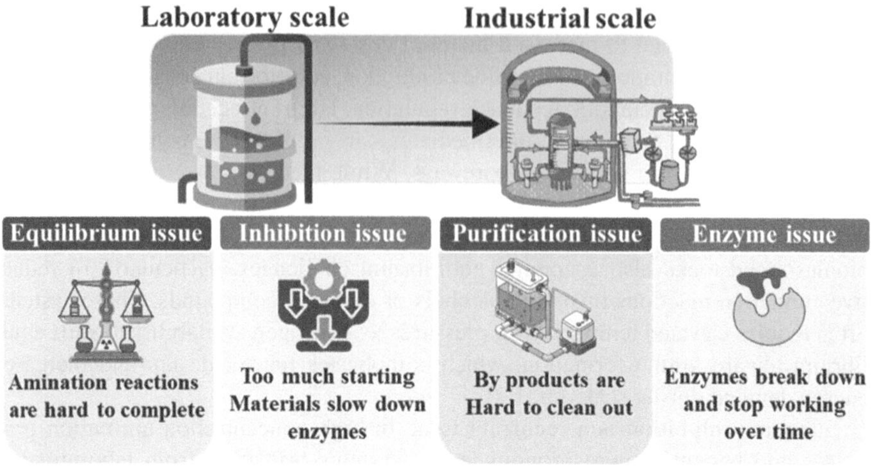

FIGURE 4.3 Key scalability challenges in bio-based amine production, including equilibrium limitations, substrate inhibition, purification difficulties, and enzyme stability.

Enzyme stability is a decisive factor governing the industrial feasibility of biocatalytic amine production. While laboratory-scale reactions often demonstrate high initial activities, long-term operation at industrial scale exposes enzymes to mechanical shear, temperature fluctuations, impurities, and inhibitory by-products. These factors contribute to gradual loss of activity and reduced process productivity over time. Industrial case studies of transaminase-based amination highlight the necessity of balancing reaction rate with enzyme lifetime. Operating conditions that maximize instantaneous conversion may accelerate enzyme deactivation, ultimately reducing cumulative product yield. As a result, industrial processes typically adopt conservative conditions that prioritize enzyme stability over peak performance [8]. Strategies such as enzyme immobilization, controlled feeding of substrates, and integration of continuous-flow systems have shown promise in enhancing enzyme productivity and operational stability. However, these approaches introduce additional complexity and cost, underscoring the need for holistic process design rather than isolated optimization of reaction kinetics [12,13]. Reaction engineering and kinetic limitations play a central role in determining the scalability of bio-based amine production processes. Equilibrium constraints, substrate inhibition, heat and mass transfer limitations, and enzyme stability issues often emerge only at larger scales, highlighting the importance of early-stage process-aware design. Addressing these challenges requires an integrated approach that balances reaction kinetics with transport phenomena and long-term operational stability. Figure 4.3 highlights the main reaction engineering challenges that arise when scaling bio-based amine production from laboratory to industrial scale.

4.5 DOWNSTREAM PROCESSING AND PURIFICATION CHALLENGES

Downstream processing represents a major cost and complexity in bio-based amine production, often outweighing challenges associated with upstream synthesis.

Fermentation-derived amines are typically produced in dilute aqueous broths containing high concentrations of cells, residual nutrients, salts, and colored organic impurities. At laboratory scale, cell removal is commonly achieved by centrifugation or filtration under well-controlled conditions. However, at industrial scale, these operations must handle large volumetric flow rates while maintaining reliability and minimizing fouling. Broth clarification becomes particularly challenging in high-cell-density fermentations used for amino acid and diamine production. Inadequate removal of biomass and colloidal impurities can adversely affect subsequent purification steps, leading to membrane fouling, reduced separation efficiency, and inconsistent product quality. Industrial amino acid processes such as L-lysine production have demonstrated that clarification efficiency strongly influences overall downstream performance and operating cost [3,4]. Decolorization is another critical step, especially for diamines derived from complex fermentation broths or protein-rich biomass feedstocks. Activated carbon treatment and ion-exchange resins are frequently employed at laboratory scale, but their large-scale implementation raises concerns related to material consumption, regeneration, and waste handling. These challenges necessitate careful selection of clarification and decolorization strategies that balance product purity with operational sustainability [16].

Salt formation is a defining characteristic of many bio-based amine purification processes. Amines are often recovered as hydrochloride or sulfate salts to facilitate isolation, improve stability, or meet product specifications. While salt formation simplifies initial recovery at small scale, it introduces additional complexity during large-scale processing due to increased ionic strength, corrosion risk, and waste salt generation. pH-controlled separation is widely used to exploit the acid–base properties of amines, enabling selective extraction or crystallization. At laboratory scale, precise pH adjustment is straightforward; however, maintaining uniform pH control in large reactors or separation units is significantly more challenging. Localized pH gradients can lead to incomplete precipitation, co-precipitation of impurities, or formation of emulsions that complicate phase separation [9]. In diamine production systems, repeated acid–base cycling is often required to achieve polymer-grade purity. Each additional pH adjustment step increases process complexity, water usage, and waste generation, emphasizing the need for simplified and integrated separation strategies during scale-up [10].

Distillation and evaporation are central to the purification of low-molecular-weight bio-based amines, particularly when solvent-free product streams are desired. At laboratory scale, rotary evaporation and short-path distillation are commonly used to concentrate amines and remove residual solvents or water. These techniques provide rapid and efficient separation but do not accurately reflect the energy demand or operational constraints of industrial systems. At industrial scale, evaporation and distillation become highly energy-intensive operations due to the high boiling points of amines and the large volumes of water associated with fermentation broths. Multistage evaporation, vacuum distillation, and heat integration are often required to achieve acceptable energy efficiency. Even minor inefficiencies at this stage can significantly impact overall process economics and environmental footprint [9]. Solvent-free purification strategies have attracted increasing interest as a means of reducing solvent consumption and waste generation. These approaches rely on careful control of temperature, pressure, and phase behavior to selectively isolate amines without auxiliary solvents.

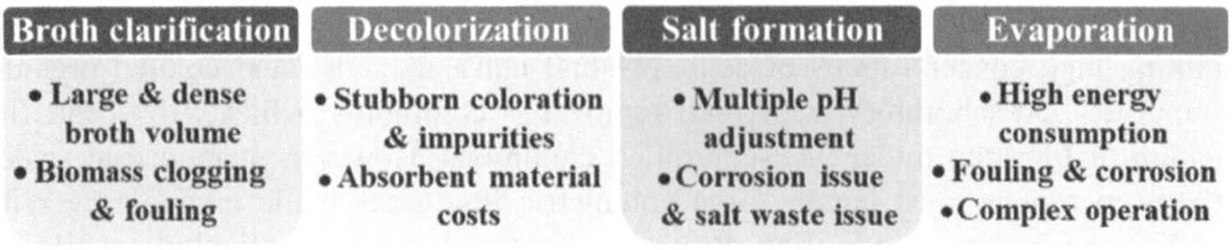

FIGURE 4.4 Major downstream processing challenges in bio-based amine production, including broth clarification, decolorization, salt formation, and energy-intensive evaporation or distillation.

While promising at pilot scale, their industrial implementation requires rigorous control of operating conditions and robust equipment design to ensure consistent product quality [13].

One of the most common pitfalls in scaling bio-based amine processes is the direct translation of laboratory evaporation protocols to industrial distillation systems. Laboratory experiments often underestimate the impact of residence time, thermal degradation, and equipment material compatibility. As a result, processes that appear straightforward at small scale may encounter severe operational challenges when implemented in large distillation columns. Industrial distillation systems must account for factors such as column hydrodynamics, tray or packing efficiency, pressure drop, and fouling behavior. In bio-based amine purification, the presence of residual salts and organic impurities can exacerbate fouling and corrosion, leading to reduced uptime and increased maintenance requirements. These issues are rarely captured in laboratory-scale studies but strongly influence long-term process viability [11]. Case studies involving cadaverine and other bio-based diamines demonstrate that successful scale-up requires early consideration of industrial distillation constraints, including material selection, heat integration, and waste stream management. Such system-level thinking is essential to avoid costly redesigns during later stages of process development [10]. Downstream processing and purification are decisive factors in the successful scale-up of bio-based amine production. Challenges associated with broth clarification, salt handling, energy-intensive separations, and translation from laboratory to industrial equipment often dominate overall process economics. Addressing these issues requires integrated downstream design strategies that prioritize simplicity, energy efficiency, and long-term operational robustness. Figure 4.4 summarizes the key downstream processing challenges encountered during scale-up of bio-based amine production.

4.6 SEPARATION OF BIO-BASED DIAMINES AND MULTIFUNCTIONAL AMINES

Separation of short-chain bio-based diamines, particularly C3–C5 species such as 1,3-diaminopropane, putrescine (1,4-diaminobutane), and cadaverine (1,5-diaminopentane), represents one of the most technically demanding steps in the overall production chain. These diamines are typically obtained from fermentation broths at relatively low concentrations and in the presence of high salt loads, residual biomass, and structurally similar nitrogen-containing impurities. Their high polarity, strong hydrogen-bonding capacity, and tendency to form stable salts complicate conventional separation approaches. At laboratory scale, separation is often achieved through multistep acid–base extraction or crystallization protocols. However, these approaches become increasingly complex at industrial scale due to high water content, overlapping boiling points, and the formation of azeotrope-like behaviors. In practice, achieving polymer-grade purity for C3–C5 diamines requires carefully designed downstream sequences that combine multiple unit operations, each contributing to cumulative yield losses [9,10]. Selectivity during diamine separation is strongly influenced by pH, ionic strength, and the presence of competing amines or amino acids. Small deviations in operating conditions can result in co-extraction or co-crystallization of impurities, reducing final product purity and increasing the need for reprocessing. While laboratory experiments often report high selectivity under tightly controlled conditions, these results are difficult to reproduce consistently at scale. Yield losses are particularly significant in processes involving repeated acid–base cycling, salt formation, and solvent exchange. Each additional separation step increases the risk of product loss through incomplete phase separation, thermal degradation, or adsorption onto process materials. From an energy perspective, evaporation and distillation steps required to concentrate diamine streams contribute disproportionately to overall process energy demand, often offsetting the sustainability advantages associated with bio-based production routes [3]. Energy demand becomes especially critical for diamines with high boiling points or strong water affinity, where deep vacuum distillation or multistage evaporation is required. These operations not only increase operating costs but also impose constraints on equipment design and heat integration strategies during scale-up [11].

Solvent extraction remains one of the most widely investigated methods for separating bio-based diamines, particularly at pilot scale. Organic solvents can provide high selectivity when combined with pH-controlled extraction schemes; however, solvent choice is critical due to amine solubility, emulsion formation, and solvent recovery requirements. At industrial scale, solvent losses, toxicity, and flammability introduce additional operational and regulatory challenges. In contrast, solvent-free routes such as direct distillation, reactive separation, or selective crystallization have gained attention as more sustainable alternatives. These methods reduce solvent consumption and waste generation but typically require higher energy input and tighter process control. Solvent-free strategies also place greater demands on equipment materials and design, particularly when handling corrosive amine salts at elevated temperatures [13]. Comparative studies indicate that while solvent extraction may offer operational flexibility at smaller scales, solvent-free approaches are often favored for long-term industrial implementation when energy efficiency and waste minimization are prioritized. The optimal separation strategy therefore depends on a balance between selectivity, energy consumption, and overall process integration [1].

Scalability and safety are critical considerations in the separation of bio-based diamines, particularly due to their high basicity, corrosiveness, and strong odor. Large-scale handling of diamine-rich streams requires careful material selection to prevent corrosion and degradation of equipment. In addition, the potential release of volatile amines poses occupational health concerns, necessitating effective containment and ventilation systems. From a process safety perspective, separation steps involving elevated temperatures, vacuum conditions, or reactive intermediates introduce additional risks during scale-up. Laboratory-scale separation protocols often underestimate these hazards, as small volumes and short operating times mask cumulative exposure and equipment stress. Industrial-scale processes must therefore be designed with conservative safety margins and robust monitoring systems to ensure stable long-term operation [16]. The scalability of diamine separation processes ultimately depends on their ability to maintain consistent performance over extended campaigns. Strategies that minimize the number of unit operations, reduce energy demand, and simplify waste handling are more likely to succeed commercially than highly selective but operationally complex alternatives. Separation of bio-based diamines and multifunctional amines remains a central bottleneck in the scale-up of sustainable amine production. Challenges associated with selectivity, yield loss, energy demand, and safety often dominate process economics and dictate industrial feasibility. Addressing these issues requires integrated separation strategies that balance purity requirements with scalability, energy efficiency, and operational robustness.

4.7 CONTINUOUS PROCESSING AND FLOW-BASED SCALE-UP

Continuous fermentation has long been considered an attractive alternative to batch and fed-batch operation to produce bio-based amines due to its potential for higher space–time yields, steady-state operation, and reduced downtime. In principle, continuous systems enable more efficient utilization of reactor volume and can simplify process control once stable operating conditions are established. However, integration of continuous fermentation with downstream separation remains challenging, particularly for amine products that are obtained at relatively low concentrations and in complex aqueous matrices. In large-scale amino acid and diamine production, maintaining long-term genetic and metabolic stability of production strains is a key concern. Continuous operation amplifies the consequences of contamination, strain drift, and subtle changes in feed composition, which can lead to gradual loss of productivity over extended campaigns. These issues have limited the widespread adoption of fully continuous fermentation in industrial amino acid manufacturing, despite its theoretical advantages [3,4]. Downstream integration further complicates continuous operation. Clarification, pH-controlled extraction, and evaporation steps must be carefully synchronized with upstream flow rates to avoid bottlenecks and product losses. As a result, many industrial processes adopt hybrid configurations, where continuous or semi-continuous upstream production is combined with batch or semi-batch downstream purification to balance operational flexibility and robustness [9].

Flow-based biocatalysis using immobilized enzymes has emerged as a promising strategy for continuous amine synthesis, particularly for transaminase- and amine-dehydrogenase-mediated reactions. Immobilization enables enzyme reuse, improves operational stability, and facilitates separation of the biocatalyst from the product stream. In laboratory and pilot-scale studies, immobilized transaminases have demonstrated enhanced reproducibility and prolonged activity compared to soluble enzyme systems [14]. From a process-engineering perspective, immobilized enzyme systems offer several advantages, including reduced enzyme consumption, simplified downstream processing, and compatibility with continuous flow reactors. However, immobilization introduces additional considerations such as mass-transfer resistance, support material cost, and mechanical stability under prolonged operation. These factors become increasingly important at industrial scale, where pressure drop, fouling, and uneven flow distribution can negatively impact performance [8]. Industrial case studies highlight that successful implementation of flow biocatalysis depends on careful matching of enzyme properties, support materials, and reactor design. While promising results have been achieved at kilogram scale, broader industrial adoption requires further demonstration of long-term stability and economic competitiveness under realistic operating conditions [12].

Flow systems offer several inherent advantages for amine synthesis, including improved heat and mass transfer, precise residence-time control, and enhanced safety when handling reactive intermediates. These features are particularly beneficial for highly exothermic or equilibrium-limited amination reactions, where rapid heat dissipation and controlled contact times can improve selectivity and yield. Despite these benefits, flow-based amine synthesis also faces notable limitations. Many bio-based amine processes involve multiphase systems, high water content, or solid formation, all of which complicate continuous operation. In addition, fouling and clogging risks increase significantly when processing fermentation-derived streams containing residual salts or biomass-related impurities. These challenges often necessitate extensive pretreatment, reducing the overall simplicity of flow-based approaches [13]. Furthermore, the economic advantages of flow systems are highly sensitive to throughput and equipment utilization. While continuous flow reactors can deliver superior performance at steady state, frequent interruptions due to maintenance or feedstock variability can negate these benefits, particularly in early-stage industrial implementations [1].

Industrial implementation of continuous and flow-based amine synthesis requires specialized equipment capable of withstanding corrosive environments, elevated temperatures, and prolonged operation. Reactor materials must be compatible with amines and associated salts, while downstream units must handle variable flow rates without compromising separation efficiency. These requirements often result in higher capital expenditure compared to conventional batch systems. Equipment selection is closely tied to process scale and product specification. Small-to-medium-scale production of specialty amines may justify the use of modular flow reactors and immobilized enzyme systems, whereas large-volume commodity amines typically favor more established batch or fed-batch configurations. As a result, continuous processing is more likely to gain traction in niche applications where product value offsets the additional complexity and investment [10]. Ultimately, the feasibility of continuous processing for bio-based amines depends on achieving a balance between operational stability, product quality, and economic performance. Incremental integration of continuous unit operations, rather than full process conversion, is increasingly viewed as a pragmatic pathway toward industrial adoption.

Continuous processing and flow-based scale-up offer compelling advantages for bio-based amine synthesis, including improved control, safety, and productivity.

However, challenges related to strain stability, enzyme lifetime, fouling, and equipment cost continue to limit widespread industrial implementation. Hybrid process designs that combine continuous and batch operations currently represent the most viable strategy for translating flow-based concepts into robust industrial processes.

4.8 PROCESS SAFETY, HANDLING, AND REGULATORY CONSTRAINTS

Process safety and regulatory compliance are critical considerations in the scale-up of bio-based amine production, as many inherent hazards associated with amines persist regardless of feedstock origin. Low-molecular-weight amines and diamines are characterized by high basicity, strong odor, and, in many cases, significant volatility. During large-scale fermentation, purification, and distillation, fugitive emissions can lead to occupational exposure and odor-related issues, necessitating robust containment, ventilation, and scrubbing systems. These challenges are often minimal at laboratory scale but become dominant concerns in industrial environments [1]. Corrosion and material compatibility represent additional safety constraints during scale-up. Amines and their salts, particularly hydrochloride and sulfate forms, are corrosive in the presence of moisture and elevated temperatures. While laboratory experiments typically use glassware or short-term stainless-steel equipment, industrial systems must withstand prolonged exposure, increasing the risk of corrosion-induced failures and product contamination. Appropriate material selection, including corrosion-resistant alloys or polymer-lined equipment, is therefore essential for long-term operational stability [9]. Large-scale purification steps such as evaporation and distillation pose some of the highest safety risks in bio-based amine processing. Elevated temperatures, reduced pressures, and the presence of residual salts can promote fouling, thermal degradation, and unstable operation in distillation columns. Laboratory-scale evaporation techniques often underestimate these hazards, leading to challenges when processes are directly translated to industrial equipment. Case studies involving bio-based diamines highlight that downstream purification frequently dictates overall process safety and reliability [10]. From a regulatory perspective, bio-based amines are generally governed by the same frameworks as petrochemical amines with identical chemical structures. Regulatory requirements related to toxicity, transport, storage, and waste management are based on chemical identity rather than production route. Consequently, bio-based origin alone does not reduce regulatory burden. However, sustainability benefits such as reduced carbon footprint may support favorable positioning within green chemistry or life cycle assessment frameworks, provided these claims are substantiated [2,13]. In addition, the use of genetically modified microorganisms or novel catalytic systems can introduce further regulatory scrutiny related to containment, waste handling, and environmental impact during upstream processing [5]. Overall, successful industrial deployment of bio-based amines requires that safety, handling, and regulatory constraints be addressed with the same rigor applied to conventional petrochemical processes, alongside careful integration of sustainability considerations. Figure 4.5 outlines the key safety and regulatory considerations involved in scaling bio-based amine production to industrial level.

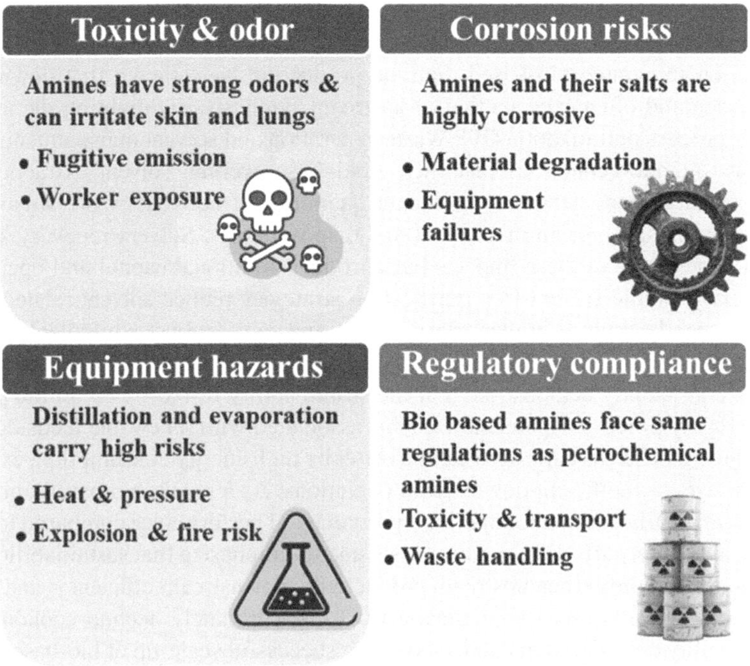

FIGURE 4.5 Safety and regulatory challenges in industrial bio-based amine production, including toxicity, corrosion, equipment hazards, and compliance requirements.

4.9 TECHNO-ECONOMIC AND SUSTAINABILITY CONSIDERATIONS

Techno-economic performance and sustainability are decisive factors in determining whether bio-based amine production routes can progress beyond pilot scale to commercial implementation. While bio-based pathways offer clear advantages in terms of renewable feedstocks and reduced reliance on fossil resources, their economic viability is strongly influenced by process complexity, energy demand, and downstream efficiency. The primary cost drivers in bio-based amine production are feedstock sourcing, upstream productivity, and downstream purification. In fermentation-based systems, raw material costs are generally lower than petrochemical feedstocks, but this advantage is often offset by dilute product streams and long processing times. Achieving high titers and yields requires extensive metabolic engineering and tightly controlled operating conditions, increasing development and operational costs. For bio-based diamines, additional expenses arise from maintaining strain stability and minimizing by-product formation during extended production campaigns [3,5]. Energy demand is a dominant contributor to operating cost and environmental footprint, particularly during downstream processing. Evaporation, distillation, and salt handling account for a substantial fraction of total energy consumption due to the high water content of fermentation broths and the relatively high boiling points of many amines. Even when upstream fermentation is optimized, energy-intensive separation steps can negate the sustainability benefits of bio-based production if not carefully integrated with heat recovery and process intensification strategies [9]. Studies on integrated bio-based diamine production have shown that downstream energy demand often exceeds that of upstream synthesis, highlighting the need for holistic process optimization [10]. Waste generation and solvent management further influence techno-economic feasibility. Acid–base cycling, solvent extraction, and decolorization steps generate significant quantities of aqueous waste streams and spent materials, increasing treatment and disposal costs. Solvent recovery systems can mitigate some of these impacts but introduce additional capital and operational complexity. While solvent-free purification strategies reduce solvent-related waste, they frequently require higher energy input and more robust equipment, creating trade-offs between waste reduction and energy efficiency [16]. Life cycle assessment studies consistently demonstrate that the sustainability of bio-based amine production is highly process-dependent. Benefits associated with renewable feedstocks and lower greenhouse gas emissions can be offset by high energy consumption, extensive water usage, or inefficient downstream operations. As a result, bio-based amines do not automatically guarantee superior environmental performance compared to petrochemical counterparts. Comparative assessments emphasize that sustainability gains are maximized only when upstream productivity, downstream efficiency, and energy integration are addressed simultaneously [1,13]. Ultimately, techno-economic and sustainability considerations underscore that successful scale-up of bio-based amine processes requires system-level optimization rather than isolated improvements. Integration of fermentation, catalysis, separation, and energy recovery is essential to balance cost competitiveness with genuine environmental benefits during industrial deployment.

4.10 INDUSTRIAL CASE STUDIES AND LESSONS LEARNED

Industrial experience with bio-based amines clearly shows that fermentation-derived systems provide the most instructive benchmarks for scale-up. Large-scale amino acid fermentation, particularly L-lysine production, demonstrates that long-term commercial success depends less on achieving maximum laboratory yields and more on robust reactor operation, reliable oxygen transfer, contamination control, and consistent product quality over extended campaigns. Decades of industrial practice have shown that maintaining process stability at scale is often more challenging than improving reaction yield, and these lessons are directly transferable to emerging bio-based amine platforms [3,4]. Fermentative production of C3–C5 diamines, such as putrescine and cadaverine, further highlights the complexity of scaling bio-based amine processes. Although engineered microorganisms can achieve promising titers at laboratory scale, pilot and industrial-scale operation often reveal limitations related to metabolic burden, by-product formation, and sensitivity to oxygen and nutrient gradients. These challenges are amplified during long production runs, where genetic and phenotypic stability become critical for maintaining productivity. Industrially relevant demonstrations of cadaverine production underscore that high upstream performance must be coupled with effective downstream purification to deliver material suitable for polymer applications [5,6,10]. A recurring bottleneck identified across industrial case studies is downstream separation, which often dominates both cost and energy consumption. Short-chain diamines are highly water-soluble and form stable salts, making their isolation from fermentation broths technically demanding. Multistep purification sequences involving pH adjustment, evaporation, and distillation are frequently required to achieve polymer-grade purity, leading to cumulative yield losses and high energy demand. Several pilot-scale processes have stalled primarily because downstream complexity outweighed improvements achieved through metabolic or catalytic optimization [9]. Biocatalytic amination routes provide additional insight into scale-up limitations. Transaminase-based processes have demonstrated high selectivity and mild reaction conditions, yet industrial translation is often constrained by reaction equilibrium, substrate inhibition, enzyme lifetime, and removal of amine donors and by-products. While kilogram-scale successes have been reported, many processes encounter diminishing returns at pilot scale when enzyme stability and downstream processing requirements are fully accounted for [7,8]. Collectively, these case studies emphasize several transferable lessons for emerging bio-based amine platforms. Early integration of upstream production with downstream purification is essential to avoid late-stage redesign. Processes must be optimized for operational robustness rather than peak laboratory performance, with particular attention to long-term stability and product specifications. Minimizing salt formation, solvent use, and energy-intensive separations significantly improves techno-economic viability. Ultimately, successful industrial deployment of bio-based amines requires system-level process design, where bioproduction, catalysis, separation, and energy management are developed as a unified framework rather than independent steps [1,2,13]. Table 4.1 summarizes the main technological pathways for producing bio-based amines, including microbial fermentation, chemocatalytic amination, and integrated bioprocess approaches. For each route, the primary processing focus and the critical scale-up constraints are highlighted. While fermentation-based systems emphasize productivity, strain robustness, and oxygen transfer, catalytic routes are more strongly influenced by catalyst stability, selectivity, and reaction severity. Across all pathways, common industrial challenges include heat and mass transfer limitations, feedstock variability, downstream purification, and maintaining consistent polymer-grade quality, underscoring the need for integrated process optimization from laboratory to commercial scale.

4.11 CONCLUSIONS

This chapter has examined the processing and scale-up challenges associated with bio-based amines, emphasizing that industrial feasibility is governed less by reaction chemistry and more by system-level process integration. While fermentation, biocatalysis, and hybrid bio-chemo routes provide viable pathways to renewable amines, their successful translation to commercial scale depends on robust operation across upstream production, reaction engineering, downstream purification, and long-term process stability. Fermentation-based amine production, particularly amino acids and C3–C5 diamines, demonstrates that large-scale bio-based manufacturing is achievable when strain performance, reactor design, and oxygen and heat transfer are carefully optimized. However, extending these platforms beyond established amino acid

TABLE 4.1
Overview of Representative Bio-Based Amine Production Routes, Key Processing Priorities, and Major Scale-Up Challenges Reported in the Literature

Sr. No.	Bio-Based Amine System/Route	Key Processing Focus	Scale-Up Challenges Highlighted	Ref.
1	Microbial production of N-containing compounds	Fermentation performance	Productivity limits, strain robustness	[17]
2	L-lysine (industrial benchmark)	Large-scale fermentation	Reactor control, oxygen transfer, consistency	[3]
3	L-lysine-HCl	Process modeling and simulation	Scale-dependent heat and mass transfer	[11]
4	Bio-based amines via heterogeneous catalysis	Chemocatalytic amination	Catalyst stability, selectivity, sustainability	[2]
5	Biomass-derived platform molecules → amines	Catalytic upgrading	Reaction severity, catalyst lifetime	[15]
6	Bio-based amines for polymers	Whole value chain	Feedstock variability, downstream cost	[1]
7	Chemical amination routes	Reaction optimization	Temperature and pressure sensitivity	[18]
8	Catalytic amination mechanisms	Catalyst behavior	Translating lab mechanisms to scale	[19]
9	Microbial bio-based diamines	Fermentation pathways	Titer, yield, by-product control	[5]
10	N-chemicals from biomass carbonyls	Chemocatalysis	Process integration limitations	[20]
11	Cadaverine → bio-polyamide	Integrated bioprocess	Downstream purification, polymer-grade quality	[10]
12	Engineered microbes for amines	Metabolic engineering	Flux imbalance, scale instability	[6]
13	Transaminase-based amination	Biocatalytic scale-up	Equilibrium limits, solvent tolerance	[7]
14	Enzyme cascades for amines	Process development	Enzyme lifetime, process robustness	[8]
15	Protein-rich biomass → amines	Feedstock processing	Separation complexity, waste streams	[16]
16	Chiral amines via transaminases	Enzymatic synthesis	Substrate inhibition, stability	[14]
17	Bio-based diamine recovery	Downstream separation	Energy-intensive purification	[9]
18	AmDH/TA systems	Process intensification	Cofactor recycling, scalability	[12]
19	Microbial N-functionalized chemicals	Host engineering	Scale-dependent productivity loss	[21]
20	Industrial amino acid fermentation	Industrial benchmarks	Capital cost, operational stability	[4]
21	Bio-based nitrogen chemicals	Critical review	Economic and sustainability trade-offs	[4]

systems reveals scale-dependent constraints, including metabolic burden, by-product formation, and sensitivity to process fluctuations. These challenges are amplified during prolonged industrial campaigns, highlighting the importance of robustness over peak laboratory productivity. Biocatalytic amination offers complementary advantages, particularly for selective and functional amine synthesis, but faces intrinsic limitations at scale. Reaction equilibrium, substrate inhibition, enzyme lifetime, and downstream removal of donors and by-products often restrict productivity and economic viability. Although immobilization and continuous-flow strategies can improve performance, their benefits must be weighed against added complexity and capital cost. Across all bio-based amine platforms, downstream processing emerges as the dominant cost and energy driver. Clarification, salt handling, evaporation, and distillation frequently outweigh upstream synthesis in terms of economic and environmental impact. Short-chain diamines are especially challenging due to their high water affinity, salt formation, and demanding purity requirements. Failure to address downstream constraints early in development is a primary reason why many processes stall at pilot scale despite promising laboratory results. From a sustainability perspective, bio-based amines do not automatically outperform petrochemical counterparts. Life cycle assessments consistently show that energy-intensive separations, high water usage, and waste generation can offset gains from renewable feedstocks unless process integration and heat recovery are implemented. Regulatory and safety requirements further reinforce that bio-based origin does not reduce the need for rigorous engineering design, material compatibility, and hazard management. Collectively, the lessons presented in this chapter underscore that successful industrial deployment of bio-based amines requires early, integrated consideration of scale-up, downstream processing, and end-use specifications. Future progress will depend on coupling advances in strain and enzyme engineering with innovations in separation technology, process intensification, and hybrid processing. By treating bio-based amine production as an interconnected system rather than a sequence of isolated steps, the gap between laboratory innovation and commercial reality can be effectively bridged.

REFERENCES

1. V. Froidevaux, C. Negrell, S. Caillol, J.P. Pascault, B. Boutevin, Biobased amines: From synthesis to polymers; present and future, *Chem. Rev.* 116 (2016) 14181–14224.
2. M. Pelckmans, T. Renders, S. Van de Vyver, B.F. Sels, Bio-based amines through sustainable heterogeneous catalysis, *Green Chem.* 19 (2017) 5303–5331.
3. F.K. do C. Félix, L.A.J. Letti, G. Vinícius de Melo Pereira, P.G.B. Bonfim, V.T. Soccol, C.R. Soccol, L-lysine production improvement: A review of the state of the art and patent landscape focusing on strain development and fermentation technologies, *Crit. Rev. Biotechnol.* 39 (2019) 1031–1055.
4. W. Leuchtenberger, K. Huthmacher, K. Drauz, Biotechnological production of amino acids and derivatives: Current status and prospects, *Appl. Microbiol. Biotechnol.* 69 (2005) 1–8.
5. W. Li, L. Guohui, D. Yu, Diamine biosynthesis: Research progress and application prospects, *Appl. Environ. Microbiol.* 86 (2020) e01972–20.
6. S. Kind, S. Neubauer, J. Becker, M. Yamamoto, M. Völkert, G. von Abendroth, O. Zelder, C. Wittmann, From zero to hero – Production of bio-based nylon from renewable resources using engineered Corynebacterium glutamicum, *Metab. Eng.* 25 (2014) 113–123.

7. C.E. Paul, M. Rodríguez-Mata, E. Busto, I. Lavandera, V. Gotor-Fernández, V. Gotor, S. García-Cerrada, J. Mendiola, Ó. de Frutos, I. Collado, Transaminases applied to the synthesis of high added-value enantiopure amines, *Org. Process Res. Dev.* 18 (2014) 788–792.
8. W. Kroutil, E.-M. Fischereder, C.S. Fuchs, H. Lechner, F.G. Mutti, D. Pressnitz, A. Rajagopalan, J.H. Sattler, R.C. Simon, E. Siirola, Asymmetric preparation of prim-, sec-, and tert-amines employing selected biocatalysts, *Org. Process Res. Dev.* 17 (2013) 751–759.
9. J.A. Lee, J.H. Ahn, I. Kim, S. Li, S.Y. Lee, Separation and purification of three, four, and five carbon diamines from fermentation broth, *Chem. Eng. Sci.* 196 (2019) 324–332.
10. H.T. Kim, K.-A. Baritugo, Y.H. Oh, S.M. Hyun, T.U. Khang, K.H. Kang, S.H. Jung, B.K. Song, K. Park, I.-K. Kim, M.O. Lee, Y. Kam, Y.T. Hwang, S.J. Park, J.C. Joo, Metabolic engineering of corynebacterium glutamicum for the high-level production of cadaverine that can be used for the synthesis of biopolyamide 510, *ACS Sustain. Chem. Eng.* 6 (2018) 5296–5305.
11. O. Anaya-Reza, T. Lopez-Arenas, Comprehensive assessment of the l-lysine production process from fermentation of sugarcane molasses, *Bioprocess Biosyst. Eng.* 40 (2017) 1033–1048.
12. H. Gröger, Biocatalytic concepts for synthesizing amine bulk chemicals: Recent approaches towards linear and cyclic aliphatic primary amines and ω-substituted derivatives thereof, *Appl. Microbiol. Biotechnol.* 103 (2019) 83–95.
13. F.G. Calvo-Flores, F.J. Martin-Martinez, Biorefineries: Achievements and challenges for a bio-based economy, *Front. Chem.* 10 (2022) 973417.
14. A. Gomm, E. O'Reilly, Transaminases for chiral amine synthesis, *Curr. Opin. Chem. Biol.* 43 (2018) 106–112.
15. N.K. Gupta, P. Reif, P. Palenicek, M. Rose, Toward renewable amines: Recent advances in the catalytic amination of biomass-derived oxygenates, *ACS Catal.* 12 (2022) 10400–10440.
16. F. De Schouwer, L. Claes, A. Vandekerkhove, J. Verduyckt, D.E. De Vos, Protein-rich biomass waste as a resource for future biorefineries: State of the art, challenges, and opportunities, *ChemSusChem.* 12 (2019) 1272–1303.
17. J. Liu, J.-Z. Xu, Z.-M. Rao, W.-G. Zhang, Industrial production of L-lysine in Corynebacterium glutamicum: Progress and prospects, *Microbiol. Res.* 262 (2022) 127101.
18. H. Wang, F. Shi, Towards economic and sustainable amination with green and renewable feedstocks, *Chin. J. Chem.* 39 (2021) 1051–1069.
19. H. Li, H. Guo, Y. Su, Y. Hiraga, Z. Fang, E.J.M. Hensen, M. Watanabe, R.L. Smith, N-formyl-stabilizing quasi-catalytic species afford rapid and selective solvent-free amination of biomass-derived feedstocks, *Nat. Commun.* 10 (2019) 699.
20. Z. He, S. Liu, F. Cardenas-Lizana, D. Bi, A. Sanna, Sustainable synthesis of nitrogen-containing chemicals from biomass-derived carbonyls: Catalytic strategies and mechanistic perspectives in multicomponent systems, *Green Chem.* 27 (2025) 10372–10401.
21. M. Mindt, T. Walter, P. Kugler, V.F. Wendisch, Microbial engineering for production of N-functionalized amino acids and amines, *Biotechnol. J.* 15 (2020) 1900451.

5 Bio-Based Amines in Polyurea and Polyurethane Synthesis

5.1 INTRODUCTION

Polyurethanes constitute one of the most versatile and widely used classes of polymeric materials, with applications spanning flexible and rigid foams, coatings, adhesives, sealants, elastomers, and advanced composites. Their widespread use across construction, automotive, electronics, packaging, and biomedical sectors is primarily attributed to the exceptional flexibility of polyurethane chemistry. By careful selection of isocyanates, polyols, amines, chain extenders, catalysts, and additives, polyurethane materials can be engineered to exhibit a wide range of mechanical, thermal, and chemical properties. This molecular-level tunability allows materials to range from soft and highly elastic foams to rigid, high-strength structural polymers. Despite these advantages, the conventional polyurethane industry remains strongly dependent on petroleum-derived raw materials, raising concerns related to fossil resource depletion, environmental impact, and long-term sustainability [1,2].

In response to growing environmental awareness, stricter regulations, and increasing demand for sustainable materials, significant research efforts have been directed toward the development of renewable alternatives for polyurethane synthesis. Bio-based building blocks derived from biomass have emerged as promising candidates to partially or fully replace fossil-based components. To date, most of the research and industrial implementation have focused on bio-based polyols, which are now commercially available and widely used. In contrast, the incorporation of bio-based amines into polyurethane systems has received comparatively limited attention, despite the critical role amines play in controlling reaction kinetics, network structure, and final material performance. This imbalance highlights an important opportunity for innovation in sustainable polyurethane design [3,4].

At the core of polyurethane chemistry lies a step-growth polymerization reaction involving isocyanate groups (–NCO) and compounds containing active hydrogen atoms. The most common and defining reaction is the addition of an isocyanate to a hydroxyl group (–OH), forming a urethane linkage (Figure 5.1). This reaction establishes the fundamental backbone of conventional polyurethane materials. Isocyanates are typically classified as aromatic or aliphatic, with aromatic isocyanates exhibiting higher reactivity due to electron-withdrawing effects, while aliphatic isocyanates offer improved resistance to ultraviolet (UV) radiation and weathering. Polyols, which form the soft segment of the polymer, vary widely in molecular weight, functionality, and chemical structure, allowing control over elasticity, flexibility, and glass transition temperature [5].

DOI: 10.1201/9781003731122-5

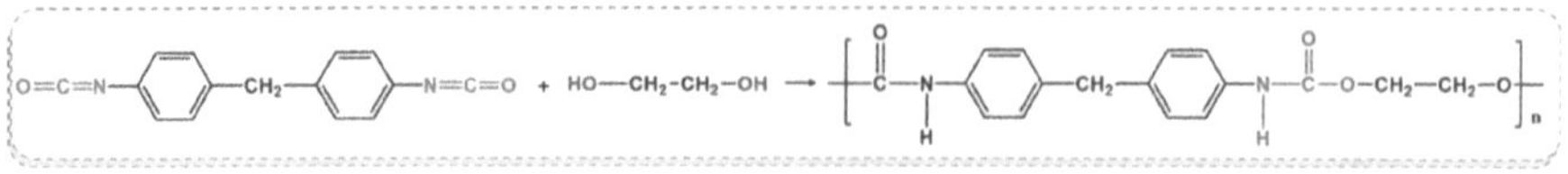

FIGURE 5.1 Reaction scheme of the synthesis of the polyurethane. (Adapted with permission [5]. Copyright 2025 American Chemical Society.)

In practical polyurethane formulations, however, the chemistry extends beyond the isocyanate–polyol reaction. Amines are essential components in many polyurethane systems and serve multiple functions, including chain extenders, crosslinkers, curing agents, and catalysts. In contrast to polyols, amines react with isocyanates to form urea linkages rather than urethane linkages. This reaction is significantly faster and more exothermic than the isocyanate–hydroxyl reaction and often proceeds without the need for a catalyst. As a result, amine–isocyanate reactions play a decisive role in early-stage network formation, cure kinetics, and heat evolution during processing [6–8].

The formation of urea linkages has profound implications for polyurethane morphology and performance. Urea groups can form strong, highly directional hydrogen bonds, which promote hard-segment aggregation and microphase separation between hard and soft domains. Compared to urethane linkages, urea linkages generally lead to increased hard-segment cohesion, higher stiffness, and enhanced thermal stability. Consequently, the incorporation of amines often results in materials with higher modulus and tensile strength, although excessive urea content can reduce elongation at break and impact resistance. Achieving an optimal balance between urethane and urea linkages is therefore a key design consideration in polyurethane formulation [9].

Amines play particularly important roles in polyurethane foams, elastomers, coatings, and adhesives. In foam systems, tertiary amines are widely used as catalysts to control the balance between gelling and blowing reactions, thereby influencing foam density, cell structure, and mechanical properties. Water, which is commonly present in foam formulations, reacts with isocyanates to generate carbon dioxide for foam expansion and produces amines as intermediates. These amines can further react with isocyanates to form urea linkages, contributing to the development of the foam's hard-segment structure. In elastomers, coatings, and adhesives, low-molecular-weight diamines and polyamines are often employed as chain extenders or curing agents, leading to tightly packed hard domains and improved mechanical performance [6,10–12].

From a mechanistic perspective, the reaction between an amine and an isocyanate proceeds through nucleophilic attack of the nitrogen atom on the electrophilic carbon of the isocyanate group, followed by rapid rearrangement to form a urea linkage. This reaction pathway is kinetically favored over the isocyanate–polyol reaction and can dominate network formation in systems containing both amines and polyols. The coexistence of these parallel reaction pathways results in complex kinetic behavior, which influences phase separation, cure profiles, and final polymer morphology [13].

The introduction of bio-based amines into polyurethane chemistry follows these same fundamental reaction principles but introduces additional structural and functional considerations. Bio-based amines are derived from renewable feedstocks such as plant oils, carbohydrates, lignin, terpenes, and amino acids. These sources often impart unique molecular features, including rigid cyclic structures, long aliphatic chains, heteroatoms, and multifunctional groups. Such characteristics can influence the intrinsic reactivity of the amine toward isocyanates, as well as its compatibility with polyols and isocyanates. In many cases, bio-based amines act not only as reactive components but also as structure-directing elements that affect microphase separation and hard-segment organization [14,15].

Beyond their influence on material performance, bio-based amines offer important sustainability advantages. Their renewable origin contributes to reduced dependence on fossil resources and can lower the life cycle greenhouse gas emissions associated with polyurethane production. Additionally, the incorporation of naturally occurring functional motifs may enhance hydrolytic or enzymatic degradation, potentially improving biodegradability or facilitating new recycling strategies. Nevertheless, challenges remain, including variability in biomass-derived feedstocks, formulation stability, and economic competitiveness with established petrochemical amines [16].

Processing considerations are particularly important when amines are incorporated into polyurethane systems. Due to their high reactivity, amines can significantly shorten pot life and increase exothermic heat release during curing. While rapid curing is advantageous for high-throughput manufacturing, it can pose challenges in large-scale processing or thick-section applications. A thorough understanding of amine–isocyanate chemistry is therefore essential for designing formulations that balance reactivity, processability, and performance [7,8].

In summary, polyurethane chemistry is governed by the interplay between isocyanates, polyols, and amines, with amines playing a decisive role in reaction kinetics, network architecture, and material properties. Their ability to form urea linkages, promote hydrogen bonding, and control phase separation makes them indispensable components of many polyurethane systems. This foundational understanding provides the framework for exploring bio-based amines as sustainable alternatives and sets the stage for subsequent sections that examine their sources, reactivity, structure–property relationships, and application-specific performance in greater detail.

5.2 FUNDAMENTALS OF POLYURETHANE CHEMISTRY AND THE ROLE OF AMINES

5.2.1 Overview of Conventional Polyurethane Synthesis

Conventional polyurethanes are synthesized through step-growth polymerization involving the reaction of isocyanates with compounds containing active hydrogen, most commonly polyols. The defining structural feature of polyurethane materials is the urethane linkage, which is formed through a nucleophilic addition reaction between the hydroxyl group of a polyol and the electrophilic carbon of an isocyanate group. This reaction proceeds without the elimination of small molecules and is therefore classified as a polyaddition process rather than a condensation reaction [17]. At the molecular level, the high reactivity of isocyanates originates from their resonance-stabilized structure. The isocyanate functional group exhibits strong polarization due to differences in electronegativity between nitrogen, carbon, and oxygen. As a result, the carbon atom carries a partial positive charge, making it highly susceptible to nucleophilic attack by oxygen- or nitrogen-containing species with active hydrogen atoms. In conventional polyurethane synthesis, the hydroxyl oxygen of the polyol acts as the nucleophile, attacking the electrophilic carbon of the isocyanate to form a tetrahedral intermediate. Subsequent proton transfer and rearrangement yield the urethane bond [18].

The reactivity of the isocyanate component plays a critical role in determining reaction kinetics and processing behavior. Aromatic isocyanates such as toluene diisocyanate and diphenylmethane diisocyanate are significantly more reactive than aliphatic isocyanates like hexamethylene diisocyanate or isophorone diisocyanate. This enhanced reactivity is attributed to electron-withdrawing aromatic rings that increase the electrophilicity of the isocyanate carbon. Consequently, aromatic isocyanates are often preferred in applications requiring rapid curing, while aliphatic isocyanates are selected for improved light stability and weather resistance despite their lower reactivity [5].

Catalysts are commonly employed to accelerate the urethane-forming reaction, particularly under ambient processing conditions where the reaction between isocyanates and polyols can be sluggish due to phase incompatibility. Tertiary amine catalysts and metal-based catalysts represent the two principal classes used in conventional polyurethane synthesis. Mechanistically, tertiary amines enhance reaction rates through dual activation pathways. In one pathway, hydrogen bonding between the amine and the hydroxyl group weakens the O–H bond, increasing the nucleophilicity of the oxygen atom. In the second pathway, the amine coordinates with the isocyanate group, increasing the electrophilicity of the carbon center and facilitating nucleophilic attack. These catalytic cycles regenerate the amine after urethane bond formation, allowing efficient turnover [19].

Overall, conventional polyurethane synthesis is governed by a complex interplay of nucleophilic addition chemistry, catalyst-mediated activation, and secondary reactions involving isocyanates. Understanding these mechanistic principles is essential for rational formulation design and provides the foundation for evaluating alternative approaches, including the incorporation of bio-based amines, which introduce new reaction pathways and structure–property relationships.

5.2.2 Chemical Role of Amines in Polyurethane and Polyurea Formation

Amines play an important supporting and structural role in polyurethane and polyurea chemistry. Their presence strongly influences reaction pathways, network development, and the balance between urethane and urea linkages in the final material. Depending on their functionality and formulation role, amines may participate directly in polymer formation or indirectly by regulating reaction rates and curing behavior. In polyurethane systems, amines are commonly used as chain extenders, crosslinkers, or catalysts, contributing to hard-segment formation and affecting phase separation and hydrogen bonding within the polymer matrix. In polyurea systems, amines serve as primary reactive components, leading to networks dominated by urea linkages and rapid network formation. Overall, amines provide an effective means to control polymer architecture, mechanical performance, and processing characteristics, making them a key component in both polyurethane and polyurea materials. Non-isocyanate polyurethanes synthesized from cyclic carbonates represent one of the most established and widely studied routes toward isocyanate-free polyurethane materials. In this approach, polymer formation proceeds through the step-growth reaction between cyclic carbonates and amines, resulting in poly(hydroxyurethane) networks (Figure 5.2) [20].

The amine component plays a central role in this chemistry, as it acts as the primary nucleophile responsible for ring opening of the cyclic carbonate and subsequent formation of urethane linkages [21]. The reaction mechanism involves nucleophilic attack of the amine nitrogen on the electrophilic carbon atom of the cyclic carbonate ring, followed by ring opening and formation of a urethane bond accompanied by a pendant hydroxyl group. Unlike conventional polyurethane synthesis, this pathway does not release small-molecule by-products and inherently introduces secondary hydroxyl functionalities along the polymer backbone. These hydroxyl groups promote extensive hydrogen bonding, influencing polymer morphology, thermal behavior, and mechanical properties. The structure and functionality of the amine therefore strongly determine the reaction rate, crosslinking density, and final network architecture [6]. Primary diamines and polyamines are most commonly employed due to their high nucleophilicity and ability to generate linear or crosslinked networks, depending on functionality. Aliphatic amines typically exhibit higher reactivity toward cyclic carbonates than aromatic amines, which is attributed to lower steric hindrance and greater electron density on the nitrogen atom. The use of multifunctional amines enables the formation of highly crosslinked non-isocyanate polyurethane networks, while difunctional amines favor more linear or elastomeric materials. Thus, careful selection of the amine structure provides a versatile tool for tailoring non-isocyanate polyurethane properties [22]. In addition to reactivity, amines significantly influence processing conditions in cyclic carbonate-based non-isocyanate polyurethane synthesis. The reaction generally requires elevated temperatures due to the lower electrophilicity of cyclic carbonates compared to isocyanates. However, amine structure, basicity, and steric effects can lower activation barriers and improve reaction efficiency [8]. Overall, cyclic carbonate–amine chemistry provides a robust and environmentally benign platform for non-isocyanate polyurethane synthesis,

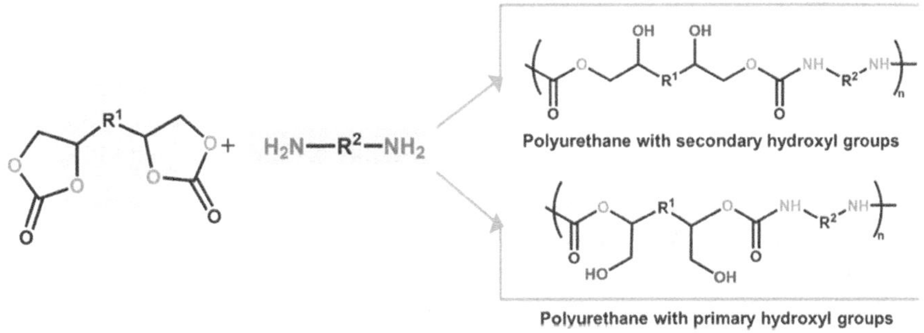

FIGURE 5.2 Synthesis of polyhydroxy urethane from cyclic carbonate and diamine. (Adapted with permission [20]. Copyright 2025 Elsevier.)

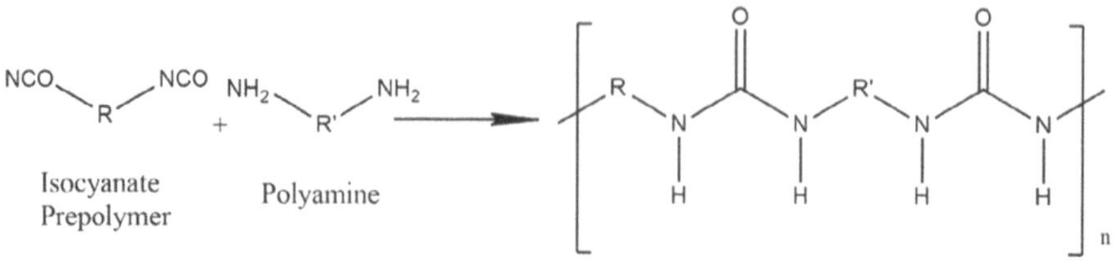

FIGURE 5.3 Polyurea formation reaction. (Adapted with permission [23]. (Copyright 2021 John Wiley & Sons Ltd.)

with amines serving as the key drivers of reaction mechanism, network formation, and structure–property relationships.

Polyurea is most synthesized via a step-growth reaction between diisocyanates and diamines, producing urea linkages (Figure 5.3). This reaction is inherently fast, autocatalytic, and does not require catalysts or elevated temperatures, making it distinct from conventional polyurethane chemistry. Notably, the amine–isocyanate reaction proceeds much faster than reactions involving alcohols or water, rendering polyurea formation largely insensitive to moisture and enabling extremely rapid curing, often within seconds in spray applications [23].

Amines play a central chemical role by defining both the reaction kinetics and the resulting polymer architecture. High-molecular-weight amine-terminated polymers serve as soft segments, contributing flexibility and elasticity, while low-molecular-weight diamines act as chain extenders, controlling molecular weight, phase separation, and mechanical performance. The chemical nature of the amine strongly influences reactivity. Aliphatic amines generally exhibit very high reactivity toward isocyanates, leading to rapid gelation, whereas aromatic amines react more slowly due to steric and electronic effects. This difference is strategically used in formulation design to balance processing time and material properties. Secondary amines, such as aspartic ester amines, are often incorporated to moderate reaction rates while maintaining urea bond formation [24].

The formation of urea groups through amine–isocyanate reactions results in a polymer network rich in intermolecular hydrogen bonding. These hydrogen bonds act as physical crosslinks, promoting phase separation between soft and hard segments and significantly enhancing mechanical strength, thermal stability, and abrasion resistance. During curing, the reorganization and stabilization of hydrogen bonds further increase hard-segment cohesion, which is directly linked to the outstanding durability and performance of polyurea materials. Overall, the chemistry described in this section highlights how the intrinsic reactivity and structural role of amines are fundamental to the rapid formation, robustness, and application versatility of polyurea systems [25,26].

5.3 BIO-BASED AMINES: SOURCES, STRUCTURES, AND REACTIVITY

5.3.1 Renewable Feedstocks and Synthesis Routes

Bio-based amines are derived from renewable carbon sources and represent an important class of sustainable nitrogen-containing building blocks for polymer chemistry. Unlike conventional amines produced from fossil-based hydrocarbons, bio-based amines originate from biomass such as plant oils, carbohydrates, lignocellulosic materials, terpenes, and amino acids. The choice of feedstock and conversion pathway strongly influences the chemical structure, purity, functionality, and reactivity of the resulting amines.

Plant oils are among the most widely explored feedstocks for bio-based amine production. Triglycerides obtained from soybean, castor, palm, or rapeseed oils provide long aliphatic chains with inherent functionality that can be chemically transformed into amines. Common synthetic routes include epoxidation of unsaturated fatty acids followed by ring opening with ammonia or amines, nitrile formation through dehydration of fatty amides and subsequent hydrogenation, and reductive amination of carbonyl-containing intermediates. These approaches yield monoamines, diamines, or polyamines with flexible hydrocarbon backbones, making them suitable for use as chain extenders, curing agents, or soft-segment modifiers in polymer systems [27].

Carbohydrate-based feedstocks such as glucose, sorbitol, and cellulose derivatives offer another important route to bio-based amines. Sugars can be converted into amines through multi-step processes involving dehydration to platform chemicals, formation of aldehydes or ketones, and subsequent reductive amination. Amination of sugar alcohols and hydrogenation of nitrile intermediates are also commonly employed. These routes often generate multifunctional amines containing hydroxyl groups in addition to amine functionalities, introducing opportunities for enhanced hydrogen bonding and network formation in polymers [28,29].

Lignin and lignin-derived aromatics represent a promising but more complex source of bio-based amines. Due to lignin's aromatic structure, amines derived from this feedstock often possess rigid backbones and aromatic character. Amination strategies include catalytic amination of lignin-derived phenols, reductive amination of aldehydes obtained from lignin depolymerization, and conversion of aromatic nitriles. Although lignin-based amines present challenges related to feedstock heterogeneity and processing, they offer access to aromatic amines that can enhance stiffness, thermal stability, and chemical resistance in polymer networks [30,31].

Amino acids and proteins provide a naturally nitrogen-rich source for bio-based amines. Decarboxylation of amino acids yields primary amines, while further chemical modification can produce diamines or polyamines. These routes are particularly attractive because nitrogen is already incorporated into the biomass, reducing the need for external nitrogen sources. However, controlling molecular structure and scalability remains a key challenge for amino-acid-derived amines [32,33].

Across all feedstocks, catalytic hydrogenation [34], borrowing-hydrogen methodologies [35,36], and reductive amination [37] are among the most important synthesis strategies. These approaches are often compatible with green chemistry principles, including high atom efficiency, reduced waste generation, and avoidance of hazardous intermediates. As a result, renewable feedstocks and synthesis routes provide a diverse and adaptable platform for producing bio-based amines tailored to specific polymer applications. The schematic presentation of different bio based amines is given in Figure 5.4.

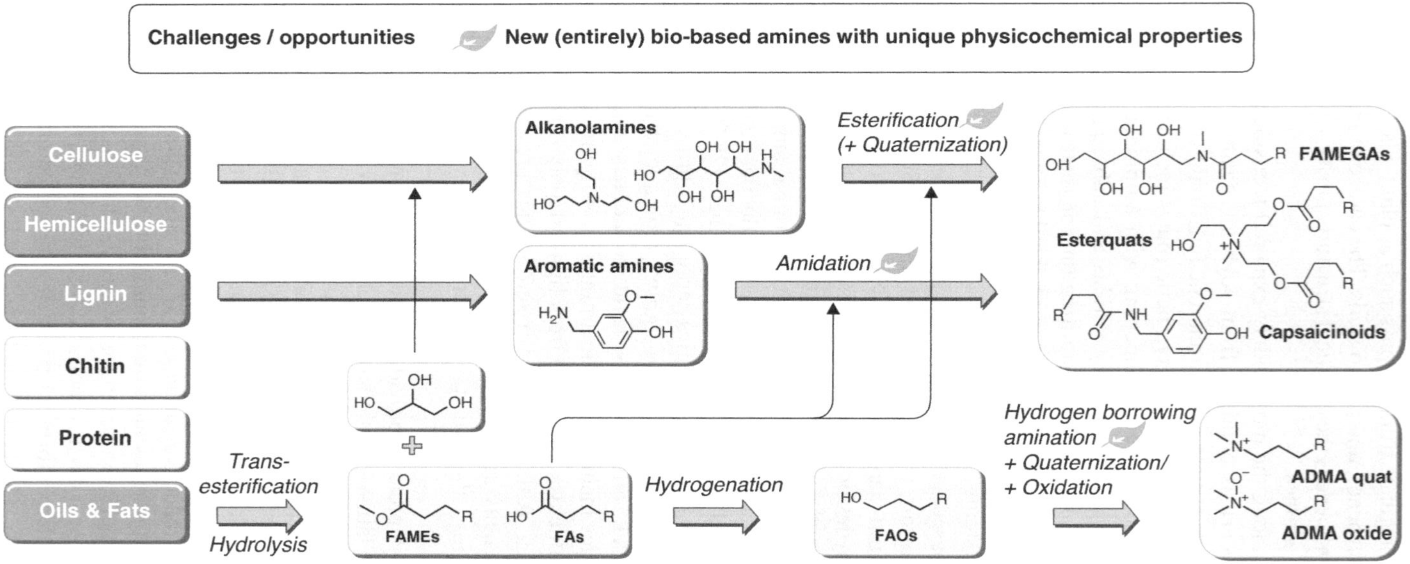

FIGURE 5.4 Schematic overview to produce many different bio-based amines. (Adapted with permission [29]. Copyright 2017 Royal Society of Chemistry.)

5.3.2 Structural Diversity of Bio-Based Amines

One of the defining advantages of bio-based amines is their exceptional structural diversity, which arises from the wide range of renewable feedstocks and chemical transformation pathways available. Bio-based amines can vary significantly in molecular weight, functionality, backbone rigidity, and substitution pattern, enabling fine control over their chemical behavior and performance in polymer systems.

Based on functionality, bio-based amines may be classified as monoamines, diamines, or polyamines [38]. Monoamines typically act as chain terminators or modifiers, while diamines and polyamines serve as chain extenders or crosslinkers. The presence of multiple amine groups increases crosslink density and enhances network formation, which is particularly important in thermoset polyurethanes and polyureas. Bio-based polyamines derived from plant oils or carbohydrates often contain long aliphatic segments between amine groups, providing flexibility while maintaining reactivity. The nature of the carbon backbone further contributes to structural diversity. Aliphatic bio-based amines, commonly derived from fatty acids or sugar alcohols, exhibit high flexibility and relatively high nucleophilicity. These amines typically show fast reaction kinetics with electrophilic species and are well suited for elastomeric and flexible polymer applications. In contrast, aromatic or cyclic bio-based amines derived from lignin or terpene-based feedstocks introduce rigidity and steric constraints, which can slow reaction rates but improve mechanical strength and thermal stability in the resulting polymers. Many bio-based amines also contain additional functional groups such as hydroxyls, ethers, esters, or unsaturated bonds. These functionalities enable secondary interactions, including hydrogen bonding and dipole–dipole interactions, which strongly influence polymer morphology and phase behavior. Hydroxyl-containing amines, for example, can participate in both amine- and alcohol-based reactions, offering dual reactivity and increased formulation flexibility. Such multifunctional structures are difficult to achieve using conventional petrochemical amines without extensive synthetic modification. Steric effects play a significant role in the reactivity of bio-based amines. Bulky substituents derived from complex biomass structures can reduce accessibility of the nitrogen lone pair, moderating reaction rates and improving processing control. This characteristic is particularly advantageous in fast-reacting systems, where overly rapid gelation can limit practical application. Secondary amines and sterically hindered primary amines derived from renewable sources are therefore often used to balance reactivity and workability. Overall, the structural diversity of bio-based amines enables precise tuning of chemical reactivity, network architecture, and final material properties. By selecting appropriate feedstocks and synthesis routes, bio-based amines can be designed to meet the specific demands of polyurethane, polyurea, and non-isocyanate polyurethane systems. This versatility positions bio-based amines as key enablers for advancing sustainable polymer chemistry while maintaining performance comparable to, or exceeding, that of conventional amines [8,16,39].

5.4 STRUCTURE–PROPERTY RELATIONSHIPS AND PERFORMANCE

The performance of polyurethane and related polymer systems is strongly governed by their molecular architecture and the interactions established during network

aqueous buffer at elevated temperature. The study showed that urethane and urea linkages are significantly more hydrolytically stable than ester bonds, and that degradation in polyurethane systems occurs primarily at ester-containing soft segments rather than at urea or urethane groups. Strong hydrogen bonding within urea-rich hard segments further limits water diffusion into the polymer network, resulting in superior hydrolytic stability compared to conventional ester-based polyurethanes [43].

Oxidative and UV stability are also influenced by amine structure. Aliphatic bio-based amines typically provide better resistance to photo-induced degradation than aromatic amines, which may undergo discoloration or embrittlement under prolonged UV exposure. Incorporation of stabilizers or protective coatings is often employed to mitigate these effects in outdoor applications. At the same time, the renewable origin of bio-based amines supports reduced environmental impact over the material life cycle, even when stabilization strategies are required. Farhadian et al. reported that polyurethane-type networks synthesized using aliphatic bio-based amines (Figure 5.5) exhibit good resistance to oxidative degradation, which was attributed to the absence of aromatic chromophores susceptible to photo-induced reactions. The study highlighted that aromatic structures are more prone to UV-induced discoloration and embrittlement, whereas aliphatic amines provide improved stability under aging conditions. The authors further noted that the use of stabilizers or protective coatings can enhance outdoor durability without diminishing the sustainability advantages of bio-based amine systems [44].

Mechanical durability under cyclic loading and long-term stress is closely related to hard-segment organization and hydrogen-bonding reversibility. Bio-based amines that promote well-distributed hard domains with strong but reversible physical interactions enable efficient stress dissipation and resistance to fatigue. This behavior is particularly advantageous in elastomeric and coating applications, where repeated deformation is common. Wang et al. demonstrated that the mechanical durability of bio-based polyurethane networks under cyclic loading is strongly linked to hard-segment organization and the reversibility of hydrogen-bonding interactions. In their study, bio-inspired polyurethanes containing well-distributed hard domains formed through multiple, reversible hydrogen bonds exhibited efficient stress dissipation during repeated deformation. These dynamic interactions acted as physical crosslinks that could reversibly break and reform under applied strain, preventing permanent network damage and delaying crack initiation. As a result, the materials retained a high fraction of their mechanical strength after repeated loading cycles, showing enhanced fatigue resistance and elastic recovery (Figure 5.6). The authors concluded that such reversible hydrogen-bonded hard segments are particularly beneficial for elastomeric and coating applications where long-term durability under repeated mechanical stress is required [45].

Overall, bio-based amines offer a powerful means to control structure–property relationships and performance in polyurethane and related polymer systems. By influencing mechanical strength, thermal behavior, chemical resistance, hydrogen bonding, and durability, these renewable amines enable the design of high-performance materials that meet both technical and sustainability objectives.

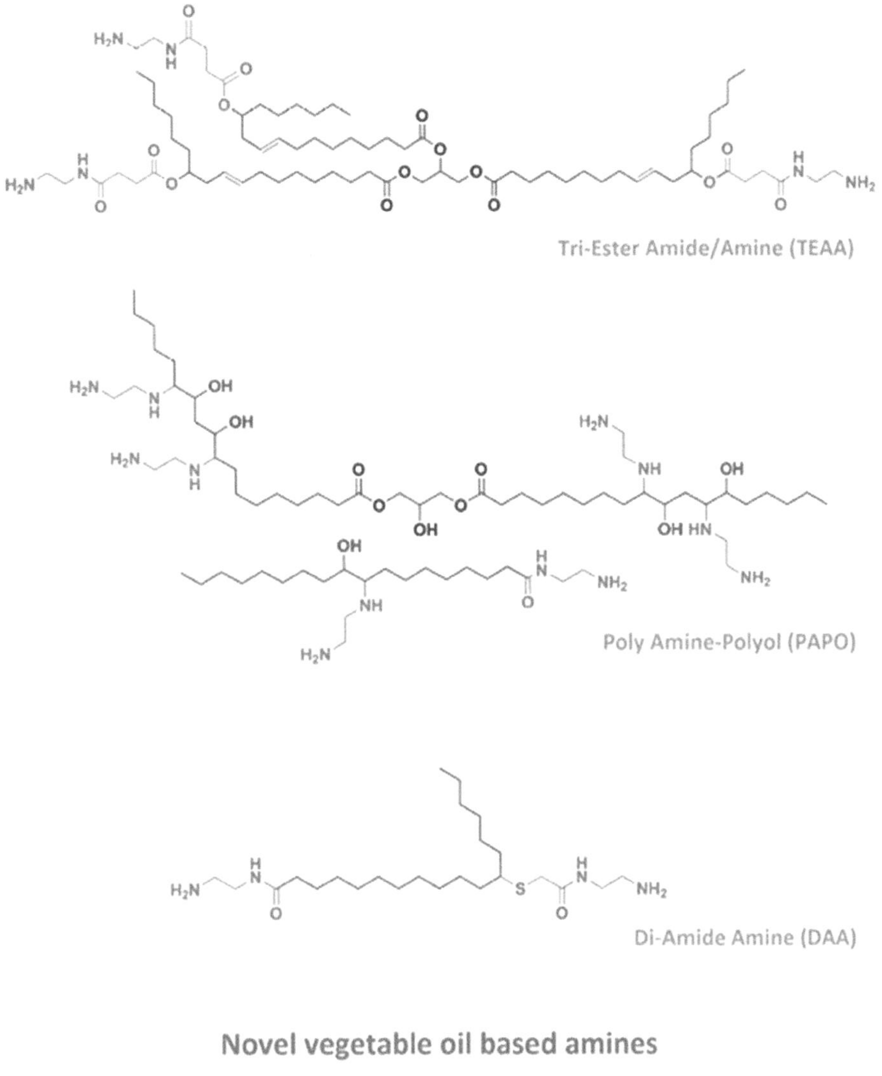

FIGURE 5.5 Structures of vegetable oil-based amines. (Adapted with permission [44]. Copyright 2018 Elsevier.)

5.5 ENVIRONMENTAL IMPACT, BIODEGRADABILITY, AND SUSTAINABILITY ASPECTS

The growing interest in sustainable polymeric materials has encouraged the replacement of petroleum-derived chain extenders and curing agents with bio-based amines in polyurethane and polyurea systems. Bio-based amines derived from renewable resources such as vegetable oils, lignin derivatives, amino acids, and fermentation products offer the potential to reduce fossil resource consumption, lower environmental impact, and improve the overall sustainability profile of these materials.

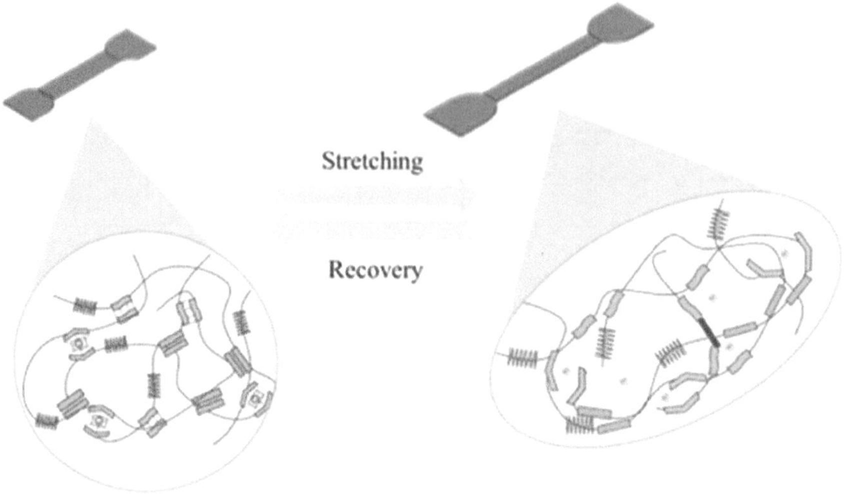

FIGURE 5.6 Schematic representation of morphological evolution during stretching and recovery. (Adapted with permission [45]. (Copyright 2025 American Chemical Society.)

In addition to contributing renewable carbon content, bio-based amines can influence polymer architecture, hydrogen bonding, and crosslink density, which in turn affect biodegradability, recyclability, and life cycle performance. Recent reviews have emphasized that sustainability in polyurethane and polyurea systems cannot be evaluated solely based on renewable content but must also consider processing energy, durability, toxicity, and end-of-life behavior [46,47]. Bio-based amines are particularly relevant because they are directly involved in forming urea or urethane linkages that determine the chemical stability and degradation characteristics of the polymer network.

5.5.1 Biodegradability and End-of-Life Behavior

Polyurethane and polyurea materials are generally resistant to biodegradation due to strong intermolecular interactions, hydrogen bonding, and crosslinked structures. However, the incorporation of bio-based amines can modify degradation pathways by introducing hydrolytically or enzymatically susceptible segments. Amines derived from fatty acids, amino alcohols, or lignin fragments often contain aliphatic chains, ester groups, or other functional moieties that increase segmental mobility and promote hydrolysis under suitable environmental conditions.

Studies have shown that polyester-based polyurethanes degrade more readily than polyether-based systems because ester bonds are susceptible to microbial and enzymatic attack [48]. When bio-based amines are used as chain extenders in these systems, they can further enhance water uptake and facilitate microbial colonization, accelerating degradation compared with fully petroleum-based analogues. Microbial degradation of polyurethane has been reported using various fungi and bacteria capable of cleaving ester and urethane linkages. For example, species of *Pseudomonas*,

Bacillus, and certain fungi have demonstrated the ability to degrade polyurethane coatings and foams under controlled conditions [48,49]. Although polyurea materials are typically more resistant due to the stability of urea bonds, the introduction of flexible aliphatic bio-based amines may increase susceptibility to environmental degradation by reducing packing density and increasing free volume. Another important development is the use of bio-based amines in non-isocyanate polyurethane systems. In these materials, cyclic carbonates react with diamines to form polyhydroxyurethanes containing pendant hydroxyl groups. These hydroxyl functionalities increase hydrophilicity and can enhance hydrolytic degradation compared with conventional polyurethane networks [50]. Such systems also eliminate isocyanates, improving environmental and occupational safety during synthesis.

End-of-life options for polyurethane and polyurea prepared with bio-based amines include mechanical recycling, chemical recycling, glycolysis, hydrolysis, and energy recovery. Chemical recycling methods that break urethane or urea linkages to recover polyols or amines are gaining attention as part of circular economy strategies [47]. The presence of bio-derived segments may also improve compatibility with emerging depolymerization and reprocessing technologies, including dynamic covalent networks and vitrimeric polyurethanes. Despite these advances, complete biodegradation in natural environments remains limited, particularly for highly crosslinked coatings and elastomers. Achieving a balance between durability during service life and degradability after disposal remains a key challenge in material design.

5.5.2 Life Cycle and Environmental Performance

Life cycle assessment (LCA) is widely used to evaluate the environmental performance of bio-based polyurethane and polyurea materials. LCA considers all stages of a material's life, including raw material production, synthesis, processing, transportation, use, and disposal. Replacing petroleum-based chain extenders with bio-based amines can significantly reduce fossil energy demand and greenhouse gas emissions, particularly when feedstocks are derived from waste biomass or nonfood resources [51]. Bio-based amines synthesized from vegetable oils, for example, often have lower embodied carbon than aromatic diamines derived from petrochemical processes. However, the overall environmental benefit depends on agricultural inputs, land use, and processing energy. Fertilizer use, irrigation, and transportation can contribute significantly to the environmental footprint of bio-based feedstocks, and these factors must be carefully considered when interpreting LCA results. Another important sustainability aspect is toxicity and worker safety. Conventional polyurethane production involves handling reactive isocyanates, which present health and environmental risks. Systems based on bio-based amines, particularly in non-isocyanate routes, can reduce exposure to hazardous intermediates and improve safety during manufacturing [50]. Durability is also a critical parameter in environmental performance. Materials with longer service life often have lower overall environmental impact because replacement and disposal are reduced. Bio-based amines can enhance mechanical strength, thermal stability, and chemical resistance through strong hydrogen bonding in urea linkages, potentially extending the lifetime of coatings, elastomers, and adhesives. This improved durability can offset some of the environmental impacts associated with raw material production. Recent studies emphasize that the most sustainable polyurethane and polyurea systems are those designed with a holistic approach that integrates renewable feedstocks, energy-efficient synthesis, long service life, and effective recycling or degradation pathways. Bio-based amines play a key role in this strategy because they influence both polymer structure and environmental performance across the entire life cycle.

5.6 RECENT ADVANCES AND APPLICATIONS

In recent years, significant progress has been made in the development of polyurethane and polyurea materials incorporating bio-based amines as chain extenders, curing agents, or crosslinkers. Advances in biomass conversion, catalytic amination, and functionalization of natural feedstocks have enabled the production of amines derived from vegetable oils, lignin derivatives, amino acids, and carbohydrate-based platform molecules. These materials are increasingly being investigated not only to reduce dependence on petroleum resources but also to tailor polymer properties such as flexibility, adhesion, thermal stability, and chemical resistance. As a result, bio-based amines are now being explored in a wide range of applications, including foams, coatings, elastomers, sealants, and structural adhesives, where both performance and sustainability are important design criteria.

Recent research has demonstrated that bio-based amines can play a central role in the synthesis of sustainable polyurethane and polyurea materials, serving as chain extenders, curing agents, or structural building blocks. In one study, biomass-derived diamines prepared from furfurylamine and isomannide were used to synthesize polyureas through a non-isocyanate route (Figure 5.7). Furfurylamine, obtained from biomass-derived furfural, was converted into a furan-based diamine, while isomannide derived from carbohydrate feedstocks was transformed into a rigid diamine. These bio-based diamines were further converted into activated dicarbamates, which

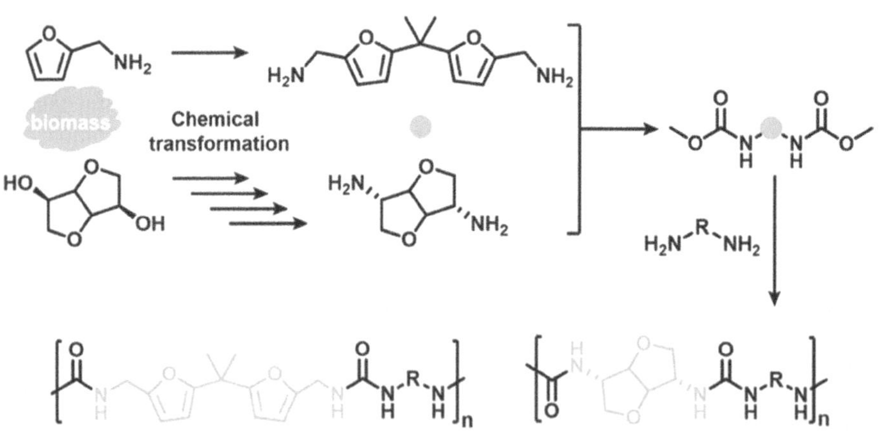

FIGURE 5.7 Synthesis of polyureas from 2-furfurylamine and isomannide. Adapted with permission [52]. (Copyright 2022 American Chemical Society.)

were then reacted with various diamines via polycondensation to form polyurea networks. This approach enabled the incorporation of renewable, rigid structural units into the polymer backbone, resulting in materials with good thermal stability and tunable glass-transition temperatures. The study also showed that the structure of the bio-based amine strongly influenced polymer properties. Furan-based and isomannide-based segments introduced rigidity and increased thermal resistance, while the choice of diamine affected molecular weight, glass-transition temperature, and mechanical performance. Thermal analysis revealed degradation temperatures above about 250°C–300°C and glass-transition temperatures over a wide range depending on the polymer composition. These results demonstrate that bio-based amines can function not only as renewable substitutes for petrochemical diamines but also as structure-directing components that allow property tailoring in polyurea systems. In terms of applications, the polyureas prepared from these bio-based amines are suitable for areas where conventional polyureas are widely used, including coatings, construction materials, automotive components, and protective systems. The strong hydrogen bonding of urea linkages provides high mechanical strength, wear resistance, and chemical durability, making such materials promising for durable coatings and structural uses [52].

A second study reported the preparation of a high-strength, recyclable, and self-healing non-isocyanate polyurethane using a bio-based diamine curing agent. In this work, a hyperbranched cyclic carbonate was first synthesized from a terpene-derived diamine through epoxidation and subsequent reaction with carbon dioxide. The resulting cyclic carbonate precursor was then reacted with a long-chain bio-based diamine (Priamine 1074), forming a crosslinked polyurethane network containing carbamate and hydroxyl groups (Figure 5.8). The reaction between cyclic carbonate and amine produced the polymer without the use of toxic isocyanates, highlighting a greener synthetic pathway. The incorporation of the bio-based diamine contributed significantly to the material properties. The flexible aliphatic structure of the diamine improved chain mobility and toughness, while the hyperbranched cyclic carbonate provided rigidity and crosslinking density. The resulting materials exhibited high tensile strength, good thermal stability, and dynamic bond exchange behavior that enabled self-healing, reprocessing, and chemical recycling. These characteristics illustrate how bio-based amines can be used not only as renewable raw materials but also as functional components that impart advanced properties such as recyclability and shape memory. In terms of applications, the polyurethane materials developed in this study demonstrated strong adhesion, solvent resistance, and mechanical durability, making them suitable for use in adhesives, coatings, packaging materials, and furniture components. The combination of high strength, recyclability, and self-healing behavior suggests that such bio-based polyurethane systems have strong potential for industrial applications where sustainability and long service life are required. Overall, these studies show that bio-based amines are emerging as important building blocks in polyurethane and polyurea chemistry. By enabling non-isocyanate synthesis routes, improving mechanical and thermal performance, and supporting recyclability and self-healing behavior, bio-based amines contribute both to material performance and environmental sustainability. Continued research in biomass-derived diamine synthesis, polymer design, and scalable processing is expected to further expand their industrial adoption in coatings, adhesives, foams, and structural materials [53].

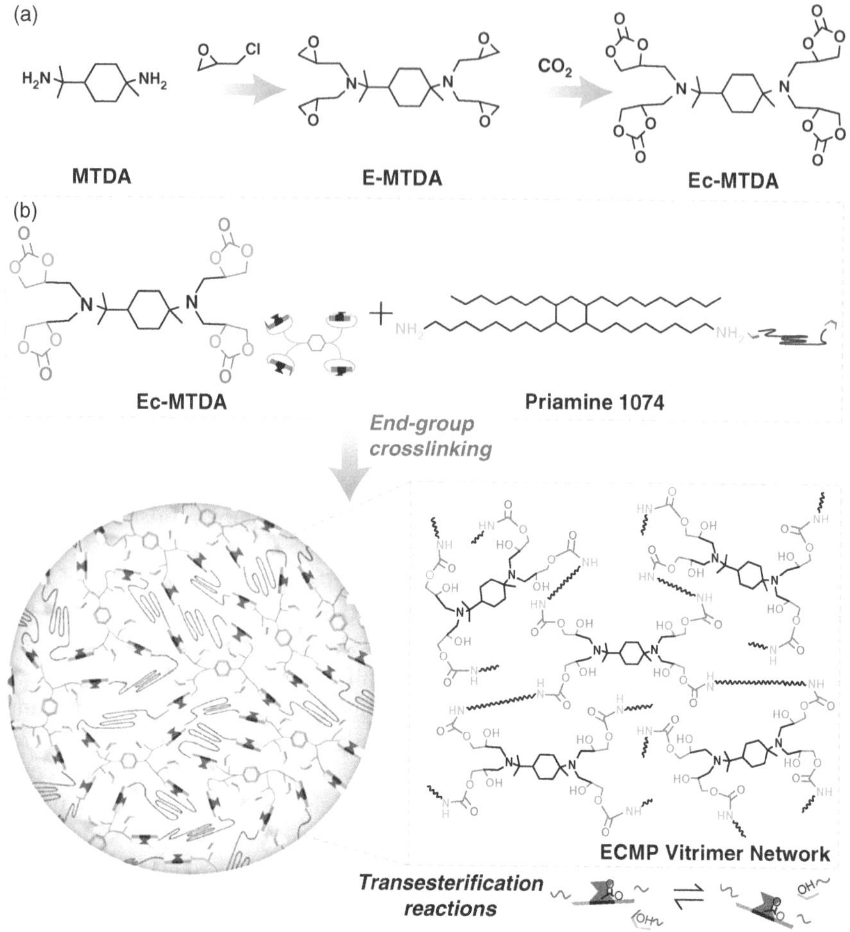

FIGURE 5.8 (a) Synthesis route of cyclic carbonate precursor. (b) Formation of polyurethane network containing carbamate and hydroxyl groups. (Adapted with permission [53]. Copyright 2023 American Chemical Society.)

5.7 COMMERCIALIZATION STATUS, CURRENT CHALLENGES, AND FUTURE OUTLOOK

The commercialization of polyurethane and polyurea systems prepared using bio-based amines is gradually progressing, particularly in coatings, adhesives, sealants, and elastomers. Research has shown that bio-based amine-cured polyurethanes can exhibit high mechanical strength, thermal stability, and strong adhesion, making them suitable for structural materials, coatings, and bonding applications. Similarly, polyureas synthesized using renewable diamines derived from biomass have demonstrated good thermal stability and mechanical performance, indicating that bio-based amines can meet many of the functional requirements of conventional materials. Despite these advances, several challenges still limit large-scale adoption. Bio-based amines are often more expensive than petrochemical counterparts, and variability in biomass feedstocks can affect consistency and processing. In addition, industrial formulations require precise curing behavior, long-term durability, and compatibility with existing processing methods, which may require further optimization when renewable amines are introduced. Looking ahead, continued advances in biomass conversion, catalytic amination, and biorefinery technologies are expected to improve the availability and cost competitiveness of bio-based amines. At the same time, increasing environmental regulations and the growing demand for low-carbon materials are likely to accelerate the development of polyurethane and polyurea systems based on renewable curing agents. With ongoing improvements in synthesis, performance, and recyclability, bio-based amines are expected to play an increasingly important role in the next generation of sustainable polymer materials.

5.8 CONCLUSION

Bio-based amines have emerged as an important class of renewable building blocks for the synthesis of polyurethane, polyurea, and non-isocyanate polyurethane materials. Unlike conventional petrochemical amines, these compounds are derived from diverse renewable resources such as plant oils, carbohydrates, lignin, terpenes, and amino acids, offering opportunities to reduce fossil resource consumption and improve the sustainability profile of polymeric materials. At the same time, the structural diversity of bio-based amines provides a versatile platform for tailoring polymer architecture, hydrogen bonding, crosslink density, and microphase separation, which directly influence mechanical, thermal, and chemical performance. Recent research demonstrates that bio-based amines can function effectively as chain extenders, curing agents, and reactive intermediates in both polyurethane and polyurea systems. Their incorporation often leads to materials with improved mechanical strength, enhanced thermal stability, and strong intermolecular interactions due to the formation of urea or hydroxyurethane linkages. Advances in non-isocyanate polyurethane chemistry further highlight the potential of bio-based amines to enable safer and more environmentally benign synthesis routes while introducing dynamic functionalities such as self-healing, recyclability, and reprocessability. These developments illustrate that renewable amines are not merely substitutes for petrochemical counterparts but also enablers of new material functionalities and design strategies. Despite these promising developments, several challenges remain before bio-based amines can be widely adopted in large-scale industrial applications. Variability in biomass feedstocks, limited commercial availability of certain diamines, higher production costs, and the need for long-term durability and performance validation continue to present obstacles. In addition, achieving an optimal balance between processability, mechanical performance, and environmental degradability remains an important research objective. Looking forward, continued progress in catalytic amination, biorefinery technologies, and green synthesis methods is expected to improve the efficiency and scalability of bio-based amine production. Integrating life cycle assessment, recyclability, and circular material design into polymer development will further enhance the sustainability of polyurethane and polyurea systems. With ongoing advances in molecular design, processing technologies, and renewable feedstock utilization, bio-based amines are likely to play a central role in the development of next-generation high-performance and environmentally responsible polymeric materials.

REFERENCES

1. M.L. Chaudhary, R. Patel, S. Chaudhari, R.K. Gupta, Impact of diverse diols and diisocyanates on thermosetting bio-based polyurethane films, *Next Mater.* 8 (2025) 100609.
2. M.L. Chaudhary, Y.N. Patel, S. Chaudhari, R. Patel, R.K. Gupta, Enhancing the mechanical strength of soy-based polyurethane adhesives through incorporation of schiff base diols, *Int. J. Adhes. Adhes.* 142 (2025) 104118.
3. D.K. Chelike, S.A. Gurusamy Thangavelu, Biodegradable isocyanate-free polyurethane films via a noncatalytic route: Facile modified polycaprolactone triol and bio-based diamine as precursors, *RSC Adv.* 13 (2023) 309–319.
4. K. Błażek, P. Kasprzyk, J. Datta, Diamine derivatives of dimerized fatty acids and bio-based polyether polyol as sustainable platforms for the synthesis of non-isocyanate polyurethanes, *Polymer (Guildf).* 205 (2020) 122768.
5. M.L. Chaudhary, R.K. Gupta, Polyurethanes: An introduction, in: M. L. Chaudhary, R. K. Gupta (Eds.), *Non-Isocyanate Polyurethanes Chem. Progress, Challenges*, American Chemical Society, Washington, DC, 2025: pp. 1–14.
6. R. Patel, A. Kumar, M.L. Chaudhary, R.K. Gupta, Design and characterization of silane-modified bio-based non-isocyanate polyurethane coatings for advanced surface applications, *Materials (Basel).* 18 (2025) 5551.
7. Y. Li, R. Shang, S. Wu, Q. Tang, Q.-E. Cao, W. Fang, Harnessing the synergism between Ni particles and an Ni–ceria interface for efficient biomass reductive amination, *Green Chem.* 27 (2025) 10019–10031.
8. V. Froidevaux, C. Negrell, S. Caillol, J.P. Pascault, B. Boutevin, Biobased amines: From synthesis to polymers; present and future, *Chem. Rev.* 116 (2016) 14181–14224.
9. K. Zhou, J. Xie, H. Shen, J. Gong, X. Pei, X. Zhang, H. Luo, Y. Yang, Q. Wang, Y. Zhang, High-performance polyurethane urea for sealing applications in extreme temperature conditions, *Chem. Eng. J.* 527 (2026) 171699.
10. P. Patel, R. Patel, M.L. Chaudhary, N. Chaudhary, R.K. Gupta, High-performance bio-based non-isocyanate polyurethane adhesive: A solvent and catalyst-free synthesis approach, *J. Polym. Environ.* 32 (2024) 5024–5035.
11. X. Mou, Z. Yang, X. Lai, J. Ding, Y. Chen, H. Li, X. Zeng, Self-healing and reprocessable biobased non-isocyanate polyurethane elastomer with dual dynamic covalent adaptive network for flexible strain sensor, *Chem. Eng. J.* 493 (2024) 152876.
12. V. Valette, N. Kébir, F.B. Tiavarison, F. Burel, L. Lecamp, Preparation of flexible biobased non-isocyanate polyurethane (NIPU) foams using the transurethanization approach, *React. Funct. Polym.* 181 (2022) 105416.
13. M.C. Pannone, C.W. Macosko, Kinetics of isocyanate amine reactions, *J. Appl. Polym. Sci.* 34 (1987) 2409–2432.
14. E.F. Aransiola, T.V Ojumu, O.O. Oyekola, T.F. Madzimbamuto, D.I.O. Ikhu-Omoregbe, A review of current technology for biodiesel production: State of the art, *Biomass Bioenergy.* 61 (2014) 276–297.
15. W. Leuchtenberger, K. Huthmacher, K. Drauz, Biotechnological production of amino acids and derivatives: Current status and prospects, *Appl. Microbiol. Biotechnol.* 69 (2005) 1–8.
16. B. Vermeeren, S. Van Praet, W. Arts, T. Narmon, Y. Zhang, C. Zhou, H.P. Steenackers, B.F. Sels, From sugars to aliphatic amines: As sweet as it sounds? Production and applications of bio-based aliphatic amines, *Chem. Soc. Rev.* 53 (2024) 11804–11849.

17. A.A. Caraculacu, S. Coseri, Isocyanates in polyaddition processes. Structure and reaction mechanisms, *Prog. Polym. Sci.* 26 (2001) 799–851.
18. M.F. Sonnenschein, *Polyurethanes: Science, Technology, Markets, and Trends*, John Wiley & Sons, Hoboken, NJ, 2021.
19. J. Dodge, Polyurethanes and polyureas, in: T.E.L. Martin, E. Rogers (Eds.), *Synthetic Methods in Step-Growth Polymers*, John Wiley Online Library, Hoboken, NJ, 2003: pp. 197–263.
20. M.L. Chaudhary, R. Patel, R.K. Gupta, Beyond isocyanates: Advances in non-isocyanate polyurethane chemistry and applications, *Polymer (Guildf).* 332 (2025) 128553.
21. M.S. Kathalewar, P.B. Joshi, A.S. Sabnis, V.C. Malshe, Non-isocyanate polyurethanes: From chemistry to applications, *RSC Adv.* 3 (2013) 4110–4129.
22. J. Guan, Y. Song, Y. Lin, X. Yin, M. Zuo, Y. Zhao, X. Tao, Q. Zheng, Progress in study of non-isocyanate polyurethane, *Ind. Eng. Chem. Res.* 50 (2011) 6517–6527.
23. B. Shojaei, M. Najafi, A. Yazdanbakhsh, M. Abtahi, C. Zhang, A review on the applications of polyurea in the construction industry, *Polym. Adv. Technol.* 32 (2021) 2797–2812.
24. S. Das, I. Yilgor, E. Yilgor, B. Inci, O. Tezgel, F.L. Beyer, G.L. Wilkes, Structure–property relationships and melt rheology of segmented, non-chain extended polyureas: Effect of soft segment molecular weight, *Polymer (Guildf).* 48 (2007) 290–301.
25. Q. Zhou, L. Cao, Q. Li, Y. Yao, Z. Ouyang, Z. Su, X. Chen, Investigation of the curing process of spray polyurea elastomer by FTIR, DSC, and DMA, *J. Appl. Polym. Sci.* 125 (2012) 3695–3701.
26. T. Mariappan, C.A. Wilkie, Thermal and fire retardant properties of polyurea, *Polimery.* 58 (2013) 371–384.
27. J.A. Melero, J. Iglesias, A. Garcia, Biomass as renewable feedstock in standard refinery units. Feasibility, opportunities and challenges, *Energy Environ. Sci.* 5 (2012) 7393–7420.
28. H.B.H.O.H. Lamborn, Economics of sucrose, *Ind. Eng. Mater.* 47 (1955) 1392–1397.
29. M. Pelckmans, T. Renders, S. Van de Vyver, B.F. Sels, Bio-based amines through sustainable heterogeneous catalysis, *Green Chem.* 19 (2017) 5303–5331.
30. D. Verboekend, Y. Liao, W. Schutyser, B.F. Sels, Alkylphenols to phenol and olefins by zeolite catalysis: A pathway to valorize raw and fossilized lignocellulose, *Green Chem.* 18 (2016) 297–306.
31. Z. Cao, J. Engelhardt, M. Dierks, M.T. Clough, G. Wang, E. Heracleous, A. Lappas, R. Rinaldi, F. Schüth, Catalysis meets nonthermal separation for the production of (alkyl) phenols and hydrocarbons from pyrolysis oil, *Angew. Chem.* 129 (2017) 2374–2379.
32. C.O. Tuck, E. Pérez, I.T. Horváth, R.A. Sheldon, M. Poliakoff, Valorization of biomass: Deriving more value from waste, *Science.* 337 (2012) 695–699.
33. E. Scott, F. Peter, J. Sanders, Biomass in the manufacture of industrial products—The use of proteins and amino acids, *Appl. Microbiol. Biotechnol.* 75 (2007) 751–762.
34. P.S. Metkar, M.A. Scialdone, K.G. Moloy, Lysinol: A renewably resourced alternative to petrochemical polyamines and aminoalcohols, *Green Chem.* 16 (2014) 4575–4586.
35. G. Guillena, D.J. Ramón, M. Yus, Hydrogen autotransfer in the N-alkylation of amines and related compounds using alcohols and amines as electrophiles, *Chem. Rev.* 110 (2010) 1611–1641.
36. K. Shimizu, S. Kanno, K. Kon, S.M.A. Hakim Siddiki, H. Tanaka, Y. Sakata, N-alkylation of ammonia and amines with alcohols catalyzed by Ni-loaded $CaSiO_3$, *Catal. Today.* 232 (2014) 134–138.
37. A.F. Abdel-Magid, K.G. Carson, B.D. Harris, C.A. Maryanoff, R.D. Shah, Reductive amination of aldehydes and ketones with sodium triacetoxyborohydride. Studies on direct and indirect reductive amination procedures, *J. Org. Chem.* 61 (1996) 3849–3862.

industries such as textiles, military equipment, and industrial manufacturing. Over subsequent decades, polyamides expanded beyond fibers into engineering plastics, supporting advancements in machinery, automotive design, and consumer products. Polyimides emerged later but quickly became associated with high-temperature and high-reliability applications. As aerospace exploration and electronic technologies advanced in the mid-twentieth century, materials capable of withstanding extreme thermal and mechanical environments were urgently needed. Polyimides addressed this challenge, enabling progress in aircraft systems, space missions, and electronic insulation technologies. The rise of microelectronics further reinforced their importance, as flexible and heat-resistant materials became essential for compact and high-performance devices. Over time, both polyamides and polyimides have undergone continuous refinement, demonstrating how material science adapts to evolving technological demands [5–7].

In recent decades, sustainability has become a defining factor in the development and evaluation of polymeric materials. Like many synthetic polymers, polyamides and polyimides have traditionally relied on nonrenewable resources. However, increasing awareness of environmental challenges such as climate change, resource depletion, and plastic waste has prompted renewed attention to their life cycle and environmental impact. Rather than focusing solely on raw material origin, modern assessments consider production energy use, durability, efficiency during service life, recyclability, and end-of-life management. Polyamides offer certain sustainability advantages due to their durability and long service life. Components made from polyamides often remain functional for extended periods, reducing the need for frequent replacement. In transportation applications, their lightweight nature contributes to lower energy consumption and reduced greenhouse gas emissions during use. Recycling technologies for polyamides, particularly in textile and engineering sectors, are advancing steadily. Mechanical reprocessing and emerging recovery strategies are supporting more circular material flows, aligning with broader sustainability goals [8–10]. Polyimides present distinct challenges because of their exceptional stability, which can complicate recycling processes. However, their use in high-value and long-lifetime applications often offsets their environmental footprint. In aerospace and electronics, the performance benefits they provide can lead to improved energy efficiency, reliability, and extended operational lifetimes. As a result, their overall contribution must be evaluated within a comprehensive life-cycle framework [11].

Looking ahead, research and industrial practice are increasingly focused on improving environmental compatibility without compromising performance. Efforts include enhancing recyclability, reducing production energy requirements, and integrating renewable content where feasible. Regulatory frameworks, industry standards, and corporate sustainability commitments are also influencing material selection and design strategies. In this context, polyamides and polyimides are being reexamined not only as high-performance materials but also as components of a more sustainable materials economy. Overall, polyamides and polyimides have shaped the technological landscape of the modern world. Their historical development illustrates the transformative power of polymer science, while their continued evolution demonstrates adaptability to new performance and environmental expectations. As industries move toward more sustainable and efficient systems, these materials remain central to innovation, bridging the demands of strength, stability, and responsibility in contemporary engineering.

6.2 CHEMISTRY AND FUNDAMENTALS

Polyamides and polyimides are two important families of nitrogen-containing polymers, each defined by a characteristic functional group in the backbone that governs their structure–property relationships. In polyamides, the defining feature is the amide linkage, in which a carbonyl group is directly bonded to a nitrogen atom. This group is polar and capable of forming strong intermolecular hydrogen bonds between adjacent chains. As a result, polyamides typically exhibit high tensile strength, good toughness, abrasion resistance, and relatively balanced mechanical performance. The presence of hydrogen bonding enhances cohesion between chains, which contributes to durability while still allowing a certain degree of flexibility depending on the molecular structure [3,12]. Polyamides are generally classified based on the nature of their backbone structure. Aliphatic polyamides are the most widely used and are common in fibers and engineering plastics due to their good processability and balanced mechanical properties. Semi-aromatic polyamides contain both aliphatic and aromatic segments, offering improved thermal stability while maintaining reasonable toughness and moldability. Aromatic polyamides, often referred to as aramids, possess rigid backbone structures that provide very high strength, stiffness, and resistance to heat, making them suitable for high-performance fibers and protective materials. This structural diversity allows polyamides to serve a wide range of applications, from consumer products to structural components. Polyimides, in contrast, are characterized by the imide functional group, which contains two carbonyl groups bonded to the same nitrogen atom in a cyclic arrangement. This ring structure is inherently rigid and thermally stable. Unlike the amide linkage in polyamides, the imide group significantly restricts rotational freedom along the polymer chain, resulting in reduced chain mobility. Consequently, polyimides exhibit exceptional thermal stability, high glass transition temperatures (T_g), excellent dimensional stability, and strong resistance to chemicals and oxidation. These properties make them suitable for demanding environments where mechanical integrity must be maintained at elevated temperatures. Polyimides are commonly classified into fully aromatic polyimides, which dominate high-temperature engineering applications due to their rigidity and superior thermal performance, and semi-aromatic or modified polyimides, which incorporate structural variations to improve flexibility, transparency, or processability. Some polyimides are designed for electronic and optoelectronic uses where mechanical flexibility combined with heat resistance is critical. Others are optimized for aerospace and industrial insulation, where long-term stability under thermal and mechanical stress is required [12].

When comparing these two polymer families, several key differences emerge. First, in terms of thermal performance, polyimides generally outperform polyamides. The rigid cyclic structure of the imide group provides superior resistance to heat and oxidative degradation, allowing polyimides to function reliably at temperatures where most polyamides would soften or degrade. Second, in terms of mechanical flexibility and processability, polyamides often have an advantage. Their hydrogen-bonded structure provides strength while still allowing easier melt processing and molding in many cases. This makes polyamides more suitable for high-volume manufacturing and consumer applications [13]. From a mechanical standpoint, polyamides typically offer excellent toughness and impact resistance, particularly in aliphatic forms, whereas polyimides tend to be stiffer and more dimensionally stable at high temperatures but may be more brittle depending on formulation. In terms of application domains, polyamides dominate textiles, automotive components, packaging films, and general engineering plastics, while polyimides are more common in aerospace systems, advanced electronics, high-temperature insulation, and specialized coatings [14,15].

In summary, although both families contain nitrogen-based functional groups and share certain structural similarities, their functional groups create distinct performance profiles. Polyamides emphasize hydrogen bonding and balanced mechanical behavior, leading to versatile and widely used materials. Polyimides emphasize rigidity and thermal endurance, resulting in polymers designed for extreme conditions. The comparison between these two families highlights how subtle differences in backbone structure can produce markedly different material properties and application landscapes.

6.3 MONOMERS AND FEEDSTOCKS

The performance and application range of polyamides and polyimides are fundamentally determined by the nature of their monomers and feedstocks. In polyamides, the essential building blocks are diamines and diacids or their derivatives, which combine to form recurring amide linkages along the polymer backbone. In polyimides, the classical system involves the reaction of diamines with dianhydrides, producing imide-containing backbones after cyclization. Although these functional groups define the polymer family, it is the detailed structure of the monomers that ultimately governs mechanical strength, thermal resistance, processability, optical behavior, and long-term stability.

Traditionally, most diamines, diacids, and dianhydrides used in both polyamides and polyimides have been derived from petrochemical sources. In polyamides, aliphatic diamines and diacids have been widely used for engineering plastics and fibers, while aromatic variants have enabled the development of high-strength aramid materials. The selection of these monomers has historically been driven by availability, cost, and the ability to achieve desirable crystallinity and hydrogen-bonding density [1]. In polyimides, aromatic diamines and aromatic dianhydrides have been particularly important because their rigid, symmetrical structures promote dense chain packing and exceptional thermal stability. As noted in recent literature, most conventional polyimide monomers originate from petroleum feedstocks, and similar reliance on fossil-based intermediates has characterized commercial polyamide production. While these systems have delivered outstanding performance, concerns related to sustainability, resource depletion, and potential toxicity of certain intermediates have motivated the search for renewable alternatives [2].

In response, considerable research efforts have focused on developing bio-based monomers suitable for both polymer families. For polyamides, renewable diacids and diamines derived from plant oils, biomass, or fermentation processes are increasingly being incorporated into engineering materials. These bio-based components can reduce carbon footprint while maintaining flexibility, toughness, and acceptable thermal performance. In polyimides, renewable feedstocks such as natural phenols, polyols, amino acids, terpenes, and fermentation-derived aromatic structures have been chemically modified to introduce amine functionalities compatible with imide formation. Although designing fully bio-based dianhydrides remains more challenging due to structural requirements for imide ring formation, partially bio-based systems have demonstrated promising results [2,16].

The relevance of monomer structure to polymer properties is particularly evident when examining structure–property relationships. In polyamides, increasing aliphatic chain length generally enhances flexibility and reduces melting temperature, whereas introducing aromatic units increases stiffness, tensile strength, and heat resistance. The density and orientation of hydrogen bonding between amide groups strongly influence crystallinity, water absorption, and mechanical robustness. Even subtle variations in chain length or symmetry can significantly alter performance.

In polyimides, the imide ring imparts intrinsic rigidity and high thermal endurance, but the nature of the diamine and dianhydride segments controls processability and functional characteristics. Fully aromatic systems produce highly stable materials with strong intermolecular interactions, often at the expense of solubility. Incorporation of bulky, asymmetric, or alicyclic segments can reduce tight chain packing, improve transparency, and enhance solubility while retaining substantial thermal resistance. Flexible segments, including those derived from renewable fatty diamines, can generate more processable polyimides and even enable thermoplastic behavior in certain systems [16].

Overall, diamines, diacids, and dianhydrides constitute the structural foundation of both polyamide and polyimide families. Whether derived from conventional petrochemical routes or renewable feedstocks, their molecular architecture directly determines the balance between rigidity and flexibility, thermal stability and processability, and crystallinity and transparency. The ongoing transition toward bio-based monomers does not merely address sustainability goals but also expands the design space for advanced materials. Ultimately, the structure–property relationship remains central: careful monomer selection and molecular design continue to define the performance envelope of next-generation polyamides and polyimides.

6.4 SYNTHESIS STRATEGIES

6.4.1 Bio-Based Polyamide Synthesis Routes

Bio-based polyamides are synthesized using the same fundamental polymerization principles as conventional nylons, but with renewable carbon sources replacing fossil-derived monomers. Bio-based polyamides are generally classified according to two principal synthetic strategies: polycondensation reactions and ring-opening polymerization (ROP) Figure 6.1 [1,17,18]. The key innovation is therefore not a new reaction mechanism, but the transformation of renewable feedstocks such as castor oil, fatty acids, carbohydrates, terpenes, and amino acids into suitable bifunctional

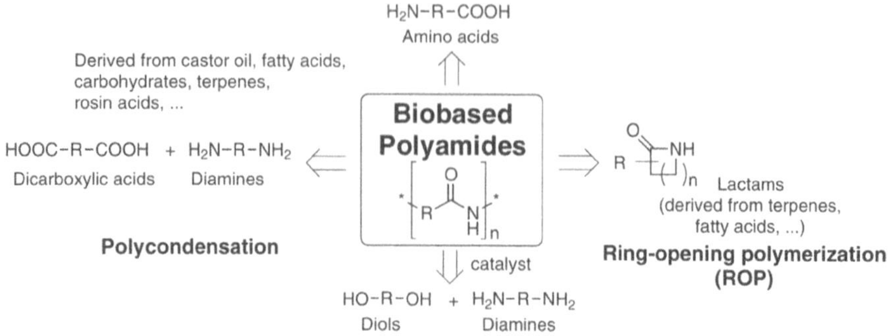

FIGURE 6.1 Synthesis ways of sustainable polyamides. (Adapted with permission [1]. Copyright 2016 WILEY-VCH Verlag GmbH & Co. KGaA, Weinheim.)

monomers. These natural resources are converted into dicarboxylic acids, diamines, α,ω-amino acids, or lactams that can be polymerized into linear polyamide backbones.

The most industrially established route is polycondensation of bio-based dicarboxylic acids and diamines (AABB-type polymerization). A central renewable platform in this context is castor oil, derived from *Ricinus communis*. Castor beans contain a high percentage of ricinoleic acid, which serves as the starting material for two important monomer pathways: the undecenoic acid route and the sebacic acid route. In the undecenoic acid pathway, ricinoleic acid is cleaved and chemically modified to yield 11-aminoundecanoic acid. This monomer undergoes self-polycondensation to form polyamide-11 (PA11), a fully bio-based polyamide that has been produced commercially for decades. PA11, marketed under the name Rilsan, is known for its excellent chemical resistance, flexibility, and durability, and is widely used in automotive fuel lines, pneumatic tubing, cable sheathing, and medical devices. Its long industrial history demonstrates that renewable feedstocks can support large-scale production of high-performance engineering plastics [19]. The second major castor oil-derived pathway involves sebacic acid, which is obtained through alkali cleavage of ricinoleic acid. Sebacic acid is an important renewable dicarboxylic acid used in the preparation of polyamides such as PA610, PA1010, and PA1012. In these AABB systems, polymerization requires strict equimolar ratios of diacid and diamine. Because slight stoichiometric imbalances can lead to chain termination and low molecular weight, industrial processes typically employ the "nylon salt" method. In this approach, the diacid and diamine are first combined at room temperature to form a precisely balanced salt, which is then polymerized at elevated temperature. This method ensures controlled molecular weight development and consistent material properties. While some diamines used in these systems are still petrochemical, efforts have been made to derive renewable diamines from sebacic acid itself, demonstrating the potential for higher overall bio content [1]. In addition to AABB polycondensation, AB-type polycondensation of bio-based α,ω-amino acids represents another important route. In this case, a single monomer contains both amine and carboxylic acid functionalities, allowing self-condensation without the need for equimolar control between two different components. The synthesis of PA11 from 11-aminoundecanoic acid is a prime example. AB-type polymerizations are generally simpler to control and can enable fully bio-based materials when the amino acid precursor is derived entirely from renewable sources. Moreover, natural amino acids and structurally modified derivatives introduce stereochemical features and functional side groups that can influence crystallinity, hydrogen bonding, and mechanical behavior [1].

A second major strategy for bio-based polyamide production is ROP of bio-based lactams. Lactams are cyclic amides that can undergo polymerization to form linear polyamide chains. Renewable lactams can be derived from fatty acids, terpenes, and other natural feedstocks. For example, ω-amino lauric acid derived from palm kernel oil can be converted into laurin lactam, which provides a bio-based route to polyamide-12. Similarly, terpenoid-based ketones can be transformed into lactam structures suitable for ROP. Compared to polycondensation, ROP can offer advantages in controlling polymer architecture and molecular weight distribution, and it can facilitate melt processing routes analogous to those used for conventional PA6. Bio-based polyamides can be classified not only by synthesis route but also by structural features, including linear, cycloaliphatic, and aromatic backbones. Renewable monomers often introduce unique structural elements such as stereocenters, long aliphatic chains, or cyclic moieties. These features can modify hydrogen-bonding density, crystallinity, melting temperature, and mechanical performance. For example, increasing the methylene segment length between amide groups generally reduces melting temperature and increases flexibility, while higher amide bond density enhances modulus and thermal resistance [1]. Overall, bio-based polyamide synthesis relies on two established polymerization strategies: polycondensation (AABB and AB types) and ROP of lactams. The central development lies in the conversion of renewable platform chemicals, particularly castor oil-derived intermediates, into functional monomers compatible with these routes, along with the increasing integration of bio-derived diamines to enhance overall biocontent. While some systems are already commercial, others remain at the research stage, aiming to balance high performance, scalability, and sustainability. The continued expansion of renewable feedstock chemistry and sustainable amine platforms, combined with mature polymerization technologies, positions bio-based polyamides as an increasingly important class of sustainable high-performance materials.

6.4.2 Bio-Based Polyimide Synthesis Routes

Bio-based polyimides derived from bio-based amines follow the same core imide-forming chemistry as conventional systems, but renewable diamines replace fossil-based aromatic amines. The sustainability contribution therefore comes primarily from the bio-derived amine component, while dianhydrides may be petrochemical, partially bio-based, or, in emerging systems, also renewable.

The most established route is the two-step poly(amic acid) (PAA) method. In the first step, a bio-based diamine reacts with a dianhydride at moderate temperature in a polar solvent to form a PAA intermediate. In the second step, thermal or chemical imidization converts this precursor into the final polyimide (Figure 6.2) [16]. This approach allows high-molecular-weight development and good control over film formation, coatings, and composite matrices. Bio-derived diamines used in this route include structures obtained from lignin derivatives, vanillin, isosorbide-based

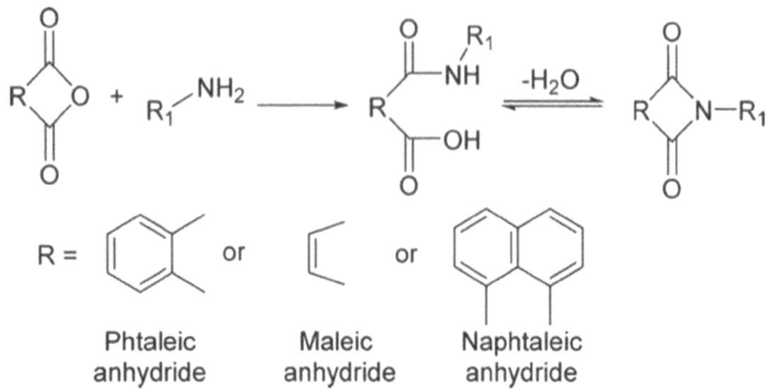

FIGURE 6.2 Synthesis of imide from an amine and a cyclic anhydride. (Adapted with permission [16]. Copyright 2016 American Chemical Society.)

systems, fatty amines, terpenes, and other biomass-derived platforms. The rigidity or flexibility of the renewable amine strongly influences T_g, solubility, transparency, and thermal stability of the final polyimide [20]. A variation of this method is the one-step high-temperature polycondensation, where bio-based diamines and dianhydrides react directly at elevated temperatures, forming the imide ring during polymerization without isolating the PAA intermediate. This approach can simplify processing but requires carefully designed monomers to avoid premature cyclization or poor molecular weight control [21].

Another important route for synthesizing bio-based polyimides is melt polymerization, particularly when renewable diamines contain alicyclic structures or flexible aliphatic segments that enhance processability. In contrast to the classical two-step solution route, melt polymerization minimizes or eliminates the need for high-boiling polar solvents, thereby reducing solvent handling, recovery requirements, and associated environmental burdens. This approach is especially attractive when the bio-based diamine exhibits sufficient thermal stability and controlled reactivity, allowing imidization to occur directly in the melt without premature degradation. However, the processing window is narrower than in solution methods. Renewable amines must tolerate elevated temperatures required for imide formation, and their volatility, oxidation resistance, and tendency toward side reactions must be carefully considered. When appropriately designed, alicyclic or long-chain bio-derived amines can lower melt viscosity, improve flow behavior, and facilitate film or composite fabrication while maintaining desirable thermal and mechanical performance. Thus, melt-based strategies offer a promising pathway toward solvent-free or reduced-solvent polyimide manufacturing, aligning with green chemistry principles while retaining compatibility with existing industrial equipment [2].

In addition to melt processing, several emerging greener technologies have been developed to reduce environmental impact and energy consumption in bio-based polyimide synthesis. Microwave-assisted polymerization accelerates both PAA formation and imidization by providing rapid, volumetric heating, significantly shortening reaction times and potentially lowering energy demand. Ionic liquid-mediated polymerization replaces conventional volatile organic solvents with low-vapor-pressure media

that can improve monomer solubility, enhance reaction control, and in some cases be recycled. Water-based or hydrothermal imidization methods represent another sustainable direction, using water as a reaction medium under controlled temperature and pressure conditions to limit solvent toxicity and simplify purification. These alternative approaches aim to reduce solvent hazards, decrease processing time, and improve overall sustainability metrics. Bio-based amines are particularly advantageous in such systems because their structural diversity allows tuning of polarity, reactivity, and solubility, enabling polymerization under milder or more environmentally compatible conditions. From a structural standpoint, renewable amines can be categorized as aromatic, alicyclic, or long-chain aliphatic. Aromatic bio-derived amines typically impart rigidity and high thermal resistance suitable for demanding electronic or aerospace applications, whereas alicyclic and fatty amines introduce flexibility, improved solubility, and in some cases enhanced optical transparency. Consequently, careful selection of the bio-based amine becomes a critical design parameter in tailoring processing strategy, sustainability profile, and final polyimide performance [2].

Overall, synthesis routes of bio-based polyimides from bio-based amines rely on established imidization chemistry, primarily the two-step PAA route and its variations. The innovation lies in incorporating renewable diamines that reduce carbon footprint while enabling tunable mechanical, thermal, and processing characteristics. As sustainable amine platforms continue to expand, they are expected to play a central role in advancing partially and fully bio-based high-performance polyimides.

6.5 STRUCTURE, PROPERTIES, AND CHARACTERIZATION OF BIO-BASED POLYAMIDES

Recent investigations show that bio-based diamines play a decisive role in tailoring the structure and performance of polyamides for advanced applications. By introducing long aliphatic chains, rigid bicyclic units, or flexible segments, renewable amines enable control over crystallinity, hydrogen bonding, thermal transitions, and mechanical behavior while maintaining high thermal stability. These design strategies expand the applicability of bio-based polyamides in engineering plastics, films, coatings, and flexible components without compromising performance.

The study by Jadhav et al. reports aromatic sulfone ether polyamides synthesized using a bio-based diamine, 4,4′-[sulfonyl-bis(4,1-phenyleneoxy-3-pentadecyl)] bisphenylamine (SPPBA) derived from cashew-nut shell liquid. The presence of long pentadecyl side chains in the diamine structure significantly influenced polymer morphology. X-ray diffraction (XRD) showed that the resulting polyamides were predominantly amorphous due to disruption of close chain packing by the bulky alkyl substituents. This structural feature enhanced solubility and processability. Thermogravimetric analysis (TGA) demonstrated high thermal stability, with no significant weight loss below about 405°C. Initial decomposition temperatures ranged from approximately 422°C–438°C, indicating that the introduction of pendant pentadecyl groups did not substantially compromise thermal resistance. Inherent viscosity values suggested moderate molecular weights, and most polymers formed good films. Solubility studies showed marked improvement in organic solvents such as

N,N-dimethylformamide, dimethylacetamide (DMAc), N-methyl-2-pyrrolidone (NMP), chloroform, and m-cresol compared to analogous polymers without pendant alkyl groups. The enhanced solubility is attributed to reduced intermolecular hydrogen bonding and decreased crystallinity. Characterization techniques included Fourier transform infrared (FT-IR) spectroscopy for confirming amide formation, XRD for morphology, TGA for thermal behavior, and viscosity measurements for molecular weight assessment. Overall, the incorporation of bio-based amines with long aliphatic side chains improved solubility and maintained high thermal stability, demonstrating the structure–property advantages of renewable diamine design [22]. Another study demonstrates the successful synthesis of bio-based polyamides using the sugar-derived diamine isosorbide(bis(propan-1-amine) (DAPIS) via melt polycondensation followed by solid-state polymerization. Structural analysis by nuclear magnetic resonance (NMR) and FT-IR confirmed efficient incorporation of the isohexide-based amine into the polymer backbone without degradation. The rigid bicyclic structure of DAPIS increased stiffness and influenced morphology, slightly reducing crystallinity and melting temperature compared to fully linear polyamides. Thermal stability remained high, with degradation occurring above 330°C. Dynamic mechanical thermal analysis showed enhanced storage modulus at low temperatures, while T_g depended mainly on molecular weight and composition. Overall, the work highlights that bio-based isohexide diamines can produce high-performance, thermally stable polyamides with tunable mechanical properties [23]. Another work demonstrates PA6 copolyamides synthesized using a novel bio-based aliphatic diamine, N^1,N^6-bis (4-aminobutyl) adipamide (BABA), derived from renewable feedstocks (Figure 6.3). Incorporation of the long-chain BABA/sebacic acid (SA) salt reduced crystallinity and shifted the crystal structure from α-form toward γ-form at higher bio-based content, as confirmed by XRD. Differential scanning colorimetry (DSC) showed decreasing melting temperature and crystallization enthalpy with increasing BABA/SA, indicating disruption of hydrogen bonding and lamellar packing. Despite this, thermal stability remained high, with degradation temperatures close to neat PA6 (onset above 350°C). Mechanically, tensile strength decreased slightly, but elongation at break increased significantly, showing enhanced flexibility due to long aliphatic segments. Overall, the bio-based diamine enabled tuning of thermal and mechanical performance while maintaining engineering-grade stability [24].

6.6 STRUCTURE, PROPERTIES, AND CHARACTERIZATION OF POLYIMIDES

Recent studies demonstrate that bio-based diamines can be strategically incorporated into polyimide backbones to enhance processability, optical transparency, and structural tunability while maintaining high thermal stability. By tailoring side chains, rigidity, and functional groups, renewable amines enable the development of high-performance polyimides for coatings, flexible electronics, and advanced film applications.

The study by Nilakshi et al. describes the synthesis of aromatic polyimides using a bio-based diamine, 4-pentadecylbenzene-1,3-diamine, derived from cashew nut shell liquid (CNSL), a renewable resource. The long pentadecyl side chain introduced into

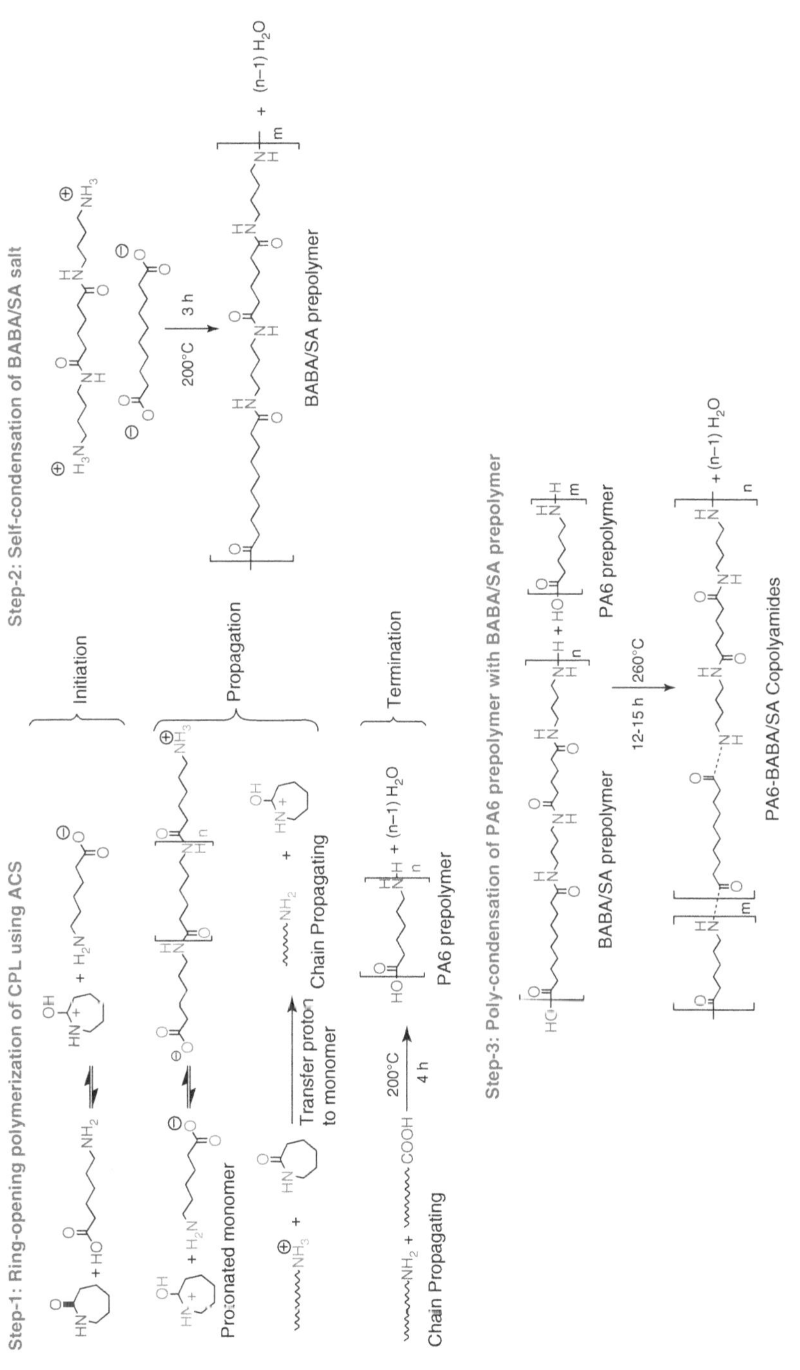

FIGURE 6.3 Synthesis of copolyimide. (Adapted with permission [24]. Copyright 2018 Wiley Periodicals, Inc.)

the aromatic diamine increased free volume and reduced chain packing efficiency, significantly improving solubility and film-forming ability. The resulting polyimides, prepared with different aromatic dianhydrides, exhibited moderate to high molecular weights and formed tough, transparent, and flexible films. TGA showed high thermal stability, with 10% weight-loss temperatures around 470°C–480°C. However, T_g (158°C–206°C) was lower than conventional fully aromatic polyimides due to increased chain mobility from the pendant alkyl group. The T_g depended on dianhydride rigidity, confirming a clear structure–property relationship. The polyimides were soluble in polar solvents such as NMP, DMAc, m-cresol, and chloroform, unlike typical insoluble aromatic polyimides. Characterization techniques included FT-IR for imide confirmation, NMR for structural verification, gel permeation chromatography for molecular weight determination, DSC and TGA for thermal properties, and ultraviolet-visible spectroscopy for evaluating charge-transfer interactions. Overall, the work demonstrates that bio-based amines with long alkyl substituents can enhance processability while maintaining high thermal performance in polyimides [25].

This study by Min et al. shows the synthesis of a furan-based bio-derived diamine prepared from t-butoxycarbonyl-protected glycine (BOC-glycine) and 2,5-furandimethanol (Figure 6.4), followed by its incorporation into polyimide structures. The authors optimized the esterification conditions using 1,3-dicyclohexylcarbodiimide (DCC) as a coupling agent and demonstrated that sufficient DCC concentration was essential for successful diamine formation. The resulting furan-containing diamine was partially substituted for 4,4′-diaminodiphenyl ether in polyimide synthesis to produce bio-based polyimide membranes. The prepared polyimides exhibited high molecular weights, good thermal stability with decomposition temperatures above 450°C, and smooth, defect-free membrane morphology as confirmed by scanning electron microscopy. Although incorporation of the furan-based diamine slightly reduced thermal stability and mechanical toughness compared to fully aromatic benzene-based systems, the materials maintained strong heat resistance and structural integrity. The results show that furan-derived bio-based amines can function as viable partial replacements for petroleum-based diamines in polyimide synthesis, offering a more sustainable route while retaining competitive thermal and structural performance [26]. The study by Chatterjee et al. reports the synthesis of partially bio-based aromatic polyimides using a renewable diamine derived from CNSL, a nonedible industrial by-product. The bio-based monomer, 4,4′-diamino-4″-pentadecyltriphenylamine, contains a long pentadecyl side chain introduced through the CNSL-derived cardanol structure. This structural modification enhances solubility and film-forming ability while maintaining the aromatic backbone necessary for thermal stability. The resulting polyimides, prepared with conventional aromatic dianhydrides, exhibited good molecular weights, excellent solubility in common organic solvents, and strong, flexible films. Thermal analysis showed high decomposition temperatures above 400°C and moderate T_g, indicating a balance between rigidity and processability. Importantly, the incorporation of the triarylamine unit enabled donor–acceptor interactions within the polymer backbone, making the materials suitable for non-volatile memory device applications. The study demonstrates that CNSL-derived bio-based amines can successfully produce high-performance polyimides with tunable thermal, optical, and electronic properties while reducing reliance on fossil-derived monomers [27].

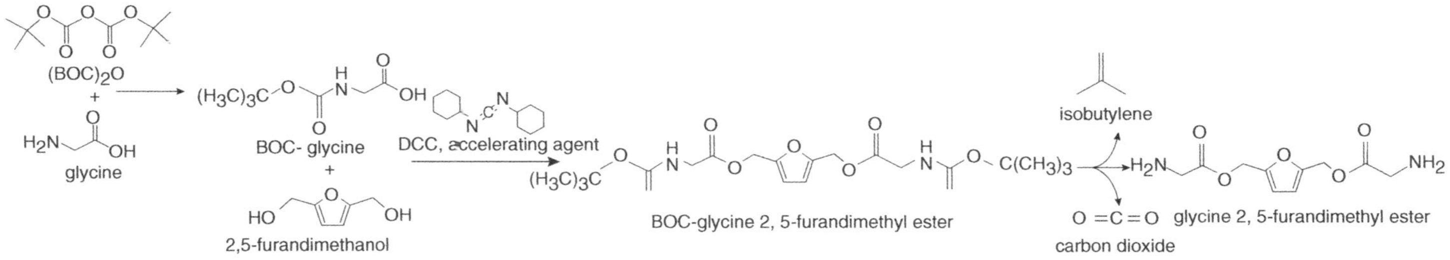

FIGURE 6.4 The synthetic route of furan diamine. (Adapted with permission [26]. Copyright 2023 MDPI. Licensed under CC BY 4.0. https://creativecommons.org/licenses/by/4.0/.)

6.7 APPLICATIONS OF BIO-BASED POLYAMIDES AND POLYIMIDES

Bio-based polyamides and polyimides are increasingly emerging as sustainable alternatives to conventional high-performance polymers across engineering and electronic sectors. By integrating renewable amines into their backbones, these materials combine reduced fossil dependency with tunable structural, thermal, and functional properties. As a result, they are expanding the application landscape of bio-derived polymers in advanced structural and optoelectronic technologies.

The study by Zhou et al. presents the development of bio-based polyesteramides synthesized from renewable diamine building blocks, emphasizing their application potential in sustainable engineering materials. The incorporation of bio-derived amines into the polyamide backbone enabled modulation of crystallinity, hydrogen-bonding interactions, and chain flexibility, resulting in materials with balanced thermal stability and mechanical performance. Although some reduction in melting temperature and stiffness was observed compared to fully petroleum-based analogues, the polymers maintained high decomposition temperatures and sufficient strength for structural applications. Enhanced ductility and processability were notable advantages, making these materials suitable for flexible components, packaging, coatings, and lightweight engineering parts. Overall, the work demonstrates that bio-based amines serve not only as sustainable alternatives but also as effective structural modifiers that allow tuning of performance while supporting reduced fossil dependency in polyamide production [28]. Moreover, the study by Chen et al. demonstrates that bio-based amines can be effectively utilized to produce high-performance polyimides suitable for advanced electronic applications. In this work, an isosorbide-derived diamine, (3*R*,6*S*)-hexahydrofuro[3,2-*b*] furan-3,6-diyl bis(4-aminobenzoate) (ISBA), synthesized from renewable starch resources, was polymerized with various dianhydrides to prepare fully bio-based or partially bio-based polyimides with biomass content up to about 53% (Figure 6.5). The rigid heterocyclic isosorbide structure in the bio-based amine significantly enhanced backbone stiffness and dimensional stability, resulting in high thermal resistance with decomposition temperatures above 400°C and T_g exceeding 250°C. At the same time, the ester linkage in ISBA helped reduce strong charge–transfer interactions, leading to improved optical transparency and low coloration, which are essential for optoelectronic devices. These combined structural advantages enabled the bio-based polyimides, particularly PI(ISBA-C95B5), to function successfully as flexible substrates and gate dielectrics in organic field-effect transistors, maintaining stable electrical performance even after thermal treatment at 200°C and repeated bending cycles. This work clearly shows that bio-based amines are not merely sustainable alternatives but structurally valuable monomers capable of delivering competitive thermal, mechanical, optical, and dielectric properties for flexible electronic applications [29]. Guo et al. further highlight the application potential of bio-based polyimides synthesized from renewable diamines. Bio-derived aromatic amines were incorporated into polyimide backbones to develop materials suitable for high-temperature and optoelectronic applications. The renewable amine structures introduced rigid aromatic frameworks along with functional groups that moderated intermolecular charge–transfer interactions. As a result, the resulting polyimides exhibited high thermal stability with decomposition temperatures

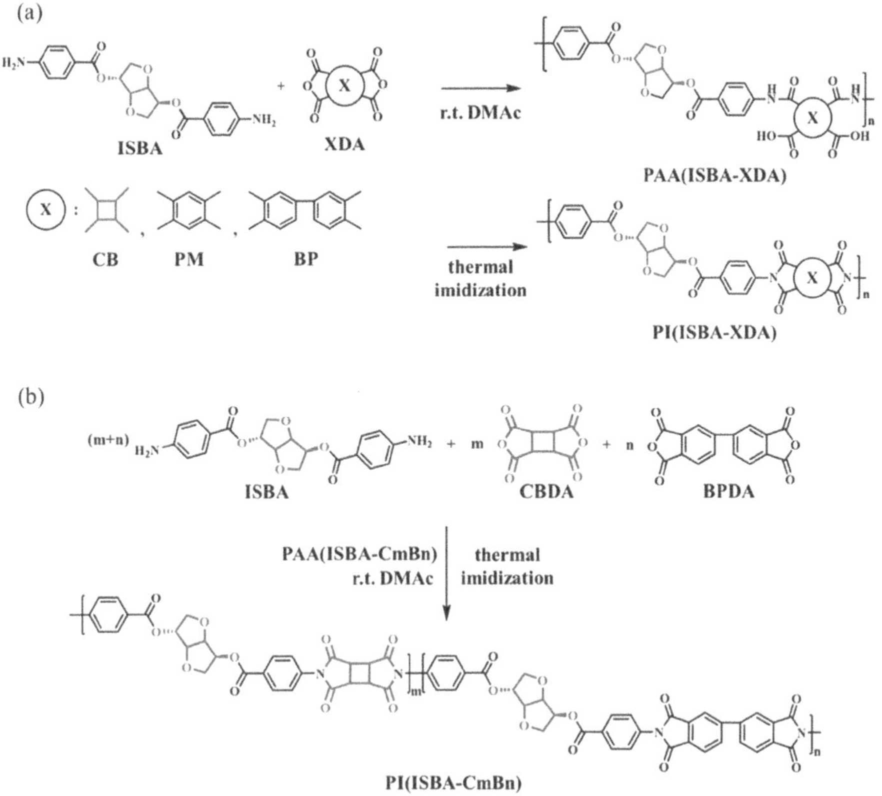

FIGURE 6.5 Synthetic routes of polyimides (a) PI(ISBA-XDA), where X represents the three studied dianhydrides, 1,2,3,4-tetracarboxylic dianhydride (CBDA), pyromellitic dianhydride (PMDA), and 3,3′,4,4′-biphenyl-tetracarboxylic dianhydride (BPDA) (b) PI(ISBA-C*m*B*n*). (Adapted with permission [29]. Copyright 2021 American Chemical Society.)

exceeding 400°C, while also maintaining good optical transparency and processability. The reduced coloration and improved solubility compared to fully petroleum-based analogues make these materials attractive for flexible electronic substrates, display technologies, and advanced coatings. Importantly, the bio-based amines contributed not only to sustainability through renewable carbon content but also to performance tuning, enabling control over T_g, dielectric properties, and mechanical flexibility. The study demonstrates that carefully designed bio-derived diamines can produce polyimides that meet the demanding requirements of modern electronic and high-performance applications while advancing material sustainability [30].

6.8 SUSTAINABILITY, CHALLENGES, AND FUTURE PERSPECTIVES

The sustainability of bio-based polyamides is primarily rooted in the use of renewable carbon feedstocks such as castor oil, fatty acids, terpenes, carbohydrates, and fermentation-derived intermediates. The review highlights that replacing fossil-derived diamines and dicarboxylic acids with biomass-derived counterparts reduces dependence on petrochemical resources and can lower the overall carbon footprint of polyamide production. Industrial examples such as polyamide-11 (PA11) demonstrate that fully bio-based engineering plastics are commercially viable and can meet demanding performance requirements. Additionally, the use of renewable long-chain monomers often results in lower water absorption, improved chemical resistance, and reduced density compared to conventional nylons, contributing to lightweighting and durability in applications. However, several challenges remain. One major issue is feedstock availability and scalability. While castor oil is a well-established platform, expanding bio-based polyamide production requires broader renewable monomer sources that do not compete with food supply or agricultural land use. Economic competitiveness is another barrier, as bio-based monomers can be more expensive than petrochemical alternatives due to smaller production volumes and more complex purification steps. Furthermore, some renewable monomers require multistep chemical transformations, which may reduce overall process efficiency and sustainability gains. The review also emphasizes that achieving high thermal performance comparable to fully aromatic petroleum-based systems can be difficult when using purely aliphatic bio-derived monomers. From a technical standpoint, maintaining consistent molecular weight, purity, and polymer quality at industrial scale is critical. Impurities in bio-based monomers may affect polymerization kinetics, mechanical properties, and long-term stability. Additionally, lifecycle assessment (LCA) data are still limited for many emerging bio-based polyamides, making it difficult to quantify environmental advantages comprehensively. Looking forward, the future of bio-based polyamides depends on expanding renewable monomer platforms, improving biotechnological production of diamines and diacids, and optimizing polymerization efficiency. The integration of biotechnology with chemical catalysis is expected to play a key role in producing high-purity monomers at competitive costs. Advances in metabolic engineering, green chemistry processing, and recyclable polymer design may further enhance sustainability. The review suggests that combining renewable feedstocks with established polymerization technologies offers a realistic pathway toward partially and fully bio-based high-performance polyamides. Overall, while bio-based polyamides have already demonstrated commercial success in selected applications, broader adoption will require addressing economic, scalability, and performance challenges. Continued innovation in renewable chemistry and process engineering is essential to fully realize their potential as sustainable alternatives to conventional nylons [1,31].

Bio-based polyimides derived from renewable amines represent a significant step toward reducing reliance on fossil-derived aromatic diamines traditionally used in high-performance polymers. The sustainability advantage primarily arises from replacing petroleum-based amines with biomass-derived counterparts such as lignin-, vanillin-, isosorbide-, terpene-, or fatty acid–based diamines. These renewable amines introduce biogenic carbon into the polymer backbone, lowering dependence on nonrenewable feedstocks and potentially reducing lifecycle carbon emissions. In addition, many bio-based amines incorporate structural diversity, such as alicyclic units, heterocycles, or flexible linkages, which can improve solubility and processability compared to fully aromatic petrochemical systems. Despite these advantages, several challenges remain. One major issue is achieving the same level of thermal and mechanical performance as conventional fully aromatic polyimides. While many bio-based diamines provide high thermal stability, their structural asymmetry or flexible linkages can reduce T_g or dimensional stability. Another challenge involves scalability and cost competitiveness. Bio-based amine synthesis often requires multistep chemical modification or purification processes, which may increase production costs and limit large-scale adoption. Additionally, maintaining consistent monomer purity is critical, as impurities can affect polymerization kinetics, molecular weight control, and long-term thermal stability. From a processing standpoint, balancing sustainability with performance remains complex. Fully aromatic systems typically provide superior thermal endurance, while bio-based amines may introduce improved transparency and reduced charge-transfer coloration but slightly lower T_g. Designing diamines that combine rigid bio-derived cores with optimized functional groups is therefore essential to match the stringent demands of aerospace, electronics, and flexible device applications. Looking forward, future research is expected to focus on developing fully bio-based polyimide systems by combining renewable diamines with bio-derived dianhydrides, thereby increasing overall biomass content. Advances in green chemistry, catalytic transformations, and biotechnological production of aromatic amines will be crucial to improving economic viability. Furthermore, LCA and recyclability considerations will likely play an increasingly important role in evaluating true sustainability. Emerging strategies such as dynamic covalent imide networks, recyclable thermosets, and low-energy processing routes may further enhance environmental performance.

Overall, bio-based polyimides derived from renewable amines demonstrate strong potential as sustainable high-performance materials. Continued innovation in monomer design, scalable synthesis, and property optimization will determine their broader industrial adoption, particularly in advanced electronics, aerospace, and high-temperature engineering applications [2].

6.9 CONCLUSION

Polyamides and polyimides have long defined the upper tier of engineering polymers, offering complementary performance profiles rooted in their distinct amide and imide functional groups. Their historical development underscores how molecular architecture governs macroscopic performance: hydrogen bonding enables the balanced toughness of polyamides, while imide ring rigidity imparts the exceptional thermal endurance of polyimides. The transition toward bio-based systems does not alter these fundamental principles but redefines the origin and design of the monomers that shape them. A key insight emerging from this chapter is that bio-based amines are not merely sustainable substitutes; they are structural design tools. Long aliphatic chains reduce crystallinity and enhance flexibility, rigid bicyclic and aromatic bio-derived diamines increase stiffness and thermal resistance, and asymmetric or functionalized renewable amines improve solubility and optical transparency. In polyamides, renewable diamines enable control over hydrogen-bonding density, crystal form, and mechanical ductility while maintaining engineering-level thermal stability. In polyimides, bio-based aromatic and heterocyclic amines allow tuning of T_g, dielectric behavior, and charge–transfer interactions, enabling applications in flexible electronics, coatings, and high-temperature substrates. Application studies confirm that bio-based polyamides can compete in structural components, lightweight engineering parts, films, and adhesives, while bio-based polyimides demonstrate suitability for advanced electronic devices, optoelectronic substrates, and aerospace-related materials. These examples illustrate that performance parity with petrochemical analogues is increasingly achievable. Nevertheless, sustainability cannot be evaluated solely through renewable content. Challenges remain in feedstock scalability, purification efficiency, molecular weight control, and LCA. Some renewable monomers require multistep chemical transformations that may offset environmental gains. Furthermore, balancing rigidity with processability, and thermal endurance with recyclability, continues to demand careful molecular design. The development of fully bio-based polyimide systems, including renewable dianhydrides, represents an important frontier. Future progress will depend on integrating biotechnology, green catalysis, and advanced polymer chemistry to produce high-purity renewable amines at scale. Innovations in dynamic covalent chemistry, recyclable thermosets, solvent-free processing, and low-energy polymerization methods may further enhance environmental performance. Most importantly, the next generation of bio-based nitrogen-containing polymers must be designed within a full life-cycle framework that considers production, service efficiency, durability, and end-of-life management. To conclude, bio-based polyamides and polyimides derived from renewable amines represent more than a sustainable alternative; they embody a strategic expansion of polymer design space. By leveraging renewable molecular diversity while preserving high-performance characteristics, these materials bridge the demands of advanced engineering and environmental responsibility. Their continued development will play a critical role in shaping a more sustainable and technologically resilient materials landscape.

REFERENCES

1. M. Winnacker, B. Rieger, Biobased polyamides: Recent advances in basic and applied research, *Macromol. Rapid Commun.* 37 (2016) 1391–1413.
2. M. Abadie, R.D. Rusu, New high-performance materials: Bio-based, eco-friendly polyimides, in: S. Diaham (Ed.), *Polyimide for Electronic and Electrical Engineering Applications*, IntechOpen, London, 2020: pp. 1–34.
3. K. Marchildon, Polyamides – Still strong after seventy years, *Macromol. React. Eng.* 5 (2011) 22–54.
4. M. Ghosh, *Polyimides: Fundamentals and Applications*, CRC Press, 2018.
5. M. Pervaiz, M. Faruq, M. Jawaid, M. Sain, Polyamides: Developments and applications towards next-generation engineered plastics, *Curr. Org. Synth.* 14 (2017) 146–155.
6. J.A. Lee, J.Y. Kim, J.H. Ahn, Y.-J. Ahn, S.Y. Lee, Current advancements in the bio-based production of polyamides, *Trends Chem.* 5 (2023) 873–891.
7. C.E. Sroog, History of the invention and development of the polyimides, in: M. Ghosh (Ed.), *Polyimides*, CRC Press, Taylor & Francis Group, Boca Raton, FL, 2018: pp. 1–6.
8. B. Zong, B. Sun, S. Cheng, X. Mu, K. Yang, J. Zhao, X. Zhang, W. Wu, Green production technology of the monomer of nylon-6: Caprolactam, *Engineering.* 3 (2017) 379–384.
9. V. Sessini, B. Haseeb, A. Boldizar, G. Lo Re, Sustainable pathway towards large scale melt processing of the new generation of renewable cellulose–polyamide composites, *RSC Adv.* 11 (2021) 637–656.

10. V. Hirschberg, D. Rodrigue, Recycling of polyamides: Processes and conditions, *J. Polym. Sci.* 61 (2023) 1937–1958.
11. E. Toto, S. Laurenzi, R.C. Pellegrini, E. Cavallini, M.G. Santonicola, Eco-friendly synthesis of high-performance polyimide materials using bio-based greener solvents: Towards sustainable technologies in space environment, *Mater. Today Sustain.* 25 (2024) 100657.
12. K.J. Saunders, Polyamides and polyimides, in: K.J. Saunders (Ed.), *Organic Polymer Chemistry*, Springer Netherlands, Dordrecht, 1988: pp. 191–223.
13. A. Mehra, M. Chakraborty, S. Nayak, Polyimide-based polymers: A new frontier in antimicrobial materials and healthcare applications, *Mater. Des.* 262 (2026) 115453.
14. Polyamide: A Versatile Polymer for Modern Industrial and Textile Applications, BTD, n.d. https://btdsco.com/en/blog/article/show/13#:~:text=Polyamide: A Versatile Polymer for Modern Industrial and Textile Applications,-Home / Event & News& text=Polyamides are one of the,environmental conditions of the application (accessed February 25, 2026).
15. Polyimide Boosts High-Temperature Performance, NASA, 2008. https://spinoff.nasa.gov/Spinoff2008/ip_5.html#:~:text=Polyimides are a class of,(step-polymerization) polyimides (accessed February 25, 2026).
16. V. Froidevaux, C. Negrell, S. Caillol, J.P. Pascault, B. Boutevin, Biobased amines: From synthesis to polymers; Present and future, *Chem. Rev.* 116 (2016) 14181–14224.
17. J.-F. Li, X.-L. Ku, Y.-D. Li, Y. Ran, J.-B. Zeng, Biobased polyamides with polyethylene-like properties through catalyst-free polycondensation, *ACS Sustain. Chem. Eng.* 13 (2025) 13492–13500.
18. G. Hua, K. Odelius, Exploiting ring-opening aminolysis–Condensation as a polymerization pathway to structurally diverse biobased polyamides, *Biomacromolecules.* 19 (2018) 1573–1581.
19. D.S. Ogunniyi, Castor oil: A vital industrial raw material, *Bioresour. Technol.* 97 (2006) 1086–1091.
20. R.-D. Rusu, M.-D. Damaceanu, M. Bruma, Copoly(1,3,4-oxadiazole-naphthylimide)s containing siloxane units in the main chain: Synthesis and properties, *High Perform. Polym.* 23 (2011) 384–393.
21. M.-D. Damaceanu, R.-D. Rusu, V.-E. Musteata, M. Bruma, Insulating polyimide films containing n-type perylenediimide moieties, *Polym. Int.* 61 (2012) 1582–1591.
22. A.S. Jadhav, S.P. Vernekar, N.N. Maldar, Synthesis and characterization of new aromatic sulfone ether polyamides containing pendant pentadecyl groups, *Polym. Int.* 32 (1993) 5–11.
23. A. Wroblewska, A. Zych, S. Thiyagarajan, D. Dudenko, D. van Es, M.R. Hansen, C. Koning, R. Duchateau, L. Jasińska-Walc, Towards sugar derived polyamides as environmentally friendly materials, *Polym. Chem.* 6 (2015) 4133–4143.
24. S.-P. Rwei, P. Ranganathan, W.-Y. Chiang, Y.-H. Lee, Synthesis and characterization of copolyamides derived from novel aliphatic bio-based diamine, *J. Appl. Polym. Sci.* 135 (2018) 46878.
25. N. V Sadavarte, M.R. Halhalli, C. V Avadhani, P.P. Wadgaonkar, Synthesis and characterization of new polyimides containing pendent pentadecyl chains, *Eur. Polym. J.* 45 (2009) 582–589.
26. Y. Zhang, L. Chen, Y. He, W. Luo, K. Li, Y. Min, Synthesis of furan-based diamine and its application in the preparation of bio-based polyimide, *Polymers (Basel).* 15 (2023) 1088.
27. D. Chatterjee, U.A. Jadhav, B.H. Javaregowda, T.D. Dongale, P.S. Patil, P.P. Wadgaonkar, Partially bio-based triarylamine-containing polyimides: Synthesis, characterization and evaluation in non-volatile memory device applications, *Eur. Polym. J.* 147 (2021) 110327.

28. X. Zhao, F. Mo, C. Zhao, S. Li, M. Wang, S. Dong, L. Zhou, Preparation, structure–property relationships, and applications of novel bio-based and biodegradable polyesteramide materials, *Npj Mater. Degrad.* 9 (2025) 71.
29. C.-K. Chen, Y.-C. Lin, L.-C. Hsu, J.-C. Ho, M. Ueda, W.-C. Chen, High performance biomass-based polyimides for flexible electronic applications, *ACS Sustain. Chem. Eng.* 9 (2021) 3278–3288.
30. G. Yu, Q. Ma, H.-Y. Chang, R. Zheng, C. Wang, L.-Y. Guo, Advancements in synthesis strategies and optoelectronic applications of bio-based photosensitive polyimides, *Int. J. Opt.* 2024 (2024) 5541865.
31. M.S.F. Khedr, Bio-based polyamide, *Phys. Sci. Rev.* 8 (2023) 827–847.

7 Bio-Based Epoxy Resins from Amines

7.1 INTRODUCTION

Epoxy resins are among the most versatile thermosetting polymers because of the intrinsic reactivity of the epoxide, or oxirane, functional group. The three-membered cyclic ether ring possesses high ring strain, making it highly susceptible to nucleophilic attack and ring-opening reactions. This chemical feature enables controlled polymerization and crosslinking when epoxies react with curing agents such as amines, anhydrides, phenols, or thiols. During curing, the initially low-viscosity resin undergoes a transformation into a highly crosslinked, three-dimensional network. The process involves both chain extension and network formation, ultimately producing a rigid, insoluble thermoset structure [1]. As crosslink density increases, the material exhibits enhanced modulus, T_g, thermal stability, and chemical resistance. These characteristics explain why epoxy thermosets are widely used in structural adhesives, protective coatings, electrical encapsulation, and high-performance fiber-reinforced composites. A major advantage of epoxy chemistry lies in its tunability. The final properties are not determined by the resin alone but by the combined selection of epoxy monomer, curing agent, stoichiometry, and additives [2]. For example, rigid aromatic epoxies cured with multifunctional amines yield high-strength, high-T_g materials suitable for aerospace composites, whereas flexible aliphatic systems produce impact-resistant coatings and adhesives. The curing mechanism also allows incorporation of modifiers, toughening agents, fillers, and reactive diluents, further expanding application possibilities. Conventionally, most commercial epoxies are glycidyl ether derivatives, particularly diglycidyl ether of bisphenol A (DGEBA), synthesized by reacting bisphenol A (BPA) with epichlorohydrin. This chemistry has dominated the market for decades due to its reliable performance, scalability, and cost efficiency [2].

The shift from petroleum-based epoxy systems to bio-based alternatives has emerged from both environmental and regulatory pressures. Traditional epoxy production depends heavily on fossil-derived feedstocks, especially BPA and propylene-derived epichlorohydrin. Concerns over BPA toxicity, VOC emissions, and carbon footprint have accelerated research into renewable replacements. One of the earliest and most practical transitions has been the development of bio-based epichlorohydrin. Instead of producing epichlorohydrin from propylene, glycerol derived from vegetable oils or biodiesel by-products can be converted into epichlorohydrin through chlorination and dehydrochlorination pathways. This approach enables partial substitution of fossil carbon without fundamentally altering epoxy manufacturing processes, allowing industry to adopt renewable content with minimal disruption. Beyond this drop-in replacement strategy, fully bio-based epoxy systems have been developed

DOI: 10.1201/9781003731122-7

from renewable building blocks such as vegetable oils, lignin derivatives, rosin, furan compounds, vanillin, and polysaccharides. For instance, epoxidized vegetable oils are prepared via peracid epoxidation of carbon–carbon double bonds, yielding non-glycidyl epoxy resins with improved flexibility and enhanced biodegradability. Aromatic bio-derived compounds, including lignin- and furan-based precursors, are particularly attractive because they can restore rigidity and thermal performance comparable to bisphenol-based systems. The transformation also extends to curing agents. Bio-based amines, poly(amidoamine)s, and organic acids derived from renewable resources are increasingly used to cure epoxies, enabling the development of fully bio-based thermoset networks. These systems often offer improved toughness, lower toxicity, and reduced environmental impact, although challenges remain in achieving identical heat resistance and long-term durability compared to conventional aromatic amine-cured systems. Overall, the evolution from petroleum-based to bio-based epoxy systems represents a strategic re-engineering of feedstocks rather than a departure from epoxy chemistry itself. The same fundamental oxirane ring-opening reactions and crosslinking mechanisms are preserved. What changes is the origin of the carbon backbone and, increasingly, the design philosophy. Modern bio-based epoxy development integrates renewable raw materials, lower-emission processing, and end-of-life considerations while striving to maintain the mechanical strength, chemical resistance, and thermal stability that have made epoxy resins indispensable in advanced materials applications [2].

Amines act as highly effective curing agents for epoxy resins due to their nucleophilic primary amine groups, which readily react with the epoxide rings of DGEBA to form a cross-linked thermoset network. Aromatic diamines such as 4,4′-diaminodiphenyl sulfone (p-DDS) and 3,3′-diaminodiphenyl sulfone (m-DDS) have been widely studied as curing agents, and their behavior has been analyzed using differential scanning calorimetry. A broad exothermic peak in the range of approximately 130°C–280°C confirms the ring-opening reaction between amine and epoxy groups. The heat of polymerization increases with amine concentration up to the stoichiometric ratio, indicating more complete network formation, whereas excess amine can reduce effective cross-link density. Differences in molecular structure also influence reactivity, as the meta isomer (m-DDS) exhibits higher reactivity than the para isomer (p-DDS), attributed to reduced resonance stabilization of the amine lone pair and enhanced nucleophilicity. Fully cured systems demonstrate improved thermal stability, while insufficient amine content leads to lower degradation temperatures and multistep decomposition due to incomplete crosslinking. Mechanically, laminates cured with p-DDS show slightly superior properties, although both systems produce rigid, high-temperature-resistant networks [3].

At the molecular level, the epoxy–amine curing reaction proceeds via a nucleophilic ring-opening mechanism in which the amine nitrogen attacks the less substituted carbon of the epoxide ring following an SN2-type pathway. The rate-determining step involves C–N bond formation and C–O bond cleavage, followed by rapid proton transfer to generate a hydroxyl group. Both primary and secondary amines participate in this reaction, and primary amines can react twice, ultimately forming tertiary amines as the network evolves. The curing process can occur through uncatalyzed, self-promoted, or alcohol-catalyzed pathways, with the alcohol-catalyzed acyclic route exhibiting the lowest activation energy and therefore dominating the reaction, particularly as curing progresses. Because each epoxy–amine reaction generates a hydroxyl group, the system becomes increasingly autocatalytic at higher conversions. Hydrogen bonding, especially interactions involving the epoxy oxygen, plays a critical role in stabilizing the transition state and lowering activation barriers. In contrast, cyclic transition-state pathways are kinetically unfavorable due to steric crowding and higher activation energies, making acyclic routes dominant. Solvent polarity further affects curing kinetics by stabilizing the charge-separated transition state more effectively than the reactants, thereby reducing activation energy and increasing reaction rates. Differences in reactivity between primary and secondary amines are influenced not only by steric and electronic factors but also significantly by solvent effects. Overall, the structure, orientation, and concentration of amine curing agents govern cure kinetics, crosslink density, thermal stability, and mechanical performance, while hydrogen-bond-assisted transition states and autocatalytic effects control the progression of epoxy thermoset network formation [4].

7.2 BIO-BASED AMINES AS EPOXY PRECURSORS AND CURING AGENTS

7.2.1 Classification of Bio-Based Amines

Bio-based amines can be systematically classified based on two primary criteria: (i) the degree of substitution at the nitrogen atom and (ii) the number of amino functional groups present within the molecule. Both classifications are essential for understanding their chemical reactivity, curing behavior, and suitability for polymer applications such as epoxies, polyamides, and polyurethanes.

From a structural perspective, bio-based amines are divided into primary, secondary, and tertiary amines according to the number of carbon substituents bonded to the nitrogen atom. Primary amines ($R–NH_2$) contain one carbon substituent and two hydrogen atoms attached to nitrogen. They are highly nucleophilic and therefore particularly reactive toward electrophilic groups such as epoxides and isocyanates. Many renewable amines derived from fatty acids, amino acids, carbohydrates, and fermentation routes fall into this category. Secondary amines (R_2NH), which contain two carbon substituents and one hydrogen attached to nitrogen, generally exhibit lower nucleophilicity compared to primary amines due to steric and electronic effects. They are often produced through modification reactions such as alkylation or condensation of primary amines and are used in tailored curing systems or as intermediates in specialty chemicals. Tertiary amines (R_3N), in which nitrogen is bonded to three carbon groups and carries no N–H bond, do not participate directly in typical addition reactions with epoxides but instead function primarily as catalysts. In epoxy and polyurethane systems, tertiary amines accelerate curing by promoting ring-opening or urethane-forming reactions without becoming incorporated into the polymer backbone [5].

In addition to substitution pattern, bio-based amines are further classified based on the number of amino functional groups per molecule. Monoamines contain a single $–NH_2$ group and are typically used as chain terminators, modifiers, or intermediates in fine chemical synthesis. Diamines possess two amino groups and are among the most important classes for polymer chemistry, as they enable step-growth polymerization and crosslinking reactions. Bio-based diamines are widely used in the production of polyamides, epoxy-curing agents, and polyurethane hard segments (Figure 7.1) [5]. Polyamines, containing more than two amino groups, introduce multiple reactive sites that facilitate the formation of highly crosslinked three-dimensional networks. These materials are particularly valuable in high-performance coatings, adhesives, and thermosetting resins where enhanced mechanical strength, chemical resistance, and thermal stability are required [6–8]

In addition to classifying bio-based amines by substitution at nitrogen (primary, secondary, and tertiary) and by functionality (mono-, di-, and polyamines), it's also common and very useful to classify them by the nature of the carbon backbone because that strongly controls reactivity, curing rate, and final polymer properties. In this framework, bio-based amines are grouped as aliphatic, aromatic, cycloaliphatic, and heterocyclic. Aliphatic bio-based amines have flexible straight or branched

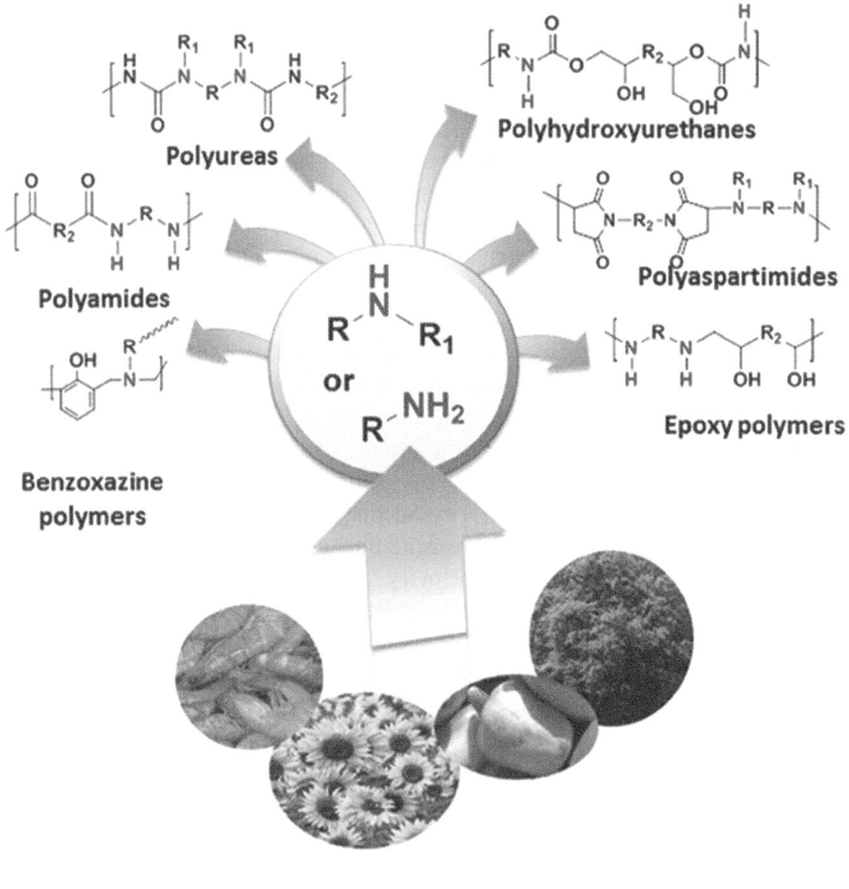

FIGURE 7.1 Applications of bio-based diamines. (Adapted with permission [5]. Copyright 2016 American Chemical Society.)

carbon chains (often originating from fatty acids, amino acids, or fermentation intermediates) and typically provide good nucleophilicity and relatively fast epoxy ring opening, often yielding networks with improved toughness and flexibility. Aromatic bio-based amines contain amine groups attached to an aromatic ring (potentially derived from lignin- or phenolic platforms such as vanillin-based chemistry); these amines tend to be less reactive than aliphatic amines due to resonance effects, but they introduce rigid segments that raise T_g, modulus, and thermal stability. Cycloaliphatic amines contain amines on saturated ring structures; they often combine better weathering and color stability than aromatic amines with higher rigidity than linear aliphatic amines, making them attractive for coatings and durable adhesive systems. Finally, heterocyclic amines incorporate nitrogen within a ring (and may include additional heteroatoms), and their behavior can vary widely depending on ring type and basicity; they are often used when specific catalytic activity, polarity, or specialized interactions (e.g., hydrogen bonding and coordination) are desired. When combined with the earlier classifications (primary/secondary/tertiary and mono/di/poly), this backbone-based grouping provides a complete, practical way to describe bio-based amines and predict how they will perform as epoxy curing agents and polymer building blocks [9].

Together, these classification frameworks provide a clear structure–function understanding of bio-based amines. The degree of substitution at nitrogen determines intrinsic reactivity and catalytic behavior, while the number of amino groups governs network formation potential and crosslink density. This dual classification is therefore critical when designing sustainable amine-based systems for advanced polymer applications.

7.2.2 Biomass Sources and Synthesis Routes

Bio-based amines are derived from multiple renewable biomass platforms, and their synthesis routes depend strongly on the chemical functionality available in the original feedstock. The document highlights four major biomass sources: lipid-based systems, carbohydrate-derived platforms, polyol intermediates such as glycerol, and lignocellulosic or aromatic biomass.

Lipid-based feedstocks, particularly vegetable oils such as soybean, rapeseed, and castor oil, are important renewable sources. These oils contain long-chain unsaturated fatty acids that can be chemically transformed into amines through multistep processes. Common routes include epoxidation of double bonds followed by ring-opening and amination, or nitrile formation through fatty acid conversion followed by catalytic hydrogenation to produce primary fatty amines. Another widely applied industrial pathway involves converting fatty acids into fatty nitriles via dehydration with ammonia, followed by hydrogenation to yield aliphatic primary amines. These lipid-derived amines are typically long-chain aliphatic amines used in surfactants, curing agents, and flexible polymer systems [10,11].

Carbohydrate-derived platforms represent another major route. Sugars obtained from starch or cellulose can be converted into platform molecules such as 5-hydroxymethylfurfural, levulinic acid, or furfural [12]. These intermediates can then undergo reductive amination to introduce amino functionalities. Reductive amination is one of the most important synthesis routes. In this process, an aldehyde or ketone group reacts with ammonia or an amine to form an imine intermediate, which is subsequently hydrogenated to produce the corresponding amine. This method is widely used because it allows selective introduction of primary, secondary, or tertiary amine groups depending on reaction conditions and nitrogen source (Figure 7.2) [13].

Polyol-based intermediates, especially glycerol derived as a by-product of biodiesel production, are also emphasized. Glycerol can be transformed into amines through catalytic amination or through intermediate formation of chlorinated or epoxidized derivatives followed by substitution reactions. Borrowing hydrogen methodologies and catalytic direct amination of alcohols are presented as greener alternatives, where alcohols are converted directly into amines using ammonia and transition metal catalysts without isolating intermediates. These routes improve atom economy and reduce waste generation [14,15].

Lignocellulosic biomass, particularly lignin, provides access to aromatic structures. Lignin depolymerization yields phenolic compounds such as vanillin, guaiacol, and syringaldehyde. These aromatic aldehydes can undergo reductive amination to form aromatic bio-based amines. Because lignin inherently contains aromatic rings, lignin-derived amines are especially attractive for producing rigid, high-performance materials such as epoxy curing agents and polyamide monomers. Nitrile hydrogenation also acts as a key route for producing diamines from dicarboxylic acids or di-nitrile intermediates, particularly in the context of bio-based polyamide production [16].

Overall, the main synthesis routes highlighted include reductive amination of aldehydes and ketones, direct amination of alcohols (including borrowing hydrogen strategies), nitrile formation followed by catalytic hydrogenation, and functional

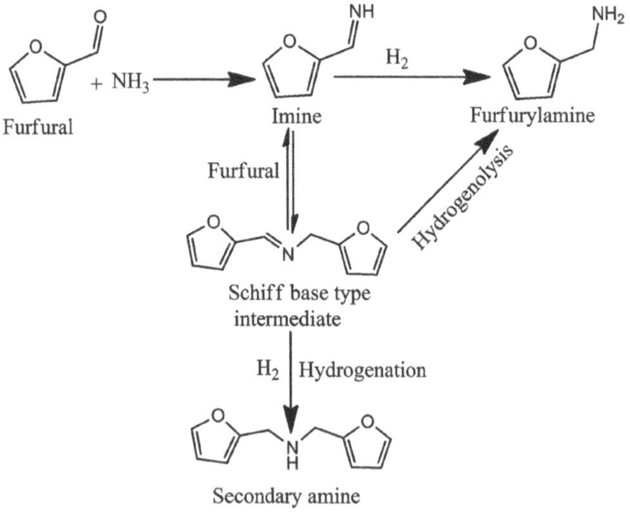

FIGURE 7.2 Proposed reaction pathway for the formation of furfurylamine under the investigated reaction conditions. (Adapted with permission [13]. Copyright 2016 Royal Society of Chemistry.)

group transformation of epoxidized or halogenated intermediates. The choice of route depends on the functional groups present in the biomass-derived precursor, desired amine type (primary, secondary, tertiary, or diamine), and intended polymer application. These pathways collectively demonstrate that bio-based amines can be produced through both biochemical and thermochemical transformations, enabling partial or complete replacement of fossil-derived amine feedstocks.

7.3 SYNTHESIS OF BIO-BASED EPOXY RESINS

A series of epoxy–amine thermosetting resins was synthesized by Mauck et al. using lignin-derived epoxy monomers and a bio-based diamine curing agent. The epoxy resins employed in the formulations were DGEGd, DGEVA, DGEBG, and DGEHQ, along with a commercial petroleum-based reference epoxy EPON 828 (DGEBA). The curing agent used for all systems was 5,5′-methylenedifurfurylamine (DFDA), a furfural-derived diamine. Liquid epoxy DGEGd and low-melting DGEVA were directly mixed with DFDA, while the higher-melting crystalline epoxies DGEBG and DGEHQ were first blended with DGEGd (25–50 wt%) to improve processability before curing. The hardener was added in stoichiometric amounts based on the epoxy equivalent weight of the resin and the amine hydrogen equivalent weight of DFDA (51.57 g/eq). The mixtures were degassed under vacuum, cast into molds, and thermally cured at 90°C for 5 h, followed by post-curing at 160°C for 2 h, producing fully crosslinked epoxy thermoset networks [17]. A magnolol-based epoxy monomer was synthesized by Bu et al. through a two-step procedure. Initially, magnolol underwent methylation under alkaline conditions using dimethyl sulfate, converting the phenolic hydroxyl groups into methoxy groups and forming the intermediate (MDBP). In the subsequent step, the allylic double bonds of MDBP were epoxidized using m-chloroperoxybenzoic acid to produce the bio-based epoxy monomer (EMDBP). Both reactions proceeded with high efficiency, yielding products above 80%, indicating that the synthetic route is feasible for scalable production. The obtained epoxy monomer was then cured with two furfurylamine-based hardeners (PDFA and BDFA), to produce fully bio-based epoxy thermosets (Figure 7.3) [18].

A series of bio-based epoxy and amine monomers were synthesized from resveratrol, a renewable phenolic compound, and subsequently used to prepare epoxy–amine thermoset networks. In this work, the authors first prepared resveratrol-derived epoxy monomers through epoxidation of resveratrol and dihydroresveratrol using epichlorohydrin under alkaline conditions, producing trans-resveratrol trisepoxide (1) and dihydroresveratrol trisepoxide (2). In addition to the epoxy monomers, a trifunctional aniline curing agent was synthesized by electrophilic aromatic substitution of resveratrol or dihydroresveratrol with 4-fluoronitrobenzene, followed by catalytic hydrogenation to convert the nitro groups into amines, yielding a trisaniline hardener (4). The synthesized epoxy monomers were then cured with either the conventional petroleum-derived hardener 4,4′-methylenedianiline (MDA) or the newly synthesized resveratrol-derived trisaniline, producing four different epoxy–amine thermoset networks. This strategy enabled the preparation of thermosetting systems containing significant renewable content while maintaining highly crosslinked network structures [19].

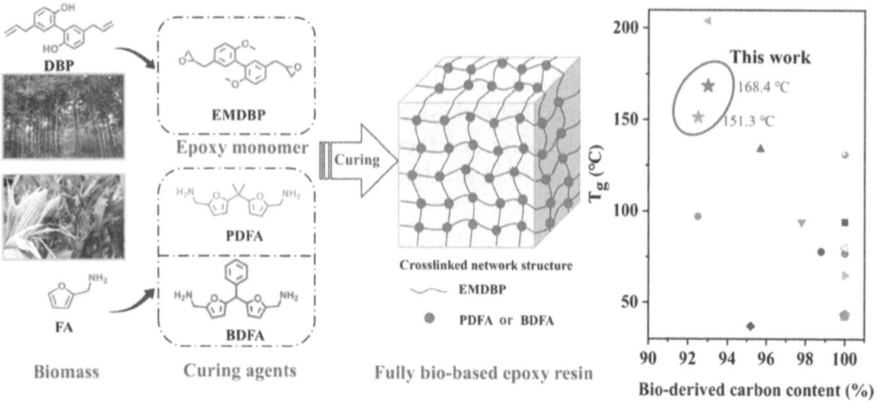

FIGURE 7.3 EMDBP cured with PDFA/BDFA) to form a fully bio-based thermoset with high T_g. (Adapted with permission [18]. Copyright 2022 Elsevier.)

7.4 STRUCTURE–PROPERTY RELATIONSHIPS

The structure–property relationships of the synthesized epoxy–amine thermosets by Mauck et al. were evaluated using spectroscopic, thermal, and thermomechanical characterization techniques. Near-infrared spectroscopy confirmed successful curing of the epoxy systems through the reduction of the oxirane absorption band and amine signals after curing, indicating that the epoxy and amine groups reacted to form a crosslinked network. Quantitative analysis showed that the extent of cure for all formulations exceeded 99%, demonstrating efficient epoxy–amine reactions and formation of highly crosslinked thermosets. Thermogravimetric analysis (TGA) revealed that the cured resins possessed good thermal stability. The initial degradation temperatures of the bio-based thermosets were generally in the range of 297°C–313°C, slightly lower than the petroleum-based EPON828-DFDA system (≈330°C). However, the bio-derived systems exhibited higher $T_{50\%}$ values (≈450°C–476°C) and significantly higher char yields (≈39–42 wt%) compared to the EPON828 system (≈22.6 wt%), indicating improved stability at high temperatures due to the presence of aromatic and furan structures in the network. Dynamic mechanical analysis further demonstrated the influence of molecular structure on thermomechanical properties. The T_g of the lignin-derived epoxy systems ranged from ≈81°C to 103°C, whereas the EPON828-based reference system exhibited a T_g of about 109°C (Figure 7.4). Although slightly lower, the bio-based resins displayed higher storage modulus values (≈2.9–3.7 GPa) compared with the reference resin (≈2.44 GPa), suggesting that the presence of methoxy-substituted aromatic structures enhanced stiffness. Crosslink density calculations showed that systems with lower molecular weight epoxy monomers formed denser networks, while higher molecular weight epoxies resulted in increased molecular weight between crosslinks and slightly lower crosslink density. Overall, the results showed clear structure–property relationships. Aromatic lignin-derived epoxies increased stiffness and char yield, while the furan-based diamine improved thermal stability. These findings confirm that the molecular structure of the epoxy monomer and curing agent strongly influences T_g, crosslink density, and mechanical performance of the bio-based thermosets [17].

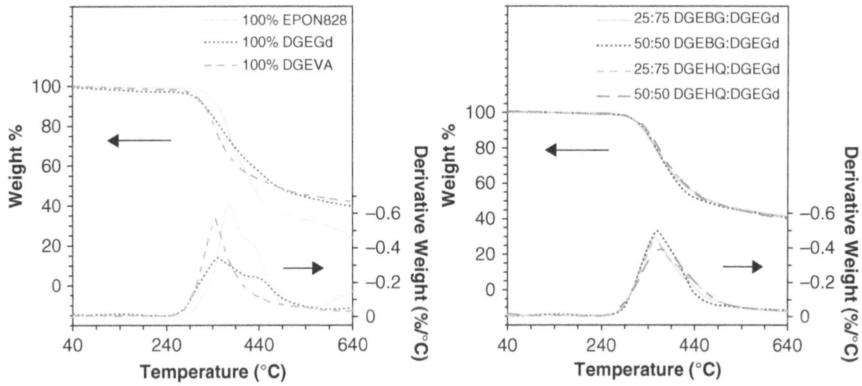

FIGURE 7.4 TGA and derivative thermogram (DTGA) of thermoset resins. (Adapted with permission [17]. Copyright 2017 Wiley.)

In another study by Bu et al., the curing behavior of the EMDBP/PDFA and EMDBP/BDFA systems was analyzed using DSC through nonisothermal and isothermal kinetic studies. The results showed that the curing reaction followed an autocatalytic mechanism at low conversion, while diffusion-controlled processes dominated at high conversion. Due to the presence of a benzene ring, BDFA exhibited higher amine reactivity than PDFA, resulting in a faster curing rate that increased with temperature. The resulting thermosets demonstrated significantly improved thermal and mechanical properties. The T_gs of the EMDBP systems reached 151.3°C and 168.4°C, representing a 71%–96% increase compared to conventional DGEBA (E51) systems. Mechanical characterization through nanoindentation and tensile testing revealed notable improvements in Young's modulus, hardness, tensile strength, and tensile modulus. TGA further indicated excellent thermal stability, with a char yield of approximately 47.5 wt% at 800°C. These results demonstrate that magnolol-derived epoxy thermosets possess enhanced heat resistance, mechanical strength, and thermal stability, making them promising alternatives to conventional petroleum-based epoxy resins for high-temperature applications [18]. The study by Garrison et al. reveals that the resveratrol-based networks exhibited significantly enhanced performance compared with conventional BPA epoxy systems. The T_g of the resveratrol-derived thermosets reached up to 285°C, which is approximately 110°C higher than typical BPA-based epoxy networks. TGA also revealed excellent thermal stability, with char yields up to 51% in nitrogen and 46% in air, substantially higher than the values observed for BPA-based systems. These improvements were attributed to the rigid stilbene structure of resveratrol and the higher crosslink density of the networks. Additionally, the materials displayed favorable moisture resistance and strong thermomechanical performance in both dry and wet conditions. Overall, the study demonstrated that resveratrol-derived epoxy and amine monomers can produce high-performance bio-based thermosets with superior heat resistance and thermal stability, highlighting their potential as sustainable alternatives to petroleum-based epoxy resins [19].

The performance of polymeric materials is strongly influenced by the molecular structure and functionality of the building blocks used in their synthesis. Variations in functional groups, molecular geometry, and chain architecture can significantly affect intermolecular interactions, crosslink density, and network formation within the material. By carefully designing molecules with different functionalities, it becomes possible to tailor properties such as mechanical strength, thermal stability, and chemical resistance. These structural modifications enable the formation of stronger intermolecular bonding and more efficient polymer networks, ultimately leading to improved material performance for specific applications. In the study by Garrison et al., the two amine curing agents differ in both functionality and molecular architecture. MDA is a difunctional aromatic diamine containing two amine groups connected through a methylene-bridged aromatic structure, whereas the resveratrol-derived hardener is a trifunctional aromatic amine containing three amine groups attached to a rigid stilbene backbone (Figure 7.5). The higher functionality and rigid aromatic structure of the resveratrol-derived amine enable the formation of denser and more rigid crosslinked networks compared with those formed using MDA. The resulting thermosets exhibited outstanding thermal and thermomechanical properties. The resveratrol-based epoxy networks showed T_gs up

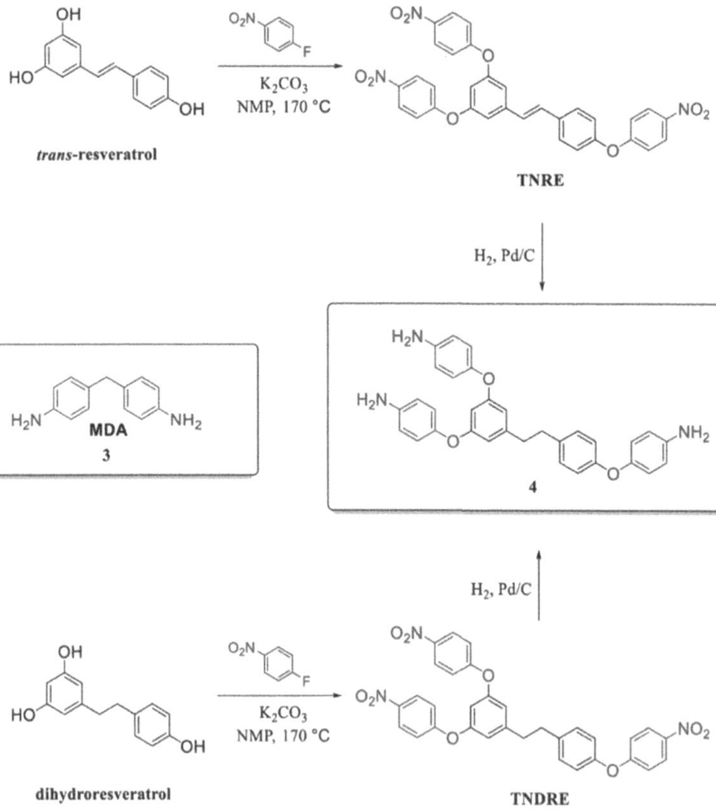

FIGURE 7.5 Different functionality amine curing agents. (Adapted with permission [19]. Copyright 2020 American Chemical Society.)

to approximately 285°C, which are significantly higher than those typically observed for conventional BPA-based epoxy systems. TGA also demonstrated excellent thermal stability, with high char yields approaching about 50%, attributed to the highly aromatic network structure and increased crosslink density. These results indicate that the incorporation of resveratrol-derived epoxy and amine monomers produces high-performance bio-based thermosets with superior heat resistance and thermal stability compared with conventional petroleum-based epoxy systems [19].

7.5 APPLICATIONS OF BIO-BASED EPOXY RESINS

Bio-based epoxy resins have gained increasing attention as sustainable alternatives to conventional petroleum-derived epoxies due to their comparable performance and reduced environmental impact. Derived from renewable resources such as lignin, vegetable oils, sugars, and natural phenolics, these materials can be tailored to achieve desirable mechanical, thermal, and chemical properties. As a result, bio-based epoxy resins are being explored for a wide range of applications. Their ability to combine sustainability with strong adhesion, chemical resistance, and thermal stability makes them promising candidates for advanced industrial applications while supporting the transition toward greener polymer technologies.

Kugler et al. investigated a solventless epoxy coating system using a bio-based cardanol-derived phenalkamine (PAA) as the curing agent. PAAs are long-chain aliphatic–aromatic amines derived from cardanol, a renewable component obtained from cashew nut shell liquid (CNSL). The bio-based amine hardener was evaluated alongside several conventional synthetic amines to understand its influence on coating performance when combined with carbon nanotubes (CNTs). The results showed that dispersing CNTs directly in the PAA hardener significantly improved electrical conductivity while maintaining high optical transparency of the coating. Coatings containing 0.35 wt% CNT predispersed in the PAA exhibited surface resistivity in the 10^7 Ω range, demonstrating effective electrostatic dissipative behavior. Additionally, the coatings displayed high gloss, good transparency, strong adhesion to glass substrates, and improved mechanical resistance, including increased cupping resistance. The PAA structure, containing both aromatic and long hydrophobic chains, also contributed to stable curing behavior and coating performance. Due to this combination of transparency, antistatic properties, and solvent-free processing, the developed bio-based epoxy coatings are particularly suitable for clear protective coatings and electrostatic dissipative surfaces, such as lampshades and sight-glasses used in chemically hazardous or explosive environments [20]. In another study, Benega et al. evaluated the use of bio-based amine curing agents derived from CNSL in epoxy resin systems for composite applications. The CNSL-based hardeners used included Ultra Lite 2009, Lite 3060, and Lite 3100, which are PAA-type curing agents derived from cardanol, a renewable phenolic compound obtained from cashew nut shells. These bio-based amines were compared with a conventional petroleum-based hardener in an epoxy infusion resin system. Structurally, PAAs contain an aromatic ring that improves chemical resistance, a long aliphatic chain that provides hydrophobic character and water resistance, and amine functional groups that promote crosslinking in the epoxy network. The results showed that epoxy systems cured with CNSL hardeners exhibited similar Young's modulus to the conventional system while demonstrating significantly lower moisture uptake, which is particularly important for marine environments. Among the tested hardeners, Lite 3060 showed mechanical properties comparable to the conventional hardener while reducing moisture absorption, whereas Lite 2009 and Lite 3100 provided much higher flexibility and impact resistance but lower stiffness. These findings indicate that CNSL-derived amine hardeners can successfully introduce renewable content into epoxy systems while maintaining adequate mechanical performance and improving moisture resistance, making them promising candidates for marine composite structures such as boat components produced by resin infusion processes [21]. In another work, Rashid et al. and authors developed bio-based epoxy vitrimers using imine (Schiff-base) hardeners prepared from lignin-derived vanillin and bio-based diamines (butanediamine (BDA) and hexanediamine (HDA)). These hardeners enabled epoxy networks that behave like conventional thermosets during use, but can be reprocessed, repaired, and chemically recycled because the imine bonds can undergo exchange reactions under heat or suitable chemical conditions. From an application standpoint, the materials are promising for recyclable composite matrices, durable coatings, and structural epoxy parts where repairability and end-of-life recovery are important (Figure 7.6). Performance-wise, the vitrimer cured with the shorter-chain hardener (V-BDA) showed higher stiffness and higher T_g than the longer-chain analogue (V-HDA), with storage modulus values around 3617 MPa vs 2872 MPa and T_g values around 75°C vs 66°C. Both systems showed flexural strengths in the ~83–88 MPa range and maintained their T_g and flexural properties after being reprocessed up to

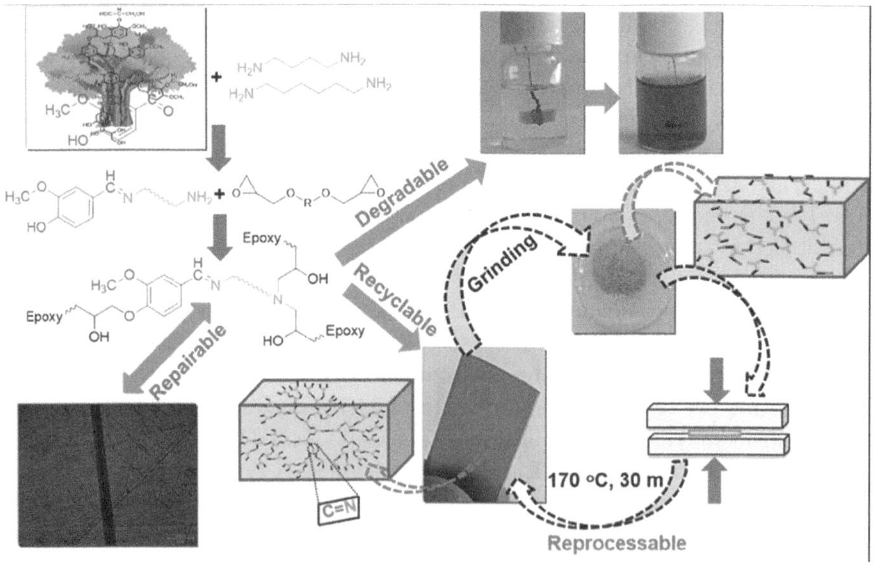

FIGURE 7.6 Recyclable reprocessable vitrimers from vanillin-derived amine hardeners. (Adapted with permission [22]. Copyright 2023 Elsevier. Licensed under CC BY 4.0. https://creativecommons.org/licenses/by/4.0/.)

three times, supporting repeated reuse in practical manufacturing. In addition, the cured networks resisted dissolution in common solvents at room temperature but could be fully degraded in an acidic solution at 50°C within about 12 h, indicating a realistic route for chemical recycling and recovery of building blocks [22].

7.6 ENVIRONMENTAL AND SUSTAINABILITY ASSESSMENT

Environmental and sustainability assessment of bio-based epoxy resins goes beyond simply stating that a resin is "renewable" and instead evaluates how the material performs across its full life cycle, from feedstock sourcing to end-of-life. A key starting point is renewable carbon content, which depends on the origin and composition of every major component in the formulation, including the epoxy monomer, curing agent, and any reactive diluents or modifiers. Feedstock choices such as lignin-derived aromatics (for rigidity and heat resistance), plant oils (for flexibility and hydrophobicity), and sugar- or fermentation-derived intermediates (for scalable platform chemistry) influence not only bio-content but also land-use considerations, supply-chain stability, and the need to avoid competition with food resources. A second major factor is VOC reduction and low-toxicity formulation design. Many bio-based epoxies are being developed as low-solvent or solvent-free systems to reduce emissions during manufacturing and application, particularly for coatings and adhesives. Using inherently lower-viscosity bio-based resins or reactive diluents can reduce or eliminate volatile solvents, while careful selection of curing agents can reduce exposure concerns. This is especially relevant because conventional epoxy systems often rely on BPA-based epoxies and aromatic amine hardeners that raise toxicity and regulatory concerns; bio-based alternatives aim to lower hazard profiles while maintaining mechanical strength, thermal stability, and chemical resistance. From a life cycle assessment (LCA) perspective, the most meaningful sustainability claims come from cradle-to-grave evaluation rather than only cradle-to-gate. LCA should consider the energy and reagent intensity of monomer synthesis, hydrogen demand, purification steps, catalyst use, and curing temperature/time, since a highly bio-based system can still have a high footprint if production is energy intensive or solvent heavy. In practice, the best-performing systems are those that combine high renewable content with efficient synthesis, scalable processing, and low-energy curing. Finally, end-of-life options and recyclability increasingly determine the true environmental advantage of epoxy materials. Conventional epoxy thermosets are difficult to recycle due to permanent crosslinking, leading to disposal by landfill, incineration, or energy recovery. Newer bio-based epoxy systems are addressing this limitation by incorporating cleavable or exchangeable linkages, enabling reprocessing, repair, and in some cases chemical recycling. Vitrimeric networks and dynamic covalent chemistries (for example, imine exchange or transesterification-type bond exchange) allow reshaping and property recovery after multiple cycles, while acid- or stimulus-triggered degradation routes can enable breakdown of the network and partial recovery of building blocks. Together, higher renewable carbon content, lower VOCs, safer chemistries, credible LCA framing, and improved end-of-life pathways form the basis for assessing whether a bio-based epoxy resin truly advances sustainability rather than only shifting impacts upstream [23–25].

7.7 COMMERCIALIZATION STATUS, TECHNICAL CHALLENGES, AND FUTURE OUTLOOK

Commercialization of bio-based epoxy resins is steadily advancing, but market penetration is still selective because epoxy users typically require proven performance, consistent quality, and established processing windows. In the current market, most "bio-based epoxies" are partially bio-based formulations, where renewable content is introduced through bio-derived epoxy monomers (for example, from plant oils or lignin-derived aromatics) and, increasingly, through bio-based curing agents, while conventional components are retained to ensure predictable viscosity, cure speed, and certification compliance. Full substitution is less common and tends to appear first in niche, high-value segments where sustainability is a purchasing driver or where unique bio-based chemistry offers a functional advantage (such as improved char yield, improved adhesion, or lower VOC processing). Scalability and cost remain central hurdles. Many bio-based building blocks are available at smaller volumes or have variable composition depending on biomass source, season, and processing method, which can complicate consistent epoxy equivalent weight control and batch-to-batch reproducibility. In addition, several bio-based epoxy routes require multistep synthesis, higher catalyst or solvent use, or more extensive purification, all of which increases cost compared with mature BPA-based supply chains. Supply chain competitiveness also depends on access to feedstocks and intermediates at industrial scale, including reliable production of bio-derived aromatics (from lignin streams), fermentation-derived intermediates, or bio-based epichlorohydrin. Logistics and regional availability matter, and there can be competition for the same biomass feedstocks from fuels, surfactants, and other polymer markets [26]. On the technical side, performance gaps relative to petro-based epoxy systems are application-dependent. Conventional DGEBA systems set a high standard for the balance of T_g, modulus, toughness, chemical resistance, and long-term reliability. Some bio-based epoxies achieve excellent stiffness or thermal properties but may suffer from brittleness, moisture sensitivity, or reduced chemical resistance; others offer improved toughness and flexibility but have lower T_g and reduced high-temperature performance. Matching "drop-in" performance is especially challenging when the bio-based structure is more aliphatic or contains flexible linkages, whereas bio-based aromatics tend to improve rigidity but can introduce viscosity and processing challenges.

Long-term durability and aging behavior are a major commercialization barrier because epoxies are frequently used in demanding conditions such as humidity, saltwater exposure, thermal cycling, UV radiation, and sustained mechanical loading. Many emerging bio-based systems still have limited long-duration field data, and dynamic or reversible chemistries (such as vitrimer networks) introduce new questions related to creep, stress relaxation, and property retention over extended service lifetimes. Manufacturers also require confidence in adhesion retention, corrosion protection, and barrier performance for coatings applications, as well as fatigue and fracture resistance for structural composites. Processing limitations also influence adoption. Bio-based resins can be more viscous, have narrower pot life windows, or require higher cure temperatures, which may conflict with established manufacturing lines for infusion, pultrusion, filament winding, or fast-cure adhesive bonding. Compatibility with fillers, pigments, and reinforcement sizing is another practical constraint, since even small changes in polarity and hydrogen bonding can affect dispersion, wet-out, and interfacial adhesion. For coatings, achieving the right balance of flow, leveling, cure speed, and hardness without solvents is particularly important, and for composites, resin flow and void control can determine whether a bio-based system is viable at scale. Looking ahead, several research and innovation pathways are shaping the outlook. One major direction is the development of bio-based aromatic building blocks that better replicate the rigidity and thermal performance of BPA-based systems while improving toxicity profiles. Another is the design of high-performance bio-based amine curing agents, including structures that deliver fast cure, high T_g, and strong chemical resistance without the hazards of traditional aromatic diamines. A third trend is the growth of reprocessable and recyclable epoxy networks, including vitrimers and other dynamic covalent systems, which can address end-of-life concerns and reduce composite waste, though these must be engineered to avoid unacceptable creep or loss of dimensional stability. Finally, manufacturing innovation is pushing toward high-solids/solvent-free processing, lower-energy cure schedules, and scalable, greener synthetic routes that reduce the cost and footprint of bio-based epoxy production [19,26,27]. Overall, near-term commercialization is likely to expand fastest in coatings, adhesives, and selected composite applications where sustainability requirements, regulatory pressures, and performance targets align. Broader adoption into mainstream structural markets will depend on improved cost competitiveness, stable large-scale supply chains, standardized specifications, and strong long-term durability data that demonstrate bio-based epoxies can match or exceed petro-based systems over real service lifetimes.

7.8 CONCLUSION

Bio-based epoxy resins derived from renewable amines are becoming practical alternatives to petroleum-based systems because they retain the same robust epoxy–amine curing chemistry while replacing fossil-derived building blocks with lignin aromatics, furan platforms, plant phenolics, and other biomass-derived intermediates. Across the examples discussed in this chapter, the key message is that performance is not inherently compromised by switching to bio-based feedstocks. Instead, properties can be deliberately tuned through molecular design of the epoxy monomer and, especially, the amine hardener. Rigid aromatic and multifunctional bio-based amines tend to increase crosslink density, stiffness, T_g, thermal stability, and char yield, while longer or more flexible aliphatic segments improve toughness and moisture resistance but may reduce T_g. Application-focused studies also show that bio-based amines can deliver real processing and durability benefits, including solvent-free or low-VOC coatings, improved moisture resistance for marine or humid environments, and functional coatings with antistatic behavior. Importantly, emerging dynamic covalent designs such as imine-containing hardeners demonstrate a pathway to repairable, reprocessable, and even chemically recyclable epoxy networks, addressing the long-standing end-of-life limitation of conventional thermosets. At the same time, broad commercialization still depends on overcoming scale-up, cost, and supply-chain consistency issues, along with closing remaining gaps in long-term aging performance and processing robustness. Continued advances in scalable bio-aromatic feedstocks, safer high-performance bio-based amines, and circular network chemistries are positioning bio-based epoxy thermosets to move from niche adoption toward wider industrial use.

REFERENCES

1. B. Bilyeu, W. Brostow, K.P. Menard, Epoxy thermosets and their applications I: Chemical structures and applications, *J. Mater. Educ.* 21(5&6) (1999) 281–286.
2. N. Karak, Overview of epoxies and their thermosets, in: N. Klarak (Ed.), *Sustainable Epoxy Thermosets Nanocomposites*, American Chemical Society, Washington, DC, 2021: p. 1–36.
3. I.K Varma, P.V. Satya Bhama, Epoxy resins: Effect of amines on curing characteristics and properties, *J. Compos. Mater.* 20 (1986) 410–422.
4. J.-E. Ehlers, N.G. Rondan, L.K. Huynh, H. Pham, M. Marks, T.N. Truong, Theoretical study on mechanisms of the epoxy–amine curing reaction, *Macromolecules.* 40 (2007) 4370–4377.
5. V. Froidevaux, C. Negrell, S. Caillol, J.P. Pascault, B. Boutevin, Biobased amines: From synthesis to polymers; Present and future, *Chem. Rev.* 116 (2016) 14181–14224.
6. X.-H. Jing, Y.-X. Liu, R. Liao, H.-J. Kang, H.-F. Tan, Y.-Y. Liu, Synthesis, characterization, and shape-memory performances of monoamine-toughened epoxy resin, *High Perform. Polym.* 28 (2016) 1082–1089.
7. G. Yang, S.-Y. Fu, J.-P. Yang, Preparation and mechanical properties of modified epoxy resins with flexible diamines, *Polymer (Guildf).* 48 (2007) 302–310.
8. J. Becker, C. Wittmann, Diamines for bio-based materials, in: C. Wittmann, J. C. Liao (Eds.), *Industrial Biotechnology*, Wiley, Hoboken, NJ, 2017: pp. 391–409.
9. B. Vermeeren, S. Van Praet, W. Arts, T. Narmon, Y. Zhang, C. Zhou, H.P. Steenackers, B.F. Sels, From sugars to aliphatic amines: as sweet as it sounds? Production and applications of bio-based aliphatic amines, *Chem. Soc. Rev.* 53 (2024) 11804–11849.
10. M.H. Mohammed, P.A. Williams, O. Tverezovskaya, Extraction of chitin from prawn shells and conversion to low molecular mass chitosan, *Food Hydrocoll.* 31 (2013) 166–171.
11. J.A. Melero, J. Iglesias, A. Garcia, Biomass as renewable feedstock in standard refinery units. Feasibility, opportunities and challenges, *Energy Environ. Sci.* 5 (2012) 7393–7420.
12. H.B. Hass, O.H. Lamborn, Economics of sucrose, *Ind. Eng. Chem.* 47 (1955) 1392–1397.
13. M. Chatterjee, T. Ishizaka, H. Kawanami, Reductive amination of furfural to furfurylamine using aqueous ammonia solution and molecular hydrogen: An environmentally friendly approach, *Green Chem.* 18 (2016) 487–496.
14. K. Shimizu, S. Kanno, K. Kon, S.M.A. Hakim Siddiki, H. Tanaka, Y. Sakata, N-alkylation of ammonia and amines with alcohols catalyzed by Ni-loaded $CaSiO_3$, *Catal. Today.* 232 (2014) 134–138.
15. M.H. Haider, N.F. Dummer, D.W. Knight, R.L. Jenkins, M. Howard, J. Moulijn, S.H. Taylor, G.J. Hutchings, Efficient green methanol synthesis from glycerol, *Nat. Chem.* 7 (2015) 1028–1032.
16. M. Pelckmans, T. Renders, S. Van de Vyver, B.F. Sels, Bio-based amines through sustainable heterogeneous catalysis, *Green Chem.* 19 (2017) 5303–5331.
17. J.R. Mauck, S.K. Yadav, J.M. Sadler, J.J. La Scala, G.R. Palmese, K.M. Schmalbach, J.F. Stanzione III, Preparation and characterization of highly bio-based epoxy amine thermosets derived from lignocellulosics, *Macromol. Chem. Phys.* 218 (2017) 1700013.

18. M. Bu, X. Zhang, T. Zhou, C. Lei, Fully bio-based epoxy resins derived from magnolol and varying furan amines: Cure kinetics, superior mechanical and thermal properties, *Eur. Polym. J.* 180 (2022) 111595.
19. M.D. Garrison, M.A. Savolainen, A.P. Chafin, J.E. Baca, A.M. Bons, B.G. Harvey, Synthesis and characterization of high-performance, bio-based epoxy–amine networks derived from resveratrol, *ACS Sustain. Chem. Eng.* 8 (2020) 14137–14149.
20. S. Kugler, K. Kowalczyk, T. Spychaj, Influence of synthetic and bio-based amine curing agents on properties of solventless epoxy varnishes and coatings with carbon nanofillers, *Prog. Org. Coat.* 109 (2017) 83–91.
21. M.A.G. Benega, R. Raja, J.I.R. Blake, A preliminary evaluation of bio-based epoxy resin hardeners for maritime application, *Proc. Eng.* 200 (2017) 186–192.
22. M.A. Rashid, M.N. Hasan, M.A. Kafi, Synthesis of novel vanillin-amine hardeners fully derived from renewable bio feedstocks and their curing with epoxy resins to produce recyclable reprocessable vitrimers, *Heliyon.* 9 (2023) e16062.
23. C. Lemesle, J. Frémiot, A. Beaugendre, M. Casetta, S. Bellayer, S. Duquesne, A.-S. Schuller, M. Jimenez, Life cycle assessment of multi-step versus one-step coating processes using oil or bio-based resins, *J. Clean. Prod.* 242 (2020) 118527.
24. A. Fitzgerald, W. Proud, A. Kandemir, R.J. Murphy, D.A. Jesson, R.S. Trask, I. Hamerton, M.L. Longana, A life cycle engineering perspective on biocomposites as a solution for a sustainable recovery, *Sustainability.* 13 (2021) 1160.
25. D. Kralisch, D. Ott, D. Gericke, Rules and benefits of Life Cycle Assessment in green chemical process and synthesis design: A tutorial review, *Green Chem.* 17 (2015) 123–145.
26. R. Auvergne, S. Caillol, G. David, B. Boutevin, J.-P. Pascault, Biobased thermosetting epoxy: present and future, *Chem. Rev.* 114 (2014) 1082–1115.
27. I. Isarn, I. Collado, A. Jiménez-Suárez, S.G. Prolongo, Analysis of bio-based epoxy resins: Impact of amine hardeners on thermal, thermomechanical, optical and electrical properties of epoxidized resveratrol with high Tg, *React. Funct. Polym.* 205 (2024) 106080.

8 Bio-Based Amines in High-Performance and Dynamic Polymer Chemistry

8.1 INTRODUCTION

Amines are among the most versatile and indispensable functional groups in modern polymer chemistry. Their nucleophilicity, tunable basicity, and ability to form covalent as well as reversible bonds make them central to the synthesis of polyamides, polyureas, epoxy thermosets, benzoxazines, and non-isocyanate polyurethanes (NIPUs) (Figure 8.1). As comprehensively reviewed by Froidevaux et al., amines constitute key building blocks in high-volume materials such as nylons, epoxy resins, and polyurethanes (PU) systems, which collectively represent multi-million-ton annual markets. The performance demands of aerospace, automotive, electronics, coatings, and structural composites continue to drive innovation in amine chemistry, particularly toward higher thermal stability, improved mechanical strength, and enhanced durability [1].

Despite their importance, the global amine industry remains heavily dependent on fossil resources. Conventional aromatic amines are typically produced via nitration–reduction sequences of benzene derivatives, processes associated with hazardous intermediates and significant environmental burden. Regulatory frameworks and sustainability targets have therefore accelerated the search for renewable nitrogen-containing chemicals. The shift toward bio-based amines is not simply a feedstock substitution but a strategic transformation in molecular design. Renewable platforms such as lignocellulosic biomass, vegetable oils, amino acids, and terpenes offer structurally diverse oxygen-rich intermediates that can be converted into primary and secondary amines via catalytic or biotechnological routes. Recent advances in catalytic amination of biomass-derived oxygenates highlight the feasibility of producing renewable amines from furfural, 5-hydroxymethylfurfural, levulinic acid, and lignin-derived aldehydes [1–3]. Reductive amination and hydrogen-borrowing methodologies enable selective incorporation of nitrogen into renewable carbon frameworks, avoiding traditional nitration chemistry. The development of heterogeneous catalysts for furanyl primary amine synthesis further enhances scalability and process sustainability. Such strategies are particularly relevant for polymer applications, where purity, functionality control, and reproducibility directly affect network formation and performance. Parallel to catalytic routes, metabolic engineering has enabled large-scale production of aliphatic diamines such as putrescine

DOI: 10.1201/9781003731122-8

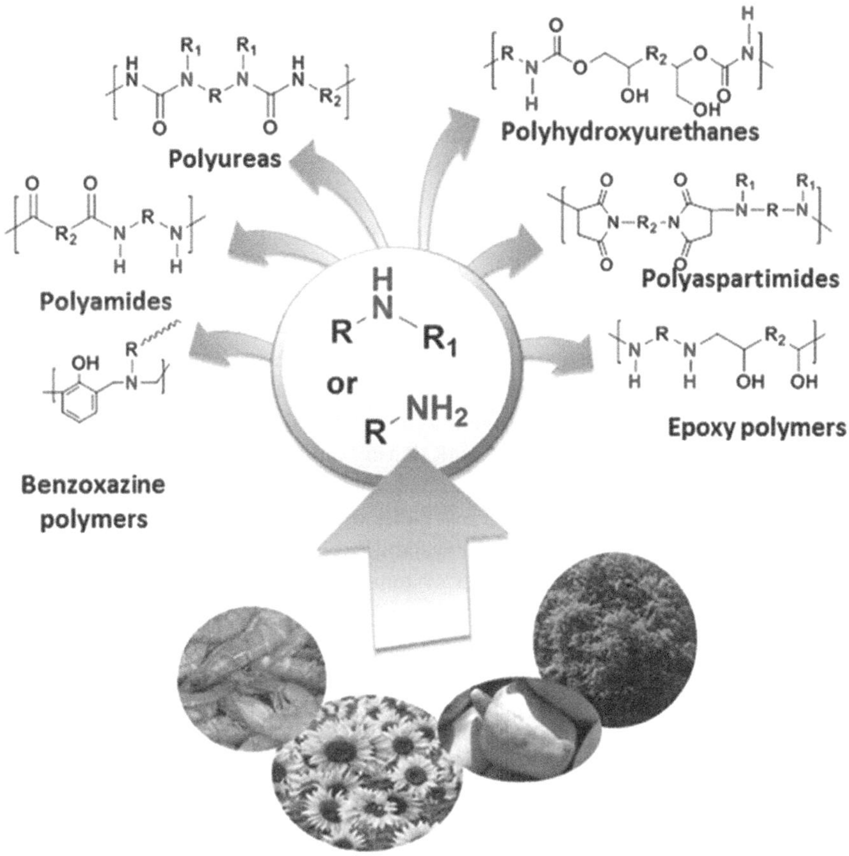

FIGURE 8.1 Bio-based amines derived from renewable resources as key precursors for various polymers. (Adapted with permission [1]. Copyright 2016 American Chemical Society.)

(1,4-diaminobutane) and cadaverine (1,5-diaminopentane) from renewable sugars. Lignin represents another crucial renewable aromatic resource. Blondiaux et al. [4] described nitration-free synthesis of 3,4-dialkoxyanilines from lignin-derived phenols via Beckmann rearrangement strategies. Complementary reductive amination approaches for lignin-derived aldehydes have also been demonstrated. These routes expand access to renewable aromatic diamines suitable for rigid thermosets and high-temperature polymers.

Amines serve multiple structural and functional roles in polymer systems. In step-growth polymerization, diamines condense with diacids to form polyamides, where hydrogen bonding between amide linkages dictates crystallinity, mechanical strength, and thermal resistance. In thermosetting systems, amines act as curing agents for epoxides and benzoxazines, forming crosslinked networks whose properties are governed by amine functionality and rigidity. For example, Zhang et al. [3] reported a renewable furan-derived diamine (propane-2,2-diyldi(furan-5,2-diyl)] dimethanamine) used to synthesize fully bio-based bis-benzoxazine thermosets.

The resulting networks exhibited glass transition temperatures (T_g) up to 72°C and char yields as high as 65%, demonstrating the strong influence of heteroaromatic amine structure on thermal performance and flame resistance. The presence of furan units contributed to enhanced rigidity and carbonization behavior, illustrating how renewable amine design can improve high-temperature stability. Similarly, Mora et al. [5] showed that β-hydroxylamines derived from lignin-based epoxides accelerate epoxy curing via hydrogen-bond-assisted autocatalysis, significantly influencing curing kinetics and crosslink density. Such hydroxyl-functional amines demonstrate how subtle molecular modifications affect network architecture and mechanical performance.

High-performance polymer systems emphasize thermal stability, mechanical strength, chemical resistance, and flame retardancy. Aromatic and heteroaromatic bio-based amines enhance rigidity and T_g, while long-chain aliphatic diamines reduce moisture uptake and improve toughness in polyamides. Control over amine structure allows fine-tuning of crosslink density and network homogeneity. Beyond static thermosets, amine chemistry also underpins dynamic and reprocessable polymer networks. Primary amines react reversibly with aldehydes to form imines (Schiff bases), enabling self-healing and recyclability. Transamination reactions allow associative bond exchange in vitrimer systems, while reversible urea and urethane exchange expands the scope of dynamic covalent networks. The structural diversity accessible from biomass-derived amines including furanyl, lignin-derived aromatic, and aliphatic motifs provides opportunities to tailor dynamic exchange kinetics without compromising mechanical integrity. Thus, bio-based amines bridge two critical domains: durable high-performance polymers and reprocessable circular materials.

This chapter examines bio-based amines as molecular design elements in high-performance and dynamic polymer chemistry, integrating advances in catalytic amination [4], biotechnological production of diamines, and structure–property investigations in benzoxazine and epoxy systems [3,5]. The discussion connects renewable feedstock conversion strategies with polymer performance metrics, emphasizing how synthetic routes influence molecular architecture and, ultimately, material behavior. This chapter first analyzes renewable sources and synthetic strategies for producing functional bio-based amines, including catalytic and fermentative pathways. It then correlates molecular structure with reactivity, curing kinetics, and network formation, highlighting how amine functionality, rigidity, and substitution patterns govern crosslink density and thermomechanical properties. Performance is evaluated across high-temperature thermosets, polyamides, and NIPU systems, with particular attention to T_g, thermal stability, and mechanical strength. The role of bio-based amines in dynamic and recyclable polymer networks is also explored, focusing on reversible covalent chemistry and topology rearrangement. Finally, sustainability aspects, life cycle considerations, and industrial translation challenges are assessed to provide a realistic perspective on large-scale implementation. By linking synthetic chemistry, catalysis, and polymer engineering, this chapter establishes a unified structure–reactivity–performance framework for designing next-generation sustainable materials, positioning bio-based amines not merely as renewable alternatives but as enabling platforms for advanced high-performance and circular polymer systems.

8.2 RENEWABLE SOURCES AND SYNTHETIC ROUTES TO BIO-BASED AMINES

8.2.1 Amino Acid-Derived Amines

Amino acids represent one of the most structurally defined and industrially accessible renewable nitrogen platforms to produce bio-based amines. Their large-scale fermentation production, inherent nitrogen functionality, stereochemical purity, and multifunctional backbone make them particularly attractive for sustainable polymer synthesis. The broader context of renewable amine development from biomass, including amino acid valorization, has been comprehensively discussed in the literature [1], where amino acids are identified as direct precursors to aliphatic and multifunctional amines relevant to polyamides, epoxy curing agents, and NIPU systems. One of the most important industrial transformations is the decarboxylation of L-lysine to produce 1,5-diaminopentane (cadaverine), a C5 diamine used in bio-based polyamides such as PA5.10. While enzymatic decarboxylation is well established, chemocatalytic strategies have also been developed. Claes et al. [6] reported an organocatalytic decarboxylation of natural amino acids mediated by α,β-unsaturated ketones such as isophorone (Figure 8.2). The mechanism proceeds via transient Schiff base formation between the amino acid and the carbonyl compound, followed by decarboxylation of the azomethine intermediate and hydrolysis to regenerate the catalyst. Under optimized conditions, neutral amino acids such as alanine, valine, leucine, and norleucine were converted into their corresponding primary amines in yields exceeding 90%. Reactivity trends correlated with side-chain electron density and steric effects,

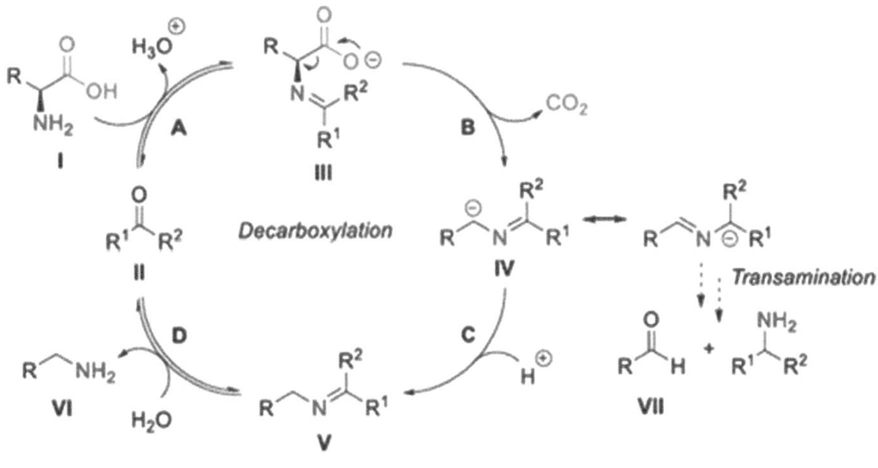

FIGURE 8.2 Proposed mechanism for the organocatalytic decarboxylation of α-amino acids: (a) condensation of the α-amino acid with a carbonyl compound to form a Schiff base; (b) decarboxylation; (c) protonation of the azomethine ylide intermediate; and (d) hydrolysis to release the amine and regenerate the catalyst. Transamination to form the corresponding aldehyde is a competing side reaction. (Adapted with permission [6]. Copyright 2019 Wiley-VCH Verlag GmbH & Co. KGaA, Weinheim.)

providing clear structure–reactivity relationships that are directly relevant to polymer curing kinetics and network formation.

Hydrogenative routes offer another direct pathway from amino acids to primary amines. A Ru-catalyzed hydrogenation–decarbonylation strategy [7] demonstrated the conversion of L-valine and other aliphatic amino acids into the corresponding primary amines through sequential hydrogenation of the carboxyl group to an amino alcohol, followed by dehydrogenation and decarbonylation. Yields up to approximately 87% were achieved under aqueous hydrogenation conditions. The process releases methane rather than CO_2 during decarbonylation, improving carbon utilization. Importantly, acidity strongly influenced selectivity toward primary amines versus over-reduced products, highlighting how reaction environment affects amine purity an essential factor in high-performance thermoset synthesis. Amino acid-derived amines also play an important role in organocatalysis and asymmetric synthesis. Zhou et al. [8] reported that vicinal diamines derived from tert-leucine efficiently catalyze asymmetric intramolecular aldol reactions, achieving yields up to 98% and enantiomeric excess values above 90%. The catalytic cycle proceeds via iminium/enamine intermediates formed between the primary amine and carbonyl substrates. This mechanistic feature is directly analogous to reversible imine chemistry in dynamic polymer networks, where primary amines enable bond exchange and self-healing behavior. The steric bulk of amino acid side chains were shown to significantly influence reactivity and selectivity, underscoring how subtle structural variations in bio-based amines affect macromolecular properties. Beyond structural polymer synthesis, amino acid-derived amines have been incorporated into functional materials. Bandyopadhyay et al. [9] synthesized conjugates based on γ-aminobutyric acid and polyamines that exhibited significant ion transport activity across phospholipid membranes. The study demonstrated that amide density, spacer length, and amine substitution pattern dramatically influence aggregation and transport behavior. These findings parallel polymer network design principles, where hydrogen-bond density and functional group spacing dictate mechanical strength, thermal stability, and dielectric properties. Analytical methodologies for characterizing amino acid-derived amines have also been well developed. A comprehensive review in Talanta [10] summarized chromatographic and spectroscopic techniques for detecting and quantifying biogenic amines. Such analytical strategies are crucial when translating renewable amine synthesis into polymer-grade materials, as trace secondary amines or residual impurities can significantly alter curing stoichiometry and crosslink density in epoxy and PU systems. Collectively, these studies demonstrate that amino acid-derived amines provide a versatile and scalable renewable platform for high-performance and dynamic polymer chemistry. Their tunable chain length, inherent chirality, multifunctionality, and compatibility with both catalytic and fermentative production routes enable precise control over polymer architecture. In polyamides, they influence crystallinity and moisture uptake; in epoxy thermosets, they modulate T_g and toughness; and in dynamic networks, primary amines derived from amino acids participate readily in imine formation and transamination processes. The integration of organocatalytic, hydrogenative, and analytical advances firmly establishes amino acids as foundational feedstocks for renewable amine chemistry in next-generation sustainable polymer materials [1,6,8–10].

8.2.2 Lignin-Derived Aromatic Amines

Lignin-derived aromatic amines have emerged as one of the most promising classes of renewable nitrogen-containing compounds for high-performance polymer chemistry. Unlike aliphatic bio-based amines derived from amino acids or fatty acids, lignin offers direct access to rigid aromatic frameworks that are essential for producing thermally stable thermosets, high-T_g epoxy networks, PUs, polyimides, and advanced functional materials. As the most abundant renewable source of aromatic carbon, lignin provides a structurally rich platform containing phenolic, methoxy, aldehyde, and alcohol functionalities, making it an ideal precursor for the synthesis of substituted anilines, aromatic diamines, and amides. In their comprehensive review, Natte et al. discussed the valorization of lignin-derived aromatic monomers as renewable feedstocks for synthesizing functional amines and other high-value chemicals (Figure 8.3) [11]. Lignin accounts for nearly 30% of natural organic carbon and represents the largest renewable reservoir of aromatic

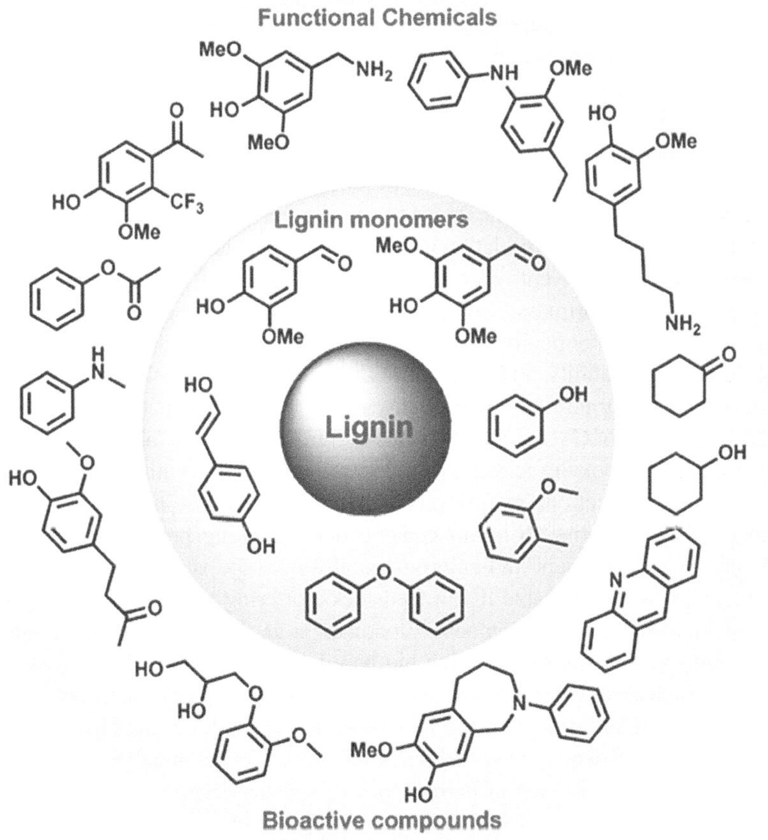

FIGURE 8.3 Lignin as a renewable platform for monomers, functional chemicals, and bioactive compounds. (Adapted with permission [11]. Copyright 2020 Wiley-VCH GmbH.)

structures. The authors highlighted several catalytic pathways for converting lignin-derived phenols and related compounds such as guaiacol, syringaldehyde, vanillin, and catechol into aromatic amines, particularly anilines. Key strategies include catalytic reductive amination, cross-coupling reactions, and hydrogenation–dehydrogenation sequences using heterogeneous catalysts such as Pd/C, Ni-based systems, and Ru catalysts. For example, Pd/C-catalyzed amination of phenol with ammonia produced aniline in yields of up to 92%, proceeding through hydrogenation to cyclohexanone, followed by amination and subsequent dehydrogenation. Advanced catalytic methodologies also enabled selective conversion of lignin model compounds such as 4-propylguaiacol and β-O-4 lignin linkages into substituted aromatic amines, which serve as valuable intermediates in pharmaceuticals, agrochemicals, and polymer synthesis. Furthermore, lignin-derived aromatic aldehydes such as vanillin and syringaldehyde were efficiently converted into N-methylated aromatic amines and diamines through reductive amination reactions with high selectivity and catalyst recyclability. Overall, this work establishes lignin as a sustainable alternative to petroleum-derived aromatic feedstocks for producing aromatic amines and other nitrogen-containing functional molecules. In a recent study, Ma et al. reported a sustainable catalytic strategy for converting lignin and lignin-derived compounds into aromatic amides using a cobalt single-atom catalyst supported on nitrogen-doped carbon (Co-SACs) [12].

Lignin, as a major component of lignocellulosic biomass, contains abundant aromatic units and represents a promising renewable alternative to fossil-derived feedstocks for producing high-value aromatic chemicals, including amides widely used in pharmaceuticals, agrochemicals, and polymer materials. The authors developed a one-pot aerobic oxidative amidation process utilizing molecular oxygen as a green oxidant and water as a solvent, enabling direct conversion of lignin model compounds and real lignin into primary, secondary, and tertiary aromatic amides. The Co-SAC catalyst, featuring atomically dispersed $Co–N_5$ active sites, demonstrated excellent catalytic activity, stability, and recyclability, achieving amide yields of up to 95% from lignin-derived substrates. Mechanistic investigations revealed that the process involves oxidative cleavage of C–O and C–C bonds in lignin structures, followed by conversion of intermediate carboxylic acids into amides through reaction with amines. Notably, the method enabled direct one-pot conversion of real pine lignin into amides with yields of up to 13 wt% relative to lignin content, demonstrating practical feasibility. This study provides an efficient and environmentally friendly pathway for producing aromatic amides from renewable lignin feedstocks, offering strong potential for sustainable polymer precursor development. Blondiaux et al. developed another efficient and sustainable approach for synthesizing bio-based aromatic amines from lignin-derived monomers such as 4-propylguaiacol and 4-propylcatechol [4]. Aromatic amines, particularly substituted anilines, are essential industrial intermediates used in the production of PUs, pharmaceuticals, dyes, and agrochemicals. Conventional synthesis typically relies on hazardous nitration of petrochemical benzene. Lignin, as the largest renewable source of aromatic compounds, provides an attractive alternative feedstock while avoiding toxic nitration processes. The authors introduced a synthetic pathway involving O-alkylation of lignin-derived phenolic monomers, followed by benzylic oxidation using sodium persulfate, Beckmann rearrangement with hydroxylamine hydrochloride to introduce the amino functionality, and final amide alcoholysis to generate the desired aromatic amines. This strategy enabled efficient conversion of 4-propylguaiacol into 3,4-dialkoxyanilines with overall yields of up to 65%, while simultaneously producing alkyl propionate esters as valuable industrial byproducts. Importantly, the process avoided nitric acid and column chromatography, improving safety and scalability. The synthesized aromatic amines serve as key intermediates for anticancer drugs such as gefitinib and erlotinib, as well as agrochemicals like diethofencarb. Green metrics analysis confirmed improved sustainability, step efficiency, and reduced environmental impact compared to traditional petroleum-based routes. In addition to synthetic and catalytic advances, understanding the environmental presence and transformation of amine-containing species is also important. Amino acids and amine compounds present in aerosols are representative reduced nitrogen-containing organic species originating from natural and anthropogenic sources, including terrestrial and marine emissions, as well as agricultural and industrial activities [13]. These compounds can also form through secondary atmospheric processes. Their chemical diversity and reactivity influence aerosol formation, physicochemical properties, cloud condensation processes, nutrient cycling, climate regulation, and potential public health impacts. Advances in high-resolution mass spectrometry and isotopic analysis have enabled detailed characterization of their compositions, abundances, origins, and atmospheric transformations. However, evolving analytical methodologies have generated datasets that are not always directly comparable. The review systematically evaluates detection methods for amino acid and amine compounds in fine aerosol particles and compares their distributions across urban, suburban, rural, and background environments. The discussion highlights that diverse emission sources and atmospheric transformation pathways are key factors driving fluctuations in composition and concentration. Overall, these studies collectively demonstrate that lignin-derived aromatic frameworks can serve as sustainable precursors for producing high-value aromatic amines and amides. Through catalytic amination, oxidative amidation, and nitration-free synthetic strategies, lignin is increasingly positioned as a viable renewable alternative to petroleum-derived aromatic feedstocks. Such developments strongly support the advancement of bio-based polymer precursors, pharmaceuticals, agrochemicals, and advanced functional materials within a sustainable chemical economy.

8.2.3 Vegetable Oil and Fatty Acid-Derived Amines

Fatty amines constitute an important class of commodity chemicals with wide applications in surfactants, corrosion inhibitors, flotation agents, lubricants, coatings, and polymer additives. Industrially, their synthesis typically begins from renewable feedstocks such as vegetable oils or animal fats. Conventional production generally follows the nitrile route, which involves hydrolysis of triglycerides to fatty acids, dehydration to fatty nitriles under high-temperature conditions, and subsequent catalytic hydrogenation to amines. Although effective at scale, this multistep process often requires harsh reaction conditions, elevated pressures, and high energy input, and may lead to mixtures of primary, secondary, and tertiary amines, thereby reducing selectivity and overall process efficiency. Recent advances have aimed to overcome these limitations through more selective catalytic and biocatalytic strategies. A highly selective cobalt–triphos

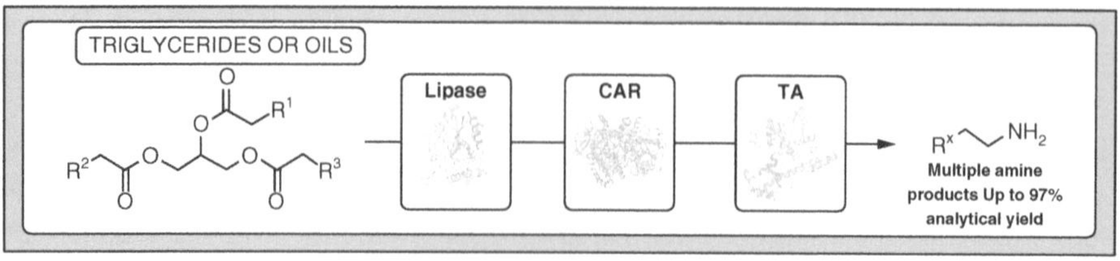

FIGURE 8.4 Enzymatic conversion of triglycerides to amines. (Adapted with permission [15]. Copyright 2021 Wiley-VCH GmbH.)

catalytic system has been developed for the direct hydrogenative amination of carboxylic acids, esters, and vegetable oils using ammonia and molecular hydrogen [14]. This homogeneous cobalt catalyst enables one-pot conversion of triglycerides into primary fatty amines under comparatively milder conditions, avoiding the formation of discrete nitrile intermediates. Mechanistic studies indicate a hydrogenative pathway involving sequential reduction steps and efficient ammonia incorporation, resulting in high selectivity toward primary amines and minimal formation of secondary products. Such catalytic systems represent an important step toward improving atom economy and reducing process complexity in fatty amine production. In parallel, a proof-of-concept biocatalytic alternative has been reported for the synthesis of primary fatty amines directly from renewable triglycerides and oils [15]. This approach integrates a lipase, a carboxylic acid reductase (CAR), and a transaminase (TA) in a one-pot enzymatic cascade. In the first step, the lipase hydrolyzes triglycerides into free fatty acids. The CAR subsequently reduces the fatty acids to the corresponding aldehydes, and the TA catalyzes transamination to generate primary fatty amines (Figure 8.4). This multienzyme system enables the direct synthesis of multiple medium- and long-chain primary fatty amines with analytical yields as high as 97%. Moreover, the methodology was successfully translated to preparative scale, where a 75mL reaction converting trilaurin into laurylamine afforded a 73% isolated yield. Compared to conventional thermochemical routes, the enzymatic cascade operates under significantly milder conditions and offers high chemo- and regioselectivity. The absence of harsh dehydration steps and high-pressure hydrogenation improves sustainability and reduces byproduct formation. Such biocatalytic strategies demonstrate strong potential for integrating renewable lipid feedstocks into circular chemical manufacturing systems. Overall, advances in both cobalt-catalyzed hydrogenative amination and enzyme-mediated cascade reactions illustrate the growing feasibility of converting vegetable oils and fatty acids directly into primary fatty amines with improved selectivity and reduced environmental impact. These developments strengthen the role of lipid-derived amines as renewable building blocks for surfactants, coatings, and high-performance polymer systems.2.4 Terpene- and Sugar-Derived Amines

8.3 STRUCTURE–REACTIVITY RELATIONSHIPS OF BIO-BASED AMINES

Understanding the structure–reactivity relationships of bio-based amines is essential for rational polymer design. Whether derived from amino acids, lignin, terpenes, or vegetable oils, the molecular architecture of an amine directly governs its nucleophilicity, steric accessibility, hydrogen-bonding ability, and thermal behavior. These parameters ultimately determine curing kinetics, crosslink density, T_g, mechanical strength, and recyclability of polymer networks. This section discusses how variations in amine structure influence polymer formation and performance.

8.3.1 Primary vs Secondary Amines

Primary amines are generally more reactive toward electrophiles than secondary amines due to lower steric hindrance and the presence of two N–H bonds. In epoxy curing, primary amines react with epoxide rings through nucleophilic ring-opening, forming β-hydroxy secondary amines. These secondary amines can further react with additional epoxide groups, increasing crosslink density. As a result, primary diamines often produce highly crosslinked, mechanically robust thermosets. Secondary amines, by contrast, exhibit reduced nucleophilicity and steric accessibility. In many systems, they react more slowly and often generate networks with lower crosslink density. However, this reduced reactivity can be advantageous in dynamic polymer systems. Secondary amines participate readily in reversible transamination and imine exchange reactions, which are central to vitrimer behavior and self-healing materials. Therefore, the choice between primary and secondary amines reflects a balance between high-performance crosslinking and dynamic bond exchange capability [16–18].

8.3.2 Aliphatic vs Aromatic Amines

The distinction between aliphatic and aromatic amines significantly affects polymer network rigidity and thermal performance. Aliphatic amines, particularly those derived from fatty acids or amino acids, are generally more nucleophilic due to the electron-donating nature of alkyl substituents. They cure epoxies rapidly at moderate temperatures and produce flexible networks with lower T_g values. The long-chain mobility of aliphatic segments reduces brittleness and improves impact resistance. Aromatic amines, such as those derived from lignin-based phenolics, exhibit lower intrinsic nucleophilicity because the nitrogen lone pair is partially delocalized into the aromatic ring. However, the rigid aromatic backbone significantly enhances thermal stability, modulus, and char formation. Epoxy systems cured with aromatic diamines typically display higher T_g values and superior thermal resistance compared to those cured with aliphatic amines. The rigidity of the aromatic structure restricts segmental motion and increases packing efficiency within the network. Thus, aliphatic amines favor toughness and flexibility, whereas aromatic amines promote stiffness and high-temperature performance. Hybrid systems combining both structures often achieve optimized property balance [19,20].

8.3.3 Chain Length, Rigidity, and Functionality

Chain length plays a critical role in determining polymer properties. Increasing the methylene content in bio-based amines reduces crosslink density per unit mass

catalyst that promotes direct condensation of carboxylic acids and amines under azeotropic conditions without the need for acid chlorides or carbodiimide activators [28]. The catalyst exhibited strong acidity, thermal stability up to 600°C, and reusability over multiple cycles while maintaining yields near 80%. Importantly, the E-factor of 9.3 and high atom economy (≈92%) demonstrate significant environmental advantages compared to conventional coupling reagents. For polyamide synthesis from bio-based diamines and renewable diacids, such heterogeneous catalysis offers a pathway to reduce waste, eliminate hazardous intermediates, and improve sustainability metrics. Beyond traditional condensation strategies, catalytic exchange methodologies provide new opportunities for incorporating amide and urea linkages into bio-based polymer systems. Han and Porco reported a zirconium(IV)-catalyzed carbamate and urea exchange process using dialkyl carbonates and carbamates [29]. The $Zr(Ot\text{-}Bu)_4$ system enables phosgene-free and isocyanate-free formation of carbamate and urea linkages with high selectivity. Although the study focused on small molecules, the underlying mechanism activation of carbonate carbonyls followed by amine exchange has strong implications for polymer chemistry. For renewable amines derived from biomass, such catalytic systems could facilitate safer polyurea and polyamide-related transformations, offering alternatives to hazardous isocyanate-based processes. The development of bio-based diamines has expanded the structural diversity available for polyamide synthesis. Long-chain aliphatic diamines derived from fatty acids introduce flexibility and reduce hydrogen bond density, resulting in lower T_g but improved moisture resistance and toughness. In contrast, shorter-chain diamines such as cadaverine increase crystallinity and tensile strength due to closer amide spacing. Sugar-derived diamines introduce rigid cyclic frameworks that reduce crystallinity but elevate T_g through restricted segmental motion. These structural variations illustrate how renewable amine architecture directly influences hydrogen bonding density, packing efficiency, and macromolecular organization. Modern catalytic systems further enhance the feasibility of converting renewable feedstocks into primary amines suitable for polyamide synthesis. Highly selective hydrogenative amination of carboxylic acids, esters, and vegetable oils using cobalt catalysts has demonstrated efficient production of primary amines under relatively mild conditions (Figure 8.5) [14]. Such advances reduce reliance on traditional nitrile routes and improve chemoselectivity toward primary amines, which are essential for step-growth polycondensation. From a structure–property perspective, polyamides derived from bio-based amines follow the same fundamental rules established in early nylon development: hydrogen bonding governs crystallinity and mechanical strength; stoichiometric precision controls molecular weight; and by-product removal determines equilibrium progression [26]. What distinguishes bio-based systems is the broader structural palette. Renewable diamines can introduce tailored flexibility, rigidity, chirality, or hydrophobicity depending on feedstock origin. This tunability allows optimization of properties for automotive components, fibers, engineering plastics, and high-performance applications. Overall, the integration of renewable diamine production, green catalytic amidation [28], ester-based polyamidation strategies, and safe catalytic exchange chemistry [29] demonstrates that high-performance polyamides from bio-based amines are not merely theoretical alternatives but technologically viable materials. By combining historical insights

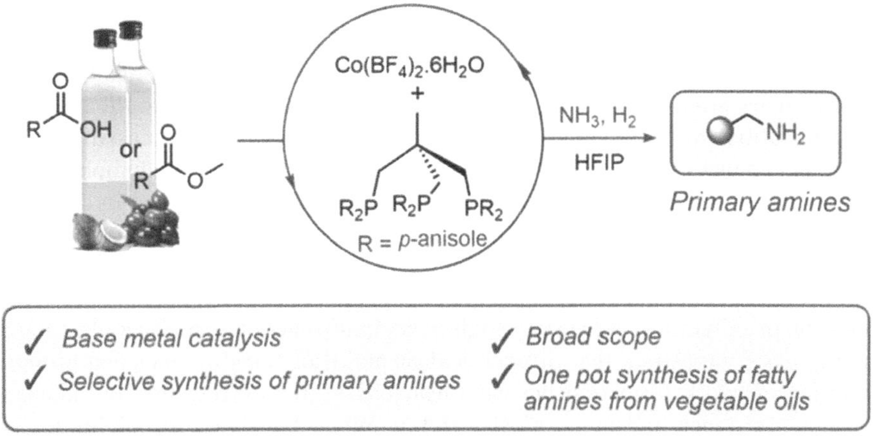

FIGURE 8.5 Base-metal catalyzed synthesis of primary amines from vegetable oils. (Adapted with permission [14]. Copyright 2025 American Chemical Society. Licensed under CC BY 4.0. https://creativecommons.org/licenses/by/4.0/.)

into condensation polymerization with modern catalytic innovation, sustainable polyamide production can achieve both environmental and performance objectives while maintaining compatibility with established industrial processing routes.

Polyimides derived from bio-based amines represent an important advancement in the development of sustainable high-performance polymers. Traditionally, polyimides are synthesized through the condensation of aromatic dianhydrides with petroleum-derived diamines, forming poly(amic acid) intermediates that undergo cyclodehydration to yield the imide ring. Replacing conventional diamines with renewable amines sourced from lignin, fatty acids, amino acids, or other biomass platforms enables the production of partially or fully bio-based polyimides while maintaining the inherent thermal stability and mechanical robustness associated with the imide structure. The rigid imide heterocycle, characterized by strong resonance stabilization and aromatic conjugation, is primarily responsible for the exceptional thermal resistance, oxidative stability, and chemical durability of these materials. The efficiency of imide formation strongly depends on the structure of the amine precursor. Mechanistic studies of phthalimide-derived systems further highlight the electronic characteristics of the imide ring and the stability of nitrogen-containing intermediates [30], providing fundamental insights into how bio-derived aromatic amines can participate effectively in polyimide formation.

The broader context of renewable amine chemistry has been outlined in comprehensive analyses of bio-based amine platforms, where lignin-derived aromatic diamines are identified as promising candidates for high-performance polymer synthesis [31]. Aromatic bio-amines derived from vanillin, guaiacol, or syringaldehyde offer structural rigidity comparable to traditional aromatic diamines such as p-phenylenediamine. Their incorporation into polyimide backbones supports high T_g and mechanical strength while increasing renewable carbon content. In contrast, fatty acid-derived aliphatic diamines introduce flexible methylene segments that reduce intermolecular charge-transfer interactions within the polymer chain, potentially lowering T_g but improving toughness and impact resistance.

The influence of amine electronic structure and steric environment on condensation polymerization has also been demonstrated in studies examining nitrogen-containing monomers in high-performance polymers [32]. These investigations show that primary aromatic amines generally exhibit higher reactivity toward anhydride functionalities than secondary or sterically hindered amines. Increased aromatic content enhances thermal resistance and rigidity, whereas extended aliphatic segments improve flexibility and processability. This balance is critical in designing bio-based polyimides, where renewable diamines may introduce additional functional groups or steric constraints that affect molecular weight development and film-forming behavior. Reactive amine functionalities can also influence crosslink density in polyimide networks. Studies on multifunctional nitrogen-containing molecules demonstrate that difunctional or multifunctional amines can promote covalent network formation beyond linear chain growth, resulting in enhanced solvent resistance and mechanical stability. When translated to polyimide systems, bio-based diamines bearing additional reactive groups may enable thermosetting polyimide architectures with improved dimensional stability at elevated temperatures.

Overall, the development of polyimides from bio-based amines requires careful consideration of amine structure, steric effects on cyclodehydration, electronic properties influencing reactivity, and the balance between rigidity and flexibility within the polymer backbone. Aromatic lignin-derived diamines appear most suitable for high-temperature applications such as aerospace films, electronic insulation, and advanced coatings, whereas aliphatic bio-diamines may enable more flexible polyimide systems for specialized applications. Continued progress in biomass-derived amine synthesis and optimized imidization strategies will be essential for expanding the role of renewable feedstocks in next-generation polyimide materials.

8.4.2 Epoxy Thermosets

The development of epoxy thermosets from bio-derived amines represents a significant advance toward replacing petroleum-based bisphenol A systems with more sustainable and high-performance alternatives. Recent research demonstrates that both epoxy monomers and amine curing agents can be sourced from lignocellulosic biomass, enabling the fabrication of fully or highly bio-based thermoset networks without compromising thermal stability or mechanical strength. A notable example is the fully bio-based epoxy system derived from syringaldehyde and a furan-based amine curing agent [33]. In this study, a syringaldehyde-based Schiff base epoxy monomer was synthesized and cured with a cellulose-derived diamine. The resulting thermoset exhibited a high T_g of approximately 204°C, substantially exceeding that of conventional petroleum-based diglycidyl ether of bisphenol A (DGEBA)/4,4′-diaminodiphenylmethane (DDM) systems. The material also demonstrated strong mechanical performance, with tensile strength around 57 MPa and storage modulus above 2 GPa. The presence of aromatic triazole and imine (C=N) structures promoted high char yield, close to 40% at 800°C under nitrogen, along with intrinsic flame retardancy reflected by a UL-94 V-0 rating and a limiting oxygen index near 40%. In addition, the Schiff base functionality contributed antibacterial behavior and improved condensed-phase charring, highlighting the multifunctional benefits of bio-amine incorporation beyond simple crosslinking.

Similarly, highly bio-derived epoxy–amine systems have been constructed using lignin-based epoxy monomers such as diglycidyl ethers of vanillyl alcohol, gastrodigenin, and bisguaiacol cured with a cellulose-derived furanyl diamine [34]. These thermosets achieved bio-derived carbon contents exceeding 97%, approaching complete biomass substitution. Thermomechanical analysis revealed T_g ranging from 80°C to 110°C depending on the specific structure, with storage moduli at room temperature between approximately 2.9 and 3.7 GPa. Crosslink density and molecular weight between crosslinks were tunable through blending strategies, allowing structural control over network architecture. Thermogravimetric analysis showed high char yields of about 39%–42%, significantly higher than conventional petroleum-based EPON828 systems. The incorporation of furan rings in the bio-amine structure contributed to improved thermal stability at elevated temperatures due to rearrangement and char-forming mechanisms. Beyond fully bio-based systems, aromatic bio-amines derived from guaiacol have also been employed to produce high-performance epoxy networks [35]. In this approach, 5-aminoguaiacol was synthesized and combined with vanillin to form a bio-based bisphenolic epoxy precursor. After glycidylation and curing with DDM, the resulting GV-EP/DDM thermoset exhibited an exceptionally high T_g of 220°C, a char yield of 44.9% at 800°C, a storage modulus of 3602 MPa at 30°C, and a Young's modulus of 4489 MPa. The incorporation of dynamic Schiff base linkages enabled reprocessability at elevated temperature and pressure, along with shape memory behavior, degradability, and enhanced acid resistance due to increased crosslink density. These findings demonstrate that aromatic bio-amines not only provide sustainability advantages but also significantly enhance rigidity, thermal resistance, and multifunctional network properties. Collectively, these studies show that epoxy thermosets derived from bio-amines offer multiple advantages, including increased T_g and modulus due to high aromatic content from lignin derivatives, improved char formation and intrinsic flame retardancy through furan and Schiff base structures, and recyclability or reprocessability enabled by dynamic imine chemistry. With bio-based carbon contents often exceeding 95%, these systems substantially reduce reliance on fossil resources while delivering competitive or superior performance. The integration of lignin-derived epoxies with cellulose-derived amines therefore represents a promising strategy for the design of next-generation sustainable thermoset materials suitable for advanced coatings, composites, electronics, and structural applications.

8.4.3 Polyurethanes and Non-Isocyanate Polyurethanes

PUs are traditionally synthesized through the polyaddition of diisocyanates and polyols; however, the toxicity of phosgene-derived isocyanates and associated environmental concerns have accelerated the search for safer and renewable alternatives. Bio-based NIPUs have emerged as a sustainable solution, primarily synthesized through the reaction of cyclic carbonates with primary amines to form poly(hydroxy urethane)s (PHUs) [36,37]. This cyclic carbonate–amine route eliminates hazardous isocyanates while introducing pendant hydroxyl groups that enhance hydrogen bonding, adhesion, and chemical resistance. A critical challenge in NIPU development has been the availability of renewable amines. Hibert et al. demonstrated a green "nitrile route" to produce bio-based aliphatic primary diamines from renewable alcohols using aerobic oxidation followed by hydrogenation [38]. These diamines were successfully polymerized with fatty acid-derived bis-cyclic carbonates to yield fully bio-based PHUs with molecular weights up to ~20,500 g/mol and thermal stability ($Td_{5\%}$) reaching 274°C. The study highlights that amine chain length directly influences crystallinity and T_g. Recent advancements further demonstrate functional tailoring of NIPUs through amine design. Bio-based amine-functionalized PHUs synthesized for home-care applications exhibited controlled water solubility, biodegradability, and tunable thermal properties [39]. The presence of hydroxyl-rich backbones enhanced hydrophilicity and environmental degradability while maintaining adequate mechanical stability. Comprehensive analyses of bio-based NIPUs show that amine structure whether aliphatic, aromatic, carbohydrate-derived, or terpene-based governs crosslink density, rigidity, and hydrogen bonding interactions. Aromatic bio-amines derived from lignin increase T_g and modulus, whereas long-chain aliphatic amines enhance flexibility. Multifunctional amines enable thermoset formation with improved solvent resistance and dimensional stability. Importantly, integration of CO_2-derived cyclic carbonates with bio-amines creates a dual sustainability advantage: carbon fixation and elimination of toxic isocyanates [36]. Compared to conventional PUs, NIPUs avoid biuret and allophanate side reactions and exhibit improved hydrolytic stability, though curing kinetics may be slower and require optimization. Overall, bio-amine-derived PUs and NIPUs represent a transformative shift in PU chemistry. Bio-amines function not only as renewable substitutes for petrochemical curing agents but also as structural design elements controlling hydrogen bonding, crosslink density, thermal behavior, and biodegradability. When combined with CO_2-based cyclic carbonates and renewable polyols, they enable the development of high-performance, environmentally benign materials suitable for coatings, adhesives, elastomers, and advanced structural applications.

8.4.4 High-Temperature and Flame-Retardant Systems

Bio-derived aromatic amines have emerged as key building blocks for developing high-temperature and flame-retardant polymer systems due to their rigid phenyl structures, high carbon content, and enhanced char-forming ability. Aromatic bio-amines derived from lignin, guaiacol, vanillin, and other phenolic biomass sources introduce restricted chain mobility and increased crosslink density when incorporated into epoxy or PU networks. This structural rigidity directly contributes to elevated T_g and improved thermal stability. For example, lignin-derived aromatic amine systems used in NIPUs have shown improved thermomechanical performance compared to aliphatic counterparts, primarily due to enhanced intermolecular interactions and hydrogen bonding density [36]. Char formation plays a critical role in flame retardancy. Aromatic bio-amines promote condensed-phase charring during thermal degradation, forming stable carbonaceous layers that act as thermal barriers and reduce heat release rates. In cyclic carbonate–amine-derived PHUs, the presence of aromatic rings and hydroxyl functionalities enhances crosslinked network integrity at elevated temperatures, increasing residue yield after thermal decomposition [38]. Furthermore, lignin-based and phenolic amine structures inherently favor carbon-rich degradation pathways, which suppress volatilization of flammable fragments and improve flame resistance [36]. Compared to aliphatic amines, aromatic bio-amines demonstrate higher decomposition onset temperatures and greater char yields due to resonance-stabilized structures and restricted segmental motion. These characteristics make them particularly suitable for coatings, structural composites, and high-temperature thermosets where thermal endurance and fire resistance are required. Overall, aromatic bio-amines serve not only as sustainable curing agents but also as molecular design tools for engineering thermally stable and flame-retardant polymer networks.

8.5 BIO-BASED AMINES IN DYNAMIC AND REPROCESSABLE POLYMER NETWORKS

The integration of bio-based amines into thermosetting polymers has enabled a transformative shift from permanently crosslinked materials toward dynamic, recyclable networks. Conventional thermosets possess outstanding mechanical strength and dimensional stability but suffer from irreversibility of covalent crosslinks, which prevents reshaping, repair, or recycling. Covalent adaptive networks (CANs) overcome these limitations by incorporating reversible covalent bonds capable of bond exchange under external stimuli while maintaining network integrity. When bio-derived aldehydes, diamines, and aromatic amines are introduced into these systems, the resulting polymers not only become reprocessable but also align with sustainability goals by reducing fossil dependence. Recent advances clearly demonstrate that bio-based imine chemistry, transamination-enabled exchange, and dynamic urethane/urea mechanisms can generate high-performance materials with closed-loop recyclability and minimal property loss.

8.5.1 Imine (Schiff Base) Chemistry in Polymer Networks

Imine (C=N) bonds formed via condensation between aldehydes and primary amines represent one of the most widely studied dynamic covalent linkages in bio-based thermosets. Because imine bonds can undergo reversible exchange (transimination) or hydrolysis under mild conditions, they enable associative bond rearrangement without destroying crosslink density. A fully bio-based imine thermoset derived from vanillin-based dialdehyde and renewable diamines demonstrated tensile strength up to 58 MPa and storage modulus of 3.07 GPa while retaining closed-loop recyclability [40]. Stress relaxation experiments revealed an activation energy of ~50 kJ/mol, confirming efficient bond exchange under moderate heating. After three thermomechanical cycles, tensile strength recovery exceeded 84%, indicating minimal degradation of network integrity. Dual dynamic systems further enhance network adaptability. A vanillin-based PU incorporating both imine and disulfide linkages exhibited tensile strength recovery above 70% after three reshaping cycles and rapid degradability within 5 h at 50°C [41]. Similarly, vanillin-derived epoxy networks containing dual S–S and C=N bonds achieved a tensile modulus of 7175 MPa and ultrafast stress relaxation (6 s at 200°C), demonstrating how combining multiple exchangeable motifs significantly improves dynamic performance [42]. These findings confirm that imine-based networks can maintain high crosslink density while enabling reshaping, welding, and chemical degradation. Imine chemistry also supports exceptional mechanical performance. A fully bio-based epoxy resin constructed from vanillin-derived monomers reached tensile strength of 104 MPa, significantly surpassing conventional DGEBA/DDM systems, while remaining recyclable through imine metathesis and acid-triggered hydrolysis [43]. Likewise, dynamic imine films formed from dialdehyde cellulose nanofibrils and plant oil-based diamines achieved tensile strength of 83.7 MPa and Young's modulus of 3.90 GPa, highlighting the synergistic role of hydrogen bonding and dynamic imine exchange [44]. These studies collectively demonstrate that imine chemistry does not compromise mechanical robustness; rather, it can enhance rigidity when aromatic bio-amines are incorporated.

8.5.2 Transamination and Vitrimer Systems

Associative bond exchange is central to vitrimer behavior, where topology rearrangement occurs without reduction in crosslink density. In bio-based imine vitrimers derived from vanillin aldehydes and renewable diamines, dynamic bond exchange enables stress relaxation, weldability, and reprocessing under hot pressing conditions (Figure 8.6) [40]. The exchange follows an associative transimination

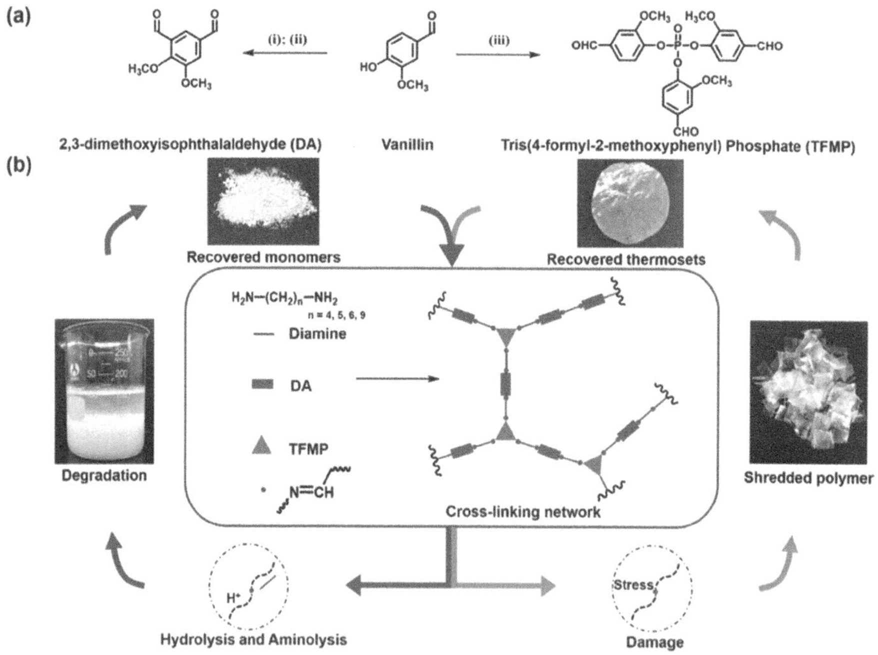

FIGURE 8.6 (a) Synthetic route of 2,3-dimethoxyisophthalaldehyde (DA) and Tris (4-formyl-2-methoxyphenyl) phosphate (TFMP) from Vanillin and (b) synthesis and reprocessing of the fully bio-based Vitrimers. (Adapted with permission [40]. Copyright 2022 American Chemical Society.)

mechanism, preserving network connectivity during rearrangement. Mechanical testing after multiple recycling cycles showed high property retention, confirming topology rearrangement without structural degradation. Bio-based PHU vitrimers incorporating hydroxyurethane and imine dual bonds further illustrate associative exchange mechanisms. These systems exhibited tensile strength up to 79 MPa and elongation at break of 296%, among the highest reported for PHUs, along with low activation energies (81–102 kJ/mol) for bond exchange [45]. The combination of hydroxyl-assisted hydrogen bonding and imine exchange promoted both toughness and reprocessability. Even after repeated reprocessing cycles, mechanical properties were largely preserved. In epoxy systems, dual dynamic bonds proportionally tuned within the network significantly accelerated stress relaxation while maintaining high modulus and thermal stability [42]. Such systems exemplify vitrimer-like behavior, where bond exchange permits permanent reshaping without sacrificing dimensional stability. Importantly, these materials demonstrate that bio-based aromatic amines can provide rigid frameworks while dynamic exchange ensures recyclability.

8.5.3 Reversible Urea and Urethane Exchange

Dynamic urethane and urea bonds offer additional routes toward adaptable PU networks. A vitrimer-like PU synthesized from castor oil, vanillin, and diamines incorporated dynamic imine bonds within a urethane matrix, enabling rapid stress relaxation and reprocessability while maintaining 86% tensile strength after three remolding cycles [46]. Hydrolysis of the imine bonds allowed rapid degradation at 60°C within 1 h, demonstrating chemical recyclability. Incorporation of imine linkages into PU backbones derived from itaconic acid and vanillin yielded dual dynamic networks capable of reshaping, degradation, and shape memory behavior [41]. The synergistic presence of disulfide and imine exchange significantly enhanced adaptability. These findings suggest that integrating reversible amine-derived linkages into urethane systems provides a viable pathway toward recyclable thermosets with maintained structural strength.

8.5.4 Stimuli-Responsive Amine-Based Polymers

Dynamic amine bonds confer responsiveness to environmental stimuli such as pH and temperature. Imine linkages readily undergo hydrolysis under acidic conditions, enabling chemical recycling and monomer recovery. Fully bio-based polyimine thermosets derived from vanillin exhibited rapid degradation in mild acid and complete recovery of aldehyde monomers [47]. Similarly, cellulose-based dynamic imine networks demonstrated complete biodegradability while maintaining strong water barrier properties [44]. In high-performance aromatic systems, guaiacol-derived amines introduced dynamic Schiff base bonds into epoxy networks with T_g of 220°C and char yield of 44.9%, while still enabling reprocessability at elevated temperatures [35]. This illustrates that dynamic behavior can coexist with high thermal stability, challenging the traditional trade-off between robustness and recyclability.

Bio-based amines play a central role in the design of dynamic and reprocessable polymer networks. Through imine exchange, transamination-enabled vitrimer behavior, reversible urethane chemistry, and acid-responsive degradation, these systems overcome the intrinsic irreversibility of conventional thermosets. Importantly, recent research demonstrates that high tensile strength (up to 104 MPa), high modulus (>7 GPa), rapid stress relaxation, and closed-loop recyclability can be simultaneously achieved. The strategic incorporation of aromatic bio-amines, dual dynamic bonds, and associative exchange mechanisms positions these materials as strong candidates for sustainable coatings, structural composites, and advanced thermosets with minimized environmental impact.

8.6 PERFORMANCE METRICS AND COMPARATIVE ANALYSIS

Bio-based amine-derived thermosets and dynamic polymer networks have progressed from proof-of-concept materials to high-performance systems capable of rivaling, and in some cases surpassing, conventional petroleum-based thermosets. A systematic comparison of mechanical, thermal, and chemical performance reveals both the strengths and current trade-offs associated with sustainable network design. Mechanical performance remains the primary benchmark for replacing conventional thermosets. Several bio-based imine and dual-dynamic systems now exhibit tensile strengths comparable to commercial epoxy and PU materials. For example, a fully bio-based epoxy resin constructed from vanillin-derived monomers reached a tensile strength of 104 MPa, significantly higher than conventional DGEBA/DDM systems (~59 MPa) [43]. Similarly, vanillin-based imine thermosets reported tensile strength up to 58 MPa and storage modulus of 3.07 GPa [40]. Dual dynamic epoxy vitrimers incorporating both disulfide (S–S) and imine (C=N) linkages achieved exceptionally high tensile modulus of 7175 MPa, demonstrating that dynamic covalent exchange does not inherently compromise stiffness when aromatic bio-based structures are incorporated [42]. In bio-based PHU vitrimers containing hydroxyurethane and imine bonds, tensile strength reached 79 MPa with elongation at break of 296%, placing them among the highest-performing PHUs reported to date [45]. Dynamic imine films derived from dialdehyde cellulose nanofibrils exhibited tensile strength of 83.7 MPa and Young's modulus of 3.90 GPa, indicating that hydrogen bonding synergistically enhances rigidity [44]. These examples demonstrate that bio-based amines can produce networks spanning rigid thermosets to tough elastomeric materials through structural control of aromatic content and crosslink density. Thermal stability is critical for structural and high-temperature applications. Bio-based imine epoxy systems derived from guaiacol amines achieved a high $T_g = 220°C$ and char yield of 44.9% at 800°C, outperforming many bisphenol A-based epoxy resins [35]. This high T_g is attributed to the rigid aromatic backbone and high crosslink density. In fully bio-based imine thermosets, T_g values ranged from 27°C to 78°C, tunable through diamine chain length and crosslink density [40]. Thermal decomposition temperatures ($Td_{5\%}$) typically exceeded 230°C, confirming good thermal stability. Dual dynamic epoxy vitrimers showed improved thermal decomposition temperatures and rapid stress relaxation at elevated temperature (200°C) without catastrophic degradation [42]. High char yields in aromatic systems contribute to flame retardancy and thermal resistance. For example, the GV-EP/DDM system exhibited significantly enhanced residue formation compared to petroleum-based counterparts [35], demonstrating the benefit of lignin-derived aromatic amines. Chemical resistance in dynamic networks depends on both crosslink density and bond stability. Aromatic imine epoxy networks demonstrated strong acid resistance when crosslink density was increased, despite containing hydrolyzable imine bonds [35]. However, controlled degradability remains a major advantage. Fully bio-based imine thermosets showed rapid degradation under mild acidic conditions, enabling monomer recovery within minutes [40]. Polyimine thermosets derived from vanillin and bio-amines could be chemically recycled via hydrolysis, while maintaining flame resistance [47]. In PU systems containing dynamic imine bonds, complete degradation occurred within 1 h at 60°C [46]. These systems balance operational stability with triggered degradability, which is critical for sustainable material design. While bio-based amine networks demonstrate impressive performance, trade-offs remain. Increasing aromatic content enhances T_g, modulus, and char yield but may reduce flexibility and increase brittleness. Conversely, incorporating long-chain aliphatic bio-amines improves elongation but lowers thermal resistance. Dual dynamic bond strategies mitigate these trade-offs by combining rigidity with adaptive exchange. However, excessive dynamic bond density may accelerate stress relaxation at elevated temperature, potentially limiting high-load structural use. From a sustainability perspective, dynamic imine networks offer acid-triggered degradation and closed-loop recyclability, but hydrolytic sensitivity may restrict long-term exposure in humid environments. Therefore, optimization of bond density, aromatic content, and network architecture is essential to balance durability with end-of-life recyclability.

8.7 CONCLUSIONS

This chapter establishes that bio-based amines are far more than renewable replacements for petrochemical amines; they function as deliberate molecular design tools for constructing high-performance and dynamic polymer systems. Across polyamides, polyimides, epoxy thermosets, PUs, and NIPUs, the structure of the amine governs curing kinetics, crosslink density, hydrogen-bonding interactions, thermal stability, and degradation pathways. Aromatic bio-amines derived from lignin, vanillin, guaiacol, and related phenolics enhance T_g, modulus, and char yield through rigid aromatic frameworks and resonance stabilization, whereas long-chain aliphatic bio-amines from fatty acids or amino acids impart flexibility, toughness, and reduced brittleness. A major advance is the integration of dynamic covalent amine chemistry into thermoset networks, where imine exchange, transimination-driven vitrimer behavior, dual dynamic motifs such as C=N and S–S bonds, and reversible urethane or urea linkages enable reshaping, welding, recycling, and controlled degradation while maintaining mechanical integrity; in several cases, tensile strengths above 100 MPa, moduli exceeding 7 GPa, and T_g values above 200°C have been achieved in largely bio-based systems. Amines occupy a central role in next-generation polymer innovation because their nucleophilicity supports condensation reactions, epoxy ring-opening, and cyclic carbonate aminolysis; their reversible bond-forming capability enables CANs with topology rearrangement; and their structural tunability allows precise control over electronic effects, steric environment, and multifunctionality, thereby defining crosslink density and exchange kinetics. From a sustainability standpoint, bio-based amines promote feedstock diversification through lignin, vegetable oils, amino acids, and sugars; reduce toxicity via isocyanate- and phosgene-free chemistries; enable circular material design through dynamic exchange and chemical recycling; deliver high-performance properties including elevated T_g and flame resistance; and support design-for-degradation strategies such as acid-triggered bond cleavage. Ultimately, sustainable polymer development must extend beyond simple biomass substitution toward molecular-level engineering of network adaptability, and bio-based amines provide the chemical versatility required to integrate durability, recyclability, and environmental responsibility within a unified materials platform, paving the way for industrially viable circular thermosets.

REFERENCES

1. V. Froidevaux, C. Negrell, S. Caillol, J.P. Pascault, B. Boutevin, Biobased amines: From synthesis to polymers; present and future, *Chem. Rev.* 116 (2016) 14181–14224.
2. N.K. Gupta, P. Reif, P. Palenicek, M. Rose, Toward renewable amines: Recent advances in the catalytic amination of biomass-derived oxygenates, *ACS Catal.* 12 (2022) 10400–10440.
3. S.-M. Zhang, J.-Q. Zhao, Y. Liu, Y.-X. Liu, C.-M. Liu, Renewable furan-derived diamine as primary amine source to prepare fully bio-based bis-benzoxazine monomer under solvent-free condition, *React. Funct. Polym.* 165 (2021) 104957.
4. E. Blondiaux, J. Bomon, M. Smoleń, N. Kaval, F. Lemière, S. Sergeyev, L. Diels, B. Sels, B.U.W. Maes, Bio-based aromatic amines from lignin-derived monomers, *ACS Sustain. Chem. Eng.* 7 (2019) 6906–6916.
5. A.-S. Mora, R. Tayouo, B. Boutevin, G. David, S. Caillol, Vanillin-derived amines for bio-based thermosets, *Green Chem.* 20 (2018) 4075–4084.
6. L. Claes, M. Janssen, D.E. De Vos, Organocatalytic decarboxylation of amino acids as a route to bio-based amines and amides, *ChemCatChem.* 11 (2019) 4297–4306.
7. J. Verduyckt, R. Coeck, D.E. De Vos, Ru-catalyzed hydrogenation–decarbonylation of amino acids to bio-based primary amines, *ACS Sustain. Chem. Eng.* 5 (2017) 3290–3295.
8. P. Zhou, L. Zhang, S. Luo, J.-P. Cheng, Asymmetric synthesis of wieland–miescher and hajos–parrish ketones catalyzed by an amino-acid-derived chiral primary amine, *J. Org. Chem.* 77 (2012) 2526–2530.
9. P. Bandyopadhyay, P. Bandyopadhyay, S.L. Regen, Ion conductors derived from biogenic amines, bile acids, and amino acids, *Bioconjug. Chem.* 13 (2002) 1314–1318.
10. R.M. Callejón, A.M. Troncoso, M.L. Morales, Determination of amino acids in grape-derived products: A review, *Talanta.* 81 (2010) 1143–1152.
11. K. Natte, A. Narani, V. Goyal, N. Sarki, R. V Jagadeesh, Synthesis of functional chemicals from lignin-derived monomers by selective organic transformations, *Adv. Synth. Catal.* 362 (2020) 5143–5169.
12. Z. Ma, Z. Chen, Z. Yuan, C. Ren, B. Zhang, Y. Cui, X. Li, R. V Jagadeesh, M. Beller, Synthesis of aromatic amides from lignin and its derivatives, *Nat. Commun.* 16 (2025) 3476.
13. Y. Xu, H.-W. Xiao, H. Xiao, H.-Y. Xiao, Amino acid and amine compounds in fine aerosol particles: A review of determination methods, abundances, origins, and atmospheric processes, *Aerosol Sci. Eng.* 10(3) (2025) 301–316.

14. F. Poovan, V.G. Chandrashekhar, D.P. Fuentes, T.H. Vuong, R. Jackstell, J. Rabeah, R.V Jagadeesh, M. Beller, A highly selective cobalt catalyst for primary amine synthesis from carboxylic acids, esters, and vegetable oils, *J. Am. Chem. Soc.* 148 (2026) 5966–5975.
15. J. Citoler, W. Finnigan, H. Bevinakatti, N.J. Turner, Direct enzymatic synthesis of fatty amines from renewable triglycerides and oils, *ChemBioChem.* 23 (2022) e202100578.
16. C. May, *Epoxy Resins: Chemistry and Technology*, 2nd Ed., Routledge, New York, 2018.
17. J.-P. Pascault, R.J.J. Williams, *Epoxy Polymers: New Materials and Innovations*, John Wiley & Sons, Hoboken, NJ, 2009.
18. R. September, Reaction kinetics of epoxy/amine model, *Macromolecules.* 27 (1994) 7589–7600.
19. Z. Karpas, Ion mobility spectrometry of aliphatic and aromatic amines, *Anal. Chem.* 61 (1989) 684–689.
20. A.P. Bhat, P.R. Gogate, Degradation of nitrogen-containing hazardous compounds using advanced oxidation processes: A review on aliphatic and aromatic amines, dyes, and pesticides, *J. Hazard. Mater.* 403 (2021) 123657.
21. A.M. Shanmugharaj, J.H. Yoon, W.J. Yang, S.H. Ryu, Synthesis, characterization, and surface wettability properties of amine functionalized graphene oxide films with varying amine chain lengths, *J. Colloid Interface Sci.* 401 (2013) 148–154.
22. E. Urbaczewski-Espuche, J. Galy, J.-F. Gerard, J.-P. Pascault, H. Sautereau, Influence of chain flexibility and crosslink density on mechanical properties of epoxy/amine networks, *Polym. Eng. Sci.* 31 (1991) 1572–1580.
23. R. Mezzenga, L. Boogh, J.-A.E. Månson, B. Pettersson, Effects of the branching architecture on the reactivity of epoxy–amine groups, *Macromolecules.* 33 (2000) 4373–4379.
24. J. Wan, Z.-Y. Bu, C.-J. Xu, B.-G. Li, H. Fan, Preparation, curing kinetics, and properties of a novel low-volatile starlike aliphatic-polyamine curing agent for epoxy resins, *Chem. Eng. J.* 171 (2011) 357–367.
25. S. Manzoor, A. Simperler, A. Korre, A theoretical study of the reaction kinetics of amines released into the atmosphere from CO_2 capture, *Int. J. Greenh. Gas Control.* 41 (2015) 219–228.
26. A. Viswanathan, Wallace Carothers: More than the inventor of nylon and neoprene, *World Pat. Inf.* 32 (2010) 300–305.
27. F. Jouffret, P.-J. Madec, Polyamide synthesis by ester aminolysis. I. A new route for nylon-6,6 synthesis, *J. Polym. Sci. Part A Polym. Chem.* 34 (1996) 2363–2370.
28. P.S. Chaudhari, S.D. Salim, R. V Sawant, K.G. Akamanchi, Sulfated tungstate: A new solid heterogeneous catalyst for amide synthesis, *Green Chem.* 12 (2010) 1707–1710.
29. C. Han, Porco, Synthesis of carbamates and ureas using Zr(IV)-catalyzed exchange processes, *Org. Lett.* 9 (2007) 1517–1520.
30. H.E. Zimmerman, D.W. Wilson, Singlet photochemistry of vinyl cyclopropenes: Regioselectivity and mechanism; mechanistic and exploratory organic photochemistry, *J. Org. Chem.* 60 (1995) 692–697.
31. J. Cong, Y. Hao, G. Boschloo, L. Kloo, Electrolytes based on TEMPO–Co tandem redox systems outperform single redox systems in dye-sensitized solar cells, *ChemSusChem.* 8 (2015) 264–268.
32. K.-L. Zhong, Z. Huang, Z. Man, L.Y. Jin, B. Yin, M. Lee, Synthesis and self-assembly of rod-coil molecules with n-shaped rod building block, *J. Polym. Sci. Part A Polym. Chem.* 48 (2010) 1415–1422.
33. H. Nabipour, X. Wang, L. Song, Y. Hu, A high performance fully bio-based epoxy thermoset from a syringaldehyde-derived epoxy monomer cured by furan-derived amine, *Green Chem.* 23 (2021) 501–510.

34. J.R. Mauck, S.K. Yadav, J.M. Sadler, J.J. La Scala, G.R. Palmese, K.M. Schmalbach, J.F. Stanzione III, Preparation and characterization of highly bio-based epoxy amine thermosets derived from lignocellulosics, *Macromol. Chem. Phys.* 218 (2017) 1700013.
35. J. Li, Z. Weng, Q. Cao, Y. Qi, B. Lu, S. Zhang, J. Wang, X. Jian, Synthesis of an aromatic amine derived from biomass and its use as a feedstock for versatile epoxy thermoset, *Chem. Eng. J.* 433 (2022) 134512.
36. K. Błażek, J. Datta, Renewable natural resources as green alternative substrates to obtain bio-based non-isocyanate polyurethanes-review, *Crit. Rev. Environ. Sci. Technol.* 49 (2019) 173–211.
37. M.L. Chaudhary, R. Patel, R.K. Gupta, Beyond isocyanates: Advances in non-isocyanate polyurethane chemistry and applications, *Polymer (Guildf).* 332 (2025) 128553.
38. G. Hibert, O. Lamarzelle, L. Maisonneuve, E. Grau, H. Cramail, Bio-based aliphatic primary amines from alcohols through the 'Nitrile route' towards non-isocyanate polyurethanes, *Eur. Polym. J.* 82 (2016) 114–121.
39. M. Aasim, A. Ayhan, R. Katırcı, A.Ş. Acar, S.A. Ali, Computing artificial neural network and genetic algorithm for the feature optimization of basal salts and cytokinin-auxin for in vitro organogenesis of royal purple (Cotinus coggygria Scop), *Ind. Crops Prod.* 199 (2023) 116718.
40. K. Hong, Q. Sun, X. Zhang, L. Fan, T. Wu, J. Du, Y. Zhu, Fully bio-based high-performance thermosets with closed-loop recyclability, *ACS Sustain. Chem. Eng.* 10 (2022) 1036–1046.
41. G. Sha, F. Kong, M. Cui, M. Quan, Y. Wang, F. Li, J. Zhu, J. Chen, Bio-based dual dynamic covalent adaptive networks based on itaconic acid with superior degradability, reprocessability, and mechanical properties, *Eur. Polym. J.* 224 (2025) 113709.
42. K. Dong, S. Tang, D. Zhao, Y. Pang, C. Zhao, Vanillin-derived bio-based epoxy resins containing dual dynamic Schiff base and disulfide bonds with reprocessability and degradability, *Polym. Degrad. Stab.* 230 (2024) 111077.
43. K. Li, S. Wang, Y. Jiang, M. Chen, W. Dong, D. Shi, High-performance, recyclable, and degradable bio-based epoxy resins based on dynamic covalent imine bonds, *J. Appl. Polym. Sci.* 142 (2025) e57033.
44. X. Zhang, P. Li, J. Zeng, J. Li, B. Wang, W. Gao, J. Xu, K. Chen, Dynamic covalent bond enabled strong Bio-based polyimide materials with thermally-driven adaptivity, healability and recycling, *Chem. Eng. J.* 465 (2023) 143017.
45. M. Fan, Y. Wang, J. Zhao, High-strength, reprocessable, and degradable bio-based non-isocyanate polyurethanes with dual dynamic covalent bonds, *J. Clean. Prod.* 513 (2025) 145732.
46. Y. Sun, D. Sheng, H. Wu, X. Tian, H. Xie, B. Shi, X. Liu, Y. Yang, Bio-based vitrimer-like polyurethane based on dynamic imine bond with high-strength, reprocessability, rapid-degradability and antibacterial ability, *Polymer (Guildf).* 233 (2021) 124208.
47. P. Li, Q. Zhang, J. Ma, Z. Liao, J. Zhang, H. Xie, S. Yang, C.-A. Xu, Y. Hu, Z. Yang, Synthesis of bio-based polyimine networks with flame-retardancy, acid-degradablility, and reprocessability, *Polymer (Guildf).* 302 (2024) 127101.

9 Bio-Based Amines in Adhesives and Coatings

9.1 INTRODUCTION

The growing demand for sustainable materials has reshaped research and industrial development across the adhesives and coatings sectors. These materials are essential in modern life and are used in construction, transportation, electronics, packaging, furniture, textiles, biomedical devices, and protective surface technologies. Adhesives provide efficient joining of similar or dissimilar substrates without the need for mechanical fastening, while coatings protect surfaces against corrosion, moisture, chemicals, ultraviolet radiation, abrasion, and biological attack, while also delivering decorative and functional value. Despite their wide utility, many conventional adhesive and coating formulations continue to rely heavily on petrochemical feedstocks, solvent-intensive processing, and non-renewable curing agents, raising concerns related to environmental impact, occupational safety, volatile organic compound (VOC) emissions, and long-term resource sustainability. As a result, there is increasing interest in replacing fossil-derived components with renewable alternatives that can reduce environmental burden without sacrificing performance [1–4].

Among the many components used in these systems, amines play a particularly important role. Amines are widely employed as curing agents, chain extenders, crosslinkers, neutralizing agents, adhesion promoters, and functional modifiers in numerous adhesive and coating chemistries. Their high reactivity toward epoxides, isocyanates, carbonyl compounds, and other electrophilic functional groups makes them central to the design of thermosetting and hybrid polymer networks. In epoxy systems, amines are commonly used to open epoxide rings and create dense crosslinked structures with strong mechanical properties and chemical resistance. In polyurethane and polyurea systems, amines participate in chain extension or react rapidly with isocyanates to form urea linkages, often contributing to toughness, rapid curing, and enhanced hydrogen bonding. They also influence viscosity, cure kinetics, substrate interaction, thermal stability, and network architecture, all of which determine the final performance of adhesives and coatings. Because of this central role, the transition from petroleum-derived amines to bio-based amines is not simply a matter of substituting one ingredient for another, but rather a significant opportunity to redesign material systems at the molecular level [5,6].

Bio-based amines are nitrogen-containing compounds derived fully or partially from renewable resources such as vegetable oils, lignin-derived aromatics, amino acids, sugars, terpenes, rosin, and other biomass-derived platform molecules. These renewable amines can be tailored to possess aliphatic, aromatic, cyclic, or multifunctional structures depending on the feedstock and synthetic route. Their chemical diversity provides opportunities to generate adhesive and coating systems with a

DOI: 10.1201/9781003731122-9

wide range of properties, including flexibility, rigidity, hydrophobicity, thermal stability, self-healing behavior, and improved interfacial adhesion. In addition to reducing reliance on fossil carbon, bio-based amines often introduce structural motifs that are not easily replicated by conventional petrochemical analogs. For example, rigid aromatic frameworks from lignin derivatives, long aliphatic chains from fatty acids, and multifunctional architectures from carbohydrate or amino acid chemistry can influence curing behavior and end-use performance in distinctive ways. This makes bio-based amines valuable not only from a sustainability standpoint but also from a materials design perspective [7].

The use of bio-based amines in adhesives and coatings is particularly relevant in the context of green chemistry and circular materials development. Traditional solvent-borne coatings and synthetic adhesives can be associated with environmental concerns such as VOC emissions, hazardous intermediates, and poor end-of-life options. The incorporation of renewable amines can contribute to lower fossil resource consumption and may support the development of formulations with reduced toxicity, lower emissions, or improved biodegradability, depending on the overall resin system. Moreover, the shift toward waterborne coatings, solvent-free adhesives, non-isocyanate polyurethanes, dynamic covalent networks, and recyclable thermosets has further increased interest in bio-derived amine chemistry. In many of these emerging systems, bio-based amines serve not merely as renewable replacements but as enabling building blocks for next-generation materials with enhanced functionality [8].

In adhesive applications, the presence and structure of amines strongly affect parameters such as curing speed, lap shear strength, tack, peel strength, flexibility, and resistance to heat, moisture, and chemicals. In coating systems, amines influence film formation, hardness, adhesion to metal or polymeric substrates, corrosion resistance, scratch resistance, gloss, and long-term durability. Therefore, selecting a suitable bio-based amine requires careful consideration of both reactivity and molecular structure. The challenge lies in achieving a balance between sustainability and performance. While many bio-based amines show promise, they may also present issues related to lower reactivity, inconsistent purity, higher viscosity, limited commercial availability, or incompatibility with existing industrial processing conditions. In some cases, renewable amines improve one set of properties while compromising others, making formulation optimization essential. These trade-offs highlight the need for a deeper understanding of structure–property relationships and formulation strategies in bio-based adhesive and coating systems [9].

Recent research has demonstrated substantial progress in this field. Bio-based amines have been explored in epoxy adhesives, polyurethane and polyurea systems, pressure-sensitive adhesives, sealants, anticorrosive coatings, self-healing coatings, and functional surface treatments. Advances in biomass valorization, catalytic amination, reductive amination, and molecular functionalization have broadened the library of renewable amines available for material synthesis. At the same time, increased regulatory pressure and consumer demand for safer and more sustainable products have encouraged both academic and industrial efforts to develop renewable alternatives to conventional amine-based additives and curing agents. However, broader commercial adoption still depends on overcoming challenges related to cost, scalability, feedstock variability, and validation of long-term performance under practical service conditions [6].

This chapter examines the emerging role of bio-based amines in adhesives and coatings, with emphasis on their chemistry, sources, synthesis, functional roles, environmental benefits, and technological challenges. It discusses how renewable amines are integrated into different formulation types and how they influence adhesion, curing, durability, and protective performance. This chapter also considers the limitations that currently restrict widespread industrial implementation and highlights ongoing research aimed at improving both functionality and sustainability. By bringing together developments in renewable amine chemistry and application-oriented materials design, this chapter aims to provide a clear perspective on how bio-based amines can contribute to the next generation of high-performance, environmentally responsible adhesives and coatings.

9.2 FUNDAMENTALS OF ADHESIVES AND COATINGS

Adhesives are materials designed to join two surfaces by creating strong interfacial contact and maintaining that bond during service. The need for adhesives arises because two solid surfaces rarely bond effectively by simple contact alone. Even when surfaces appear smooth, they contain microscopic irregularities and asperities that greatly reduce the true area of contact. As a result, only a very small fraction of the apparent surface is close enough for attractive forces to operate. An adhesive is therefore typically applied in a liquid, softened, or highly deformable state so that it can spread over the substrate, flow into surface irregularities, and establish intimate contact across a much larger area. This initial contact is essential because adhesion originates from short-range intermolecular and interatomic interactions, including van der Waals forces, hydrogen bonding, acid–base interactions, and in some systems ionic or covalent bonding. Since these forces act only over extremely small distances, proper wetting of the substrate is one of the most important requirements for effective adhesion. Good wetting allows the adhesive to spread rather than remain as droplets, which improves interfacial contact and promotes stronger bonding. This behavior depends on the balance between the surface tension of the adhesive and the surface energy of the substrate, as well as the viscosity of the adhesive during application. An adhesive with suitable flow behavior can conform to the substrate topography, fill microscopic voids, and create a continuous bondline, while excessive viscosity may prevent sufficient spreading and penetration. At the same time, the adhesive cannot remain fluid indefinitely. After wetting and contact formation, it must develop cohesion through solvent evaporation, cooling, polymerization, or crosslinking so that it becomes a stable material capable of transmitting stress between the joined surfaces [10]. The overall performance of an adhesive depends not only on interfacial bonding but also on the mechanical behavior of the adhesive layer during loading and debonding. A useful adhesive must possess a balanced combination of strength, toughness, elasticity, and resistance to creep. If the material is too brittle, it may crack under impact, thermal stress, or mechanical deformation, whereas if it is too soft, it may deform excessively or fail under continuous load. For this reason, many effective adhesives are polymeric materials with molecular architectures that provide both

stronger substrate interaction, or improved compatibility with fillers and nanoparticles [18,19]. In advanced coatings, amine chemistry is often used to build multifunctionality into the surface, for example, by enabling coupling with bio-inspired interfacial layers such as polydopamine-based coatings [20]. The schematic illustration of polydopamine coating method is given in Figure 9.1. Amines may also enhance adhesion to metals, glass, wood, and polymeric substrates through polar interactions and specific interfacial bonding. More broadly, the presence of amine groups can influence film formation, crosslinking kinetics, moisture sensitivity, and surface energy, all of which affect the protective and functional behavior of the coating. Thus, in both adhesives and coatings, amines are far more than simple additives. They are key reactive and interfacial components that control curing, network formation, adhesion strength, durability, and the overall balance of properties needed for practical applications.

9.4 RENEWABLE SOURCES AND SYNTHESIS OF BIO-BASED AMINES

Bio-based amines originate from renewable feedstocks that already contain abundant oxygenated or nitrogen-containing functionalities, making them attractive alternatives to fossil-derived amine precursors. The article identifies several major biomass platform families for amine production: lignocellulosic carbohydrates, lignin, oils and fats, chitin, and proteins (Figure 9.2). Lignocellulose is the most abundant biomass resource and is composed of cellulose, hemicellulose, and lignin [21]. Cellulose and hemicellulose mainly provide polyols, sugars, furans, levulinic acid, and other oxygenated intermediates, while lignin serves as the largest renewable source of aromatic phenolics. Oils and fats supply glycerol, fatty acids, fatty esters, and fatty alcohols. Chitin is a particularly valuable feedstock because it is a naturally abundant nitrogen-containing polysaccharide based on N-acetyl-D-glucosamine units, while proteins provide amino acids that already contain biogenic nitrogen in

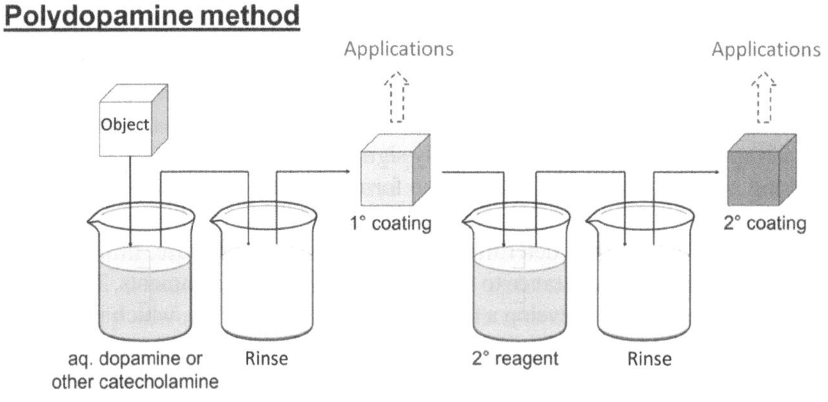

FIGURE 9.1 Polydopamine coating method. (Adapted with permission [20]. Copyright 2018 American Chemical Society.)

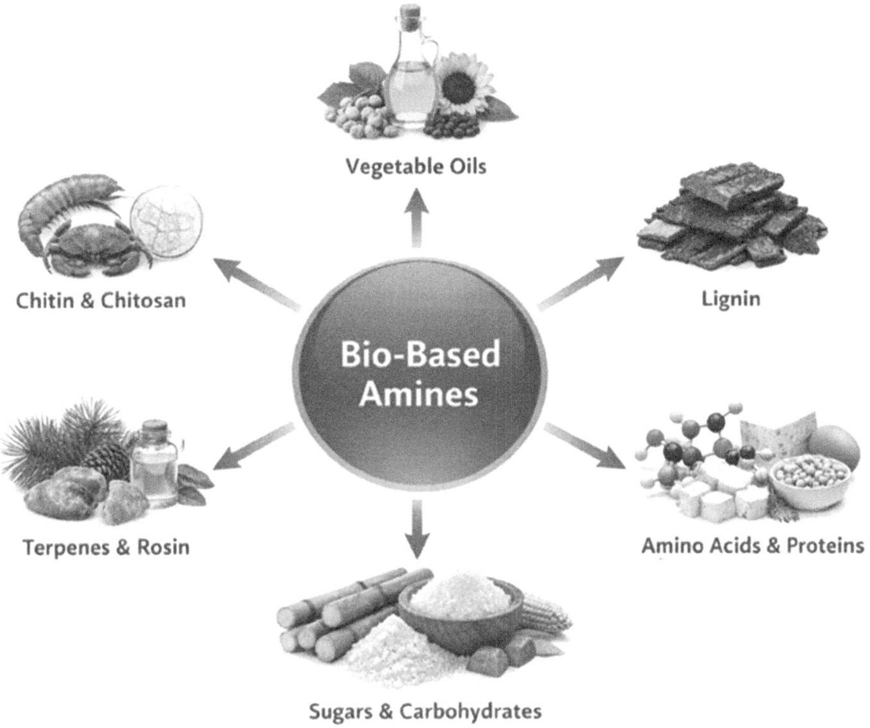

FIGURE 9.2 Renewable sources for bio-based amines.

their structure. In most cases, these macromolecular feedstocks must first be deconstructed into smaller platform molecules before amination. Typical intermediate pools therefore include alcohols, carbonyl compounds, phenolics, fatty alcohols, glucosamine derivatives, and amino acids, each of which can be transformed into different classes of amines [7].

From a synthesis perspective, the conversion of renewable substrates into amines mainly involves transforming C–O, C=O, or C=C bonds into C–N bonds by atom-economical catalytic routes. Four broad strategies are emphasized. The first is reductive amination of carbonyl compounds, in which aldehydes or ketones react with ammonia or amines to form imine-type intermediates that are then hydrogenated to the desired amines [22]. This route is especially suitable for biomass-derived sugars, furfurals, glycolaldehyde, glyceraldehyde, levulinic acid derivatives, and lignin-derived aldehydes such as vanillin. The second is hydrogen borrowing amination of alcohols [23], where a biomass-derived alcohol is first dehydrogenated to a carbonyl intermediate, then condensed with an amine source, and finally rehydrogenated to the amine product. This pathway is highly relevant because alcohol functionalities are widespread in biomass-derived molecules such as methanol, ethanol, ethylene glycol, higher diols, glycerol, sugar alcohols, and fatty alcohols. A third route is hydroamination of unsaturated compounds, where amines add across C=C or C≡C bonds; this can also be preceded by dehydration of alcohols to generate the needed unsaturation [24]. The fourth route is the use of inherent nitrogen functionality already present in chitin- and protein-derived feedstocks. In these cases, instead of introducing nitrogen from ammonia, the chemistry often involves selective defunctionalization, decarboxylation, cyclization, hydrogenation, or deacetylation to convert naturally embedded nitrogen into valuable amine structures [7].

Carbohydrate-derived feedstocks form one of the most diverse renewable sources for bio-based amine synthesis. Cellulose and hemicellulose can be hydrolyzed, hydrogenolyzed, or dehydrated to generate alcohols such as ethylene glycol and higher diols, as well as carbonyl compounds such as glycolaldehyde, glyceraldehyde, furfural, and 5-hydroxymethylfurfural. Alcohol intermediates are frequently aminated through hydrogen borrowing, whereas sugar aldehydes and ketones are more suitable for reductive amination because their carbonyl groups are already activated toward nucleophilic attack. Carbohydrate dehydration products such as furfural, 5-hydroxymethylfurfural, and levulinic acid are especially attractive because they can be converted into furfurylamines, bis(aminomethyl)furans, pyrrolidones, and other cyclic amines. The article also highlights that carbohydrate-derived alcohol routes are often more selective in upstream production but require harsher amination conditions, whereas carbonyl routes involve more reactive intermediates that are easier to aminate but can be harder to produce selectively [25,26].

Lignin represents the principal renewable aromatic source for bio-based amines. After depolymerization, lignin yields phenolic monomers such as guaiacols, syringols, alkylphenols, and aromatic aldehydes. These can be transformed into amines either at the aromatic ring or at side-chain carbonyl groups. Phenolic substrates may undergo reductive amination through partially reduced cyclohexanone-type intermediates, formal hydrogen borrowing-type pathways, or direct ammonolysis under acid-catalyzed conditions. Aromatic aldehydes derived from lignin, such as vanillin, are more straightforward substrates for reductive amination and can furnish aromatic amines like vanillylamine. The main challenge for lignin is that depolymerization rarely gives a single clean monomer stream, and methoxy-substituted aromatics often require further demethoxylation or careful catalytic tuning. Even so, lignin is highly valuable because it offers access to renewable aromatic amines that are otherwise difficult to obtain from carbohydrate-based resources [27].

Oils and fats provide a different type of amine platform based on long aliphatic chains. After hydrolysis or transesterification, triglycerides yield glycerol along with fatty acids or fatty acid methyl esters. Glycerol can be aminated through hydrogen borrowing, dehydration-reductive amination, or dehydration-hydroamination pathways to produce amino alcohols, diamines, and cyclic amines. Fatty acids and fatty esters can be hydrogenated to fatty alcohols, which can then be aminated directly, especially with dimethylamine, to form long-chain tertiary amines. These products are especially important in surfactants, quaternary ammonium salts, and related amphiphilic materials. The article emphasizes that oils and fats are therefore not only carbon sources for simple amines but also provide routes to highly functional long-chain amines with unique physicochemical properties [28,29].

Chitin and proteins are especially interesting because they contain renewable nitrogen within their own structure. Chitin can be deacetylated to chitosan or hydrolyzed to glucosamine and N-acetylglucosamine derivatives, which can then be converted into nitrogen-containing platform chemicals such as 3-acetamido-5-acetylfuran, N-acetyl monoethanolamine, or reduced amino polyols. Protein-derived amino acids can be obtained either by fermentation or by hydrolysis of protein-rich side streams. These amino acids can then be transformed into amines and related compounds through decarboxylation, hydrogenation, cyclization, or nitrile-based downstream chemistry. Examples discussed include conversion of glutamic acid into pyrrolidones and diamine precursors, lysine into lysinol or caprolactam-related intermediates, and valine into isobutylamine. These nitrogen-retaining feedstocks are particularly important because they offer a route to amines that are less dependent on externally supplied ammonia from the Haber–Bosch process [30–32].

Overall, the synthesis of bio-based amines is built around the selective catalytic upgrading of renewable platform molecules into nitrogen-containing chemicals using routes with high atom economy and minimal waste. The most important synthesis logic is feedstock-dependent: oxygen-rich biomass such as carbohydrates, lignin, and oils is upgraded mainly by introducing nitrogen through reductive amination, hydrogen borrowing, or hydroamination, whereas nitrogen-rich biomass such as chitin and proteins can be upgraded by preserving and reshaping naturally embedded nitrogen functionality. This makes renewable amine chemistry unusually broad, spanning aliphatic amines, aromatic amines, amino alcohols, diamines, heterocycles, amides, and surfactant-like amine derivatives from a diverse set of sustainable resources.

9.5 BIO-BASED AMINES IN ADHESIVES

Bio-based amines are increasingly being explored in adhesive systems as renewable reactive components that can contribute to crosslinking, interfacial interactions, and overall bond performance. Their use has attracted growing attention as researchers seek adhesive materials that combine functional efficiency with reduced reliance on petrochemical feedstocks.

An amine-functionalized biomass adhesive was prepared by first introducing amino groups into cellulose through reaction with ethylenediamine, followed by crosslinking with bio-based citric acid and further incorporation of polyamide 651 to construct a multi-crosslinked AC:CA:PA651 adhesive network. Although the amine source was not fully bio-based in a strict sense, the adhesive system still represented a renewable-material approach because cellulose and citric acid formed the main bio-based framework, while the amine functionality was used to create reactive sites and strengthen the network structure. The amine-assisted design resulted in substantial improvements in adhesive structure and performance. Spectroscopic analyses confirmed successful amination of cellulose and subsequent amidation reactions with citric acid and polyamide 651, while microscopic observations showed that the adhesive surface became denser and more compact after crosslinking. Thermal analysis also indicated improved thermal stability and higher char formation, consistent with the formation of a stronger crosslinked network. Under optimized conditions of 160°C, an AC:CA:PA651 ratio of 0.5:2:1, and 50% solid content, the plywood adhesive exhibited a dry strength of 4.5 ± 0.12 MPa and a wet strength of 3.54 ± 0.01 MPa, while retaining 3.11 ± 0.15 MPa after 3 h boiling water treatment. Even after the severe 4+4+1 water resistance test, the bond strength remained 2.81 MPa, which was far above the Class I plywood requirement of 0.7 MPa. The adhesive also showed broad substrate compatibility, giving lap shear strengths of 7.11 MPa for wood, 5.24 MPa for bamboo, 3.98 MPa for steel, 3.25 MPa for aluminum, 3.33 MPa for copper, 2.11 MPa for glass, 0.74 MPa for poly tetrafluoroethylene, and 1.13 MPa for polyvinyl chloride. Overall, the results demonstrated that amine-functional modification played a central role in improving crosslink density, cohesion, water resistance, and adhesion performance, while also suggesting that future replacement of petrochemical amines with fully renewable amine sources could further enhance the sustainability of this approach [33].

A completely bio-based epoxy structural adhesive was formulated by combining a rigid epoxy monomer (DGEDP)-methyl, with a flexible cardanol-derived epoxy resin, NC-514, and later incorporating cellulose nanocrystals (CNCs) as a bio-based filler. To achieve a fully renewable system, the conventional petroleum-derived hardener isophorone diamine (IPDA) was ultimately replaced with bis-(furfurylamine), a bio-based amine hardener (Figure 9.3). The final formulation was therefore built entirely from bio-based resin components, filler, and curing agent, making it a notable example of a fully bio-based epoxy adhesive system. The results showed that balancing the rigid and flexible epoxy components was essential for achieving both strength and toughness. A 60:40 blend of DGEDP-methyl and NC-514 gave the most favorable performance because it produced a nanoscale mixed morphology and a marked improvement in fracture toughness and adhesive strength compared with the neat rigid epoxy. This optimized resin composition displayed about a twofold increase in both fracture toughness and lap shear strength on aluminum relative to the unmodified DGEDP-methyl system. However, adding the flexible NC-514 slowed curing, so CNCs were introduced to adjust rheology and recover cure speed. The CNC filler induced shear-thinning behavior, created a yield stress at 1 wt % and above that helped prevent dripping or sagging after application, and also accelerated

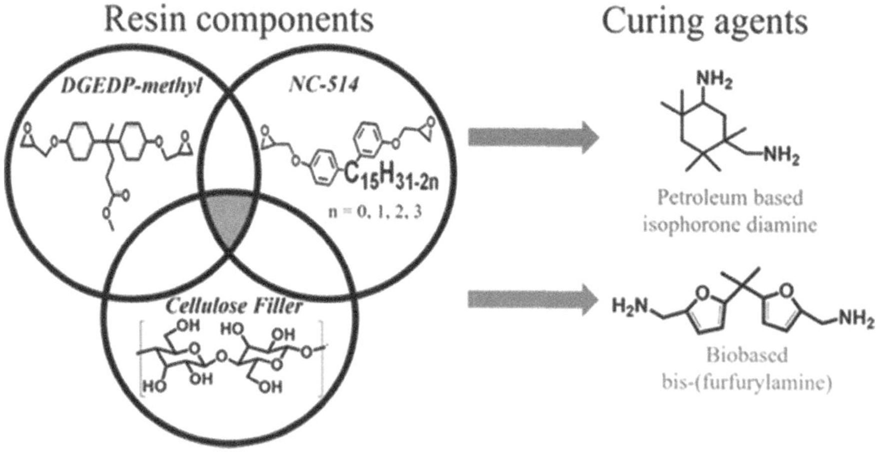

FIGURE 9.3 Structure of resin components and the use of bio based bis-(furfurylamine) as a curing agent to replace petroleum-based IPDA. (Adapted with permission [34]. Copyright 2019 American Chemical Society.)

curing through the catalytic effect of hydroxyl and carboxyl groups on the CNC surface. At 1 wt % CNC, the adhesive strength increased by about 1.5 times relative to the CNC-free 40 wt % NC-514 formulation. Replacement of the petroleum-based hardener with bis-(furfurylamine) did not significantly alter the curing rate, but it had a major effect on fracture behavior and final adhesive performance. While the IPDA-cured material still failed in a brittle manner, the bio-based amine-cured adhesive exhibited stable crack growth and much greater deformation before failure, indicating a transition from brittle to ductile fracture. This change greatly enhanced energy dissipation during loading and resulted in a dramatic increase in lap shear strength. The final optimized adhesive, consisting of DGEDP-methyl, NC-514, 1 wt % CNC, and bis-(furfurylamine), achieved adhesive strengths above 20 MPa on aluminum, representing nearly a 700% increase compared with the neat DGEDP-methyl system cured with IPDA. Overall, the article demonstrated that a bio-based amine hardener can play a critical role not only in making the formulation fully renewable but also in improving fracture resistance and adhesive strength, making such systems highly promising for sustainable structural adhesive applications [34].

An itaconic acid-based polyamidoamine-epichlorohydrin resin was prepared by reacting diethylenetriamine with renewable itaconic acid, followed by epichlorohydrin modification to generate a reactive amine-containing curing resin. Although the amine precursor itself was not bio-based, the use of bio-derived itaconic acid made the resin partially renewable and showed how bio-based components can be incorporated into amine curing systems for more sustainable adhesive formulations. The resulting resin markedly improved the performance of soy flour adhesive, particularly its resistance to water. The cured adhesive reached a wet bond strength of about 0.95 MPa, which was comparable to that of commercial polyamidoamine-epichlorohydrin modified soy adhesives and sufficient to meet the requirement for interior plywood applications. This enhancement was associated with effective crosslinking between the reactive resin and soy protein, which generated a denser and more compact network structure during curing. The higher degree of crosslinking reduced the water sensitivity of the adhesive and increased the water-insoluble fraction, indicating better structural stability after exposure to moisture. Microscopic observations also supported this behavior by showing a more uniform and compact fractured surface after modification, consistent with improved cohesion and reduced defect formation. Overall, even though the amine source was not fully bio-based, the adhesive performance demonstrated that partial replacement with renewable feedstock could still deliver strong practical benefits and could serve as a transition toward future fully bio-based amine-based curing systems [35].

Although the use of truly bio-based amines in adhesive formulations remains limited in the current literature, the available studies clearly show that amine chemistry plays a highly valuable role in improving crosslinking, cohesion, and water resistance in renewable adhesive systems. Most reported examples still rely on bio-based backbones such as cellulose, lignin, soy protein, or organic acids while using conventional amines to introduce reactive functionality. This gap highlights an important opportunity for future research, as replacing petrochemical amines with fully renewable amine counterparts could further improve the sustainability profile of adhesive materials without sacrificing performance. Therefore, the development of fully bio-based amine-containing adhesives represents a promising and worthwhile direction for next-generation green adhesive technologies.

9.6 BIO-BASED AMINES IN COATINGS

Bio-based amines have emerged as promising components in coating technologies because they can contribute both to network formation and to surface performance while reducing reliance on fossil-derived chemicals. Depending on their structure and reactivity, these renewable amines can be incorporated into protective, functional, and advanced coating systems through different formulation and processing routes. As illustrated in Figure 9.4, coating fabrication can involve a wide range of techniques, and the integration of bio-based amines into these systems offers new opportunities to tailor adhesion, durability, wettability, and surface functionality [14].

Cardanol-derived amines were synthesized by direct amination of two epoxidized cardanol monomers, NC-514 and GX-2551, using aqueous ammonia under microwave irradiation, yielding the amine hardeners NC-A and GX-A. Among the two, GX-A was highlighted as the more significant product because its precursor route avoided phenol, making it a fully bio-based amine, whereas NC-A remained only partially

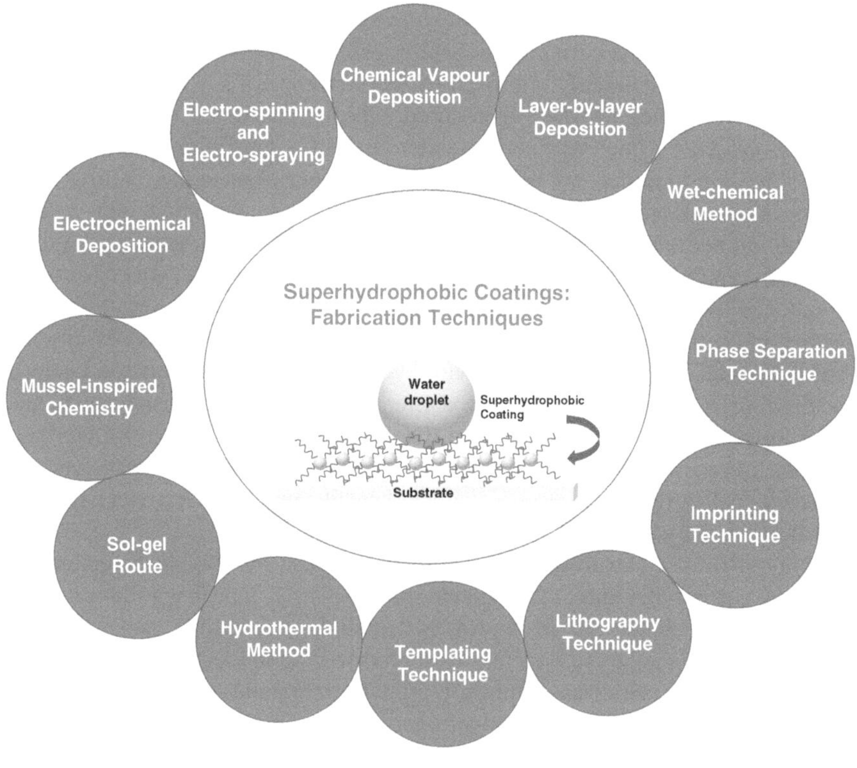

FIGURE 9.4 Fabrication strategies of superhydrophobic coatings. (Adapted with permission [14]. Copyright 2020 Elsevier.)

bio-based. These amines were then used as curing agents for diglycidyl ether of bisphenol A (DGEBA)-based epoxy thermosets and, although the work was not presented as a direct coating-performance study, such bio-based amine hardeners could be used in coating formulations based on epoxy-amine networks. The results showed that both cardanol-based amines were effective curing agents and produced thermosets with good thermal and thermomechanical performance. GX-A was particularly attractive because it combined full bio-based origin with lower viscosity, which is beneficial for processing and formulation. The cured GX-A thermoset showed slightly better performance than the NC-A system, with a glass transition temperature of about 58°C, a 5% weight-loss decomposition temperature of about 340°C, and a tan delta peak around 78°C, while the NC-A thermoset showed corresponding values of about 51°C, 332°C, and 64°C. Dynamic mechanical analysis further indicated that the GX-A network had a somewhat higher crosslink density and rigidity, while both systems maintained high stiffness in the glassy state. Overall, the work demonstrated that bio-based amines derived from cardanol can function as promising renewable hardeners for epoxy thermosets, and these materials could be valuable candidates for sustainable adhesive and coating applications where lower toxicity, reduced petrochemical dependence, and good thermal performance are desired [36].

Bio-based amines were directly used in this work as crosslinkers for epoxy-coating formulations. The coating systems were prepared from DGEBA epoxy resin and cured with a series of phenalkamine hardeners, where the bio-based set consisted of cardanol-derived phenalkamines labeled PK1 to PK4. Among them, PK1 represented a fully cardanol-based phenalkamine, while the others were commercial phenalkamine formulations containing cardanol-derived components together with additional amine functionality. Although these bio-based amines were not synthesized within the work itself, their use was central to the formulation of renewable epoxy coatings and allowed direct comparison with conventional fossil-based amine crosslinkers. The results showed that the cardanol-derived phenalkamines provided clear advantages in coating performance, particularly under low-temperature curing conditions. These bio-based amine-cured coatings generally exhibited faster curing behavior and better film formation at room temperature and low temperature than the fossil-based amine systems. They also developed superior surface hardness, adhesion, flexibility, and impact resistance, demonstrating that the cardanol-based amines were highly effective curing agents for protective epoxy coatings. In addition, the bio-based phenalkamine systems showed improved hydrophobicity and moisture tolerance, which are important for coating durability and practical application in humid or variable environments. Overall, the article demonstrated that bio-based phenalkamine crosslinkers derived from cardanol can serve as strong renewable alternatives to conventional petroleum-based amines in epoxy coatings, offering both sustainability benefits and competitive, and in some cases improved, coating performance [37].

A fully bio-based coating system was developed using acrylated epoxidized soybean oil (AESO) as the renewable acrylate component and Priamine 1074 as the key bio-based amine hardener. The soybean oil derivative was first acrylated to obtain AESO, and the coating network was then formed through aza-Michael addition between the acrylate groups of AESO and the secondary amine groups of Priamine (Figure 9.5). To further tailor surface properties, amine-functional silane agents such as

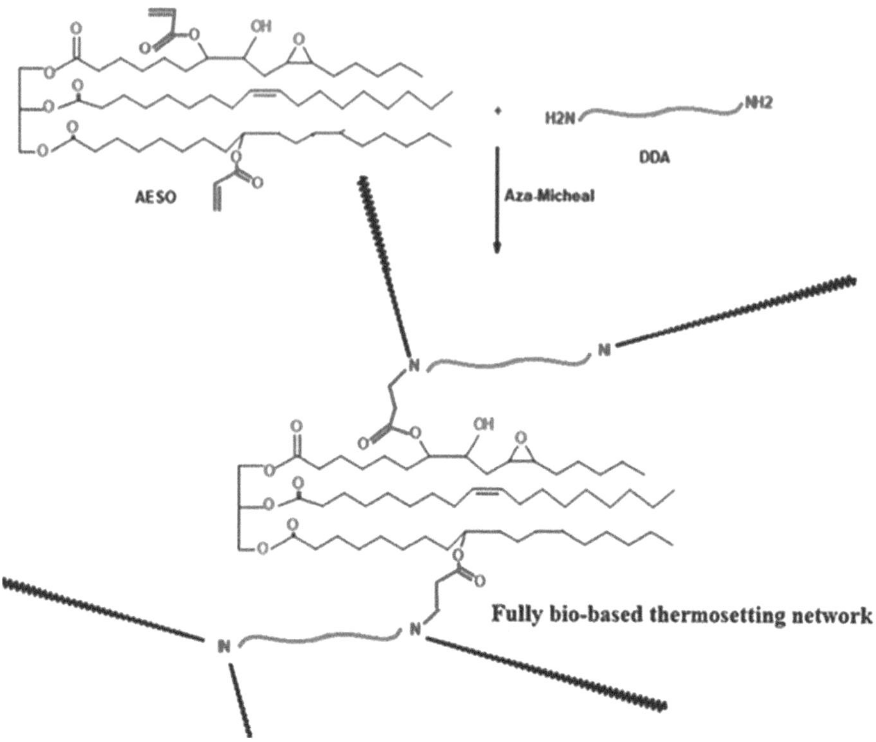

FIGURE 9.5 Mechanism of aza-Michael addition reaction between acrylate groups of AESO and the secondary amine groups of priamine. (Adapted with permission [38]. Copyright 2023 American Chemical Society.)

3-aminopropyltriethoxysiloxane, *N*-(2-aminoethyl)-3-aminopropyltrimethoxysilane, and amine-modified castor oil were incorporated together with hydrophobic nanosilica. In this formulation, the use of Priamine was especially important because it served as a truly bio-based amine building block rather than a conventional petrochemical curing agent. The results showed that the combination of bio-based AESO and bio-based Priamine produced a crosslinked coating with strong water-repellent characteristics, while the additional silane agents and nanosilica further enhanced surface roughness and hydrophobicity. Among the application routes examined, the two-step filming and two-step spraying methods gave better performance than the one-step spraying process, which did not provide the desired contact angle or mechanical properties. Structural analysis confirmed successful acrylation of soybean oil and supported the formation of the aza-Michael-crosslinked network. Overall, the work demonstrated that a bio-based amine can be used directly in coating design to create renewable thermosetting coatings with promising hydrophobic and protective properties, making this system a strong example of bio-based amine use in sustainable coating applications [38]. Overall, bio-based amines represent a promising route for the development of sustainable coating systems, as they can serve not only as renewable curing agents or reactive building blocks but also as key contributors to coating adhesion, durability, hydrophobicity, and protective performance.

9.7 SUSTAINABILITY BENEFITS, CHALLENGES, AND LIMITATIONS

Bio-based amines offer a meaningful sustainability advantage in adhesives and coatings because they can replace fossil-derived curing agents, crosslinkers, or reactive modifiers with components obtained from renewable resources such as cardanol, vegetable oils, lignin-derived molecules, carbohydrates, amino acids, and fermentation-derived intermediates. Since amines are central to the curing and network formation of epoxy, polyurethane, polyurea, and related thermosetting systems, their replacement can significantly improve the renewable carbon content of the overall formulation. More broadly, the development of bio-based amines aligns with green chemistry goals by encouraging the use of biomass feedstocks and expanding the range of renewable nitrogen-containing building blocks available for polymer materials. Reviews of the field have emphasized that bio-based amine chemistry is especially important because nitrogen-containing renewable molecules are far less naturally abundant than oxygen-rich biomass intermediates, making their development strategically valuable for sustainable materials design [6]. In addition to reducing petrochemical dependence, bio-based amines can also contribute positively to material performance. Their molecular structures often introduce useful features such as long aliphatic chains for flexibility and hydrophobicity, aromatic groups for rigidity and thermal resistance, and multifunctional amine sites for dense crosslinking. In adhesive systems, these features can strengthen interfacial bonding, cohesion, and moisture resistance, while in coatings they can improve cure behavior, hardness, barrier performance, and surface durability. For example, cardanol-derived phenalkamines have been reported as effective bio-based crosslinkers for epoxy coatings, showing fast reaction rates, higher room-temperature conversion, and competitive coating performance relative to conventional fossil-based amines. This suggests that bio-based amines are not only environmentally attractive substitutes but also valuable design tools for tailoring performance [37]. Another important benefit is their compatibility with lower-emission and high-solids material systems. Many current adhesive and coating technologies are moving toward solvent-free, waterborne, or low-VOC formulations, and bio-based amines can support this transition when they are designed with suitable reactivity and viscosity. In coatings, recent reviews have noted that sustainability is no longer evaluated only by the origin of the resin but also by the broader formulation package, including solvents, additives, fillers, and processing routes. Therefore, bio-based amines become particularly valuable when they contribute both renewable content and practical processing advantages such as ambient curing, lower toxicity potential, or improved formulation efficiency [39].

Despite these advantages, important challenges still limit the wider use of bio-based amines in adhesives and coatings. One major issue is synthesis complexity. Unlike oxygenated biomass platforms, naturally available amines are relatively scarce, so many renewable amines must be prepared through additional conversion steps such as reductive amination, hydrogen borrowing, amination of epoxides, or downstream transformation of amino-acid or lignin-derived intermediates. These extra steps can increase cost, reduce atom efficiency, and complicate purification. As a result, many reported systems are only partially bio-based, where the backbone may be renewable but the amine component remains fossil-derived. This gap is particularly visible in the adhesive literature, where numerous bio-based adhesive systems rely on renewable cellulose, soy, lignin, or citric acid frameworks but still use conventional amines to create reactive functionality. Performance-related limitations also remain significant. In adhesives, common weaknesses of bio-based systems include poor water resistance, lower bond durability, susceptibility to microbial attack, viscosity instability, and difficulty maintaining performance under harsh service conditions. In coatings, challenges often include lower long-term weatherability, reduced chemical resistance in some formulations, inconsistent curing behavior, and difficulties in balancing hardness with flexibility. These problems are not caused by bio-based amines alone, but they can affect the success of replacing conventional amines in practical formulations. The broader coatings literature also points out that true sustainability requires more than renewable feedstock content, because a coating may still depend on fossil-based solvents, auxiliary additives, or energy-intensive processing even when the amine or resin is partly bio-based. Economic and industrial implementation issues further complicate commercialization. Petrochemical amines are widely available, relatively consistent in quality, and supported by mature supply chains, whereas many bio-based amines are still produced at smaller scale and may suffer from feedstock variability or limited standardization. In industry, this creates uncertainty in cost, shelf life, cure reproducibility, and regulatory positioning. The coatings field has also pointed out the need for clearer standards to define and quantify bio-based content, since partially renewable systems are sometimes grouped together with fully renewable ones. For wider industrial acceptance, future work will need to improve synthesis efficiency, reduce cost, expand feedstock flexibility, and validate long-term performance through life cycle and service-life studies rather than relying only on renewable content claims [6,39].

9.8 FUTURE OUTLOOK AND CONCLUSION

Bio-based amines are positioned to become increasingly important in adhesive and coating technologies because they offer a rare combination of renewable origin, high chemical reactivity, and broad structural diversity. Their use already shows clear potential for improving crosslinking, interfacial bonding, curing behavior, moisture resistance, hydrophobicity, corrosion protection, and mechanical durability. Unlike simple fossil-to-renewable substitution, bio-based amines also create opportunities to redesign material performance at the molecular level. Feedstocks such as lignin, vegetable oils, carbohydrates, amino acids, chitin, and other biomass-derived intermediates can be transformed into amines with tailored rigidity, flexibility, functionality, and hydrophobic character, making them attractive for a wide range of adhesive and coating applications. Even with this promise, wider implementation still depends on overcoming several scientific and practical barriers. Many currently reported systems remain only partially renewable, with bio-based backbones combined with petrochemical amines, especially in adhesive formulations. Future progress will therefore require greater focus on truly renewable amine curing agents, crosslinkers, and reactive modifiers with controlled structure, reliable purity, and scalable production. Expanding the range of bio-based amines from lignin-derived aromatics, amino-acid-derived diamines, furan-based intermediates, terpene-derived amines, and nitrogen-rich biomass sources could significantly strengthen the field. At the same time, a deeper understanding of structure–property relationships will be necessary so that amine design can be matched more precisely with the needs of structural adhesives, flexible coatings, protective barrier systems, and multifunctional surfaces. The next major advances are likely to come from integrating bio-based amines into high-value material platforms such as recyclable thermosets, self-healing adhesives, dynamic covalent networks, anticorrosive coatings, antimicrobial coatings, flame-retardant coatings, and waterborne or solvent-free systems. Industrial adoption, however, will require more than laboratory-scale success. Challenges related to synthesis complexity, feedstock variability, cost, storage stability, processability, and long-term durability must still be addressed in a consistent way. Stronger life cycle assessment, techno-economic analysis, and standardized long-term performance testing will also be essential to confirm that the environmental advantages are meaningful under real manufacturing and service conditions. With continued advances in biomass valorization, catalytic amination, and formulation engineering, bio-based amines are likely to play a central role in the development of adhesive and coating materials that are not only more sustainable but also more functional, durable, and application-specific.

REFERENCES

1. S. Ebnesajjad, *Introduction and Adhesion Theories*, in: S.P. Ebnesajjad (Ed.), *Handbook of Adhesives and Surface Preparation*. Plastics Design Library Series, William Andrew Publishing, Oxford, 2011: pp. 3–13.
2. C.S.P. Borges, A. Akhavan-Safar, P. Tsokanas, R.J.C. Carbas, E.A.S. Marques, L.F.M. da Silva, From fundamental concepts to recent developments in the adhesive bonding technology: A general view, *Discov. Mech. Eng.* 2 (2023) 8.
3. S.F. Kistler, P.M. Schweizer, Coating science and technology: An overview, in: S.F. Kistler, P.M. Schweizer (Eds.), *Liquid Film Coating: Scientific Principles and Their Technological Implications*, Springer Netherlands, Dordrecht, 1997: pp. 3–15.
4. L. Quintino, Overview of coating technologies, in: R.M. Miranda (Ed.), *Surface Modification by Solid State Processing*, Woodhead Publishing, Cambridge, 2014: pp. 1–24.
5. P. Patel, R. Patel, M.L. Chaudhary, N. Chaudhary, R.K. Gupta, High-performance bio-based non-isocyanate polyurethane adhesive: A solvent and catalyst-free synthesis approach, *J. Polym. Environ.* 32 (2024) 5024–5035.
6. V. Froidevaux, C. Negrell, S. Caillol, J.P. Pascault, B. Boutevin, Biobased amines: From synthesis to polymers; present and future, *Chem. Rev.* 116 (2016) 14181–14224.
7. M. Pelckmans, T. Renders, S. Van de Vyver, B.F. Sels, Bio-based amines through sustainable heterogeneous catalysis, *Green Chem.* 19 (2017) 5303–5331.
8. Y. Xie, S. Gao, D. Zhang, C. Wang, F. Chu, Bio-based polymeric materials synthesized from renewable resources: A mini-review, *Resour. Chem. Mater.* 2 (2023) 223–230.
9. L.A. Heinrich, Future opportunities for bio-based adhesives – advantages beyond renewability, *Green Chem.* 21 (2019) 1866–1888.
10. K.W. Allen, Adhesion and adhesives – some fundamentals, *Stud. Conserv.* 29 (1984) 5–12.
11. C. Gay, Some fundamentals of adhesion in synthetic adhesives, *Biofouling.* 19 (2003) 53–57.
12. S. Gharde, G. Sharma, B. Kandasubramanian, Hot-melt adhesives: Fundamentals, formulations, and applications: A critical review, in: K.L. Mittal (Ed.), *Progress in Adhesion and Adhesives*, Wiley, Hoboken, NJ, 2021: pp. 1–28.

13. R. Patel, M.L. Chaudhary, R.K. Gupta, Recent advances in bio-based adhesives: sustainable materials, synthesis strategies, and emerging applications, *J. Adhes. Sci. Technol.* (2025) 1–50.
14. A. Hooda, M.S. Goyat, J.K. Pandey, A. Kumar, R. Gupta, A review on fundamentals, constraints and fabrication techniques of superhydrophobic coatings, *Prog. Org. Coatings.* 142 (2020) 105557.
15. H. Beyrami, M. Golshan, A. Zardehi-Tabriz, M. Salami-Kalajahi, Smart Coatings: Fundamentals, preparation approaches, and applications, *Adv. Mater. Technol.* 10 (2025) e00574.
16. U. Panchal, M.L. Chaudhary, P. Patel, J. Patel, R.K. Gupta, Soybean-based bio-adhesives: Role of diamine on the adhesive properties, *ACS Omega.* 9 (2024) 10738–10747.
17. P. Vijayalakshmi, T.C. Rao, V. Kale, R.S. Balakrishna, R. Subbarao, Reactive polyamides from cycloaliphatic C21 dicarboxylic acid and their evaluation as epoxy curing agents, *Polymer (Guildf).* 33 (1992) 3252–3256.
18. M.L. Alfieri, L. Panzella, A. Napolitano, Multifunctional coatings hinging on the catechol/amine interplay, *European J. Org. Chem.* 27 (2024) e202301002.
19. O. Björneholm, G. Öhrwall, A.N. de Brito, H. Ågren, V. Carravetta, Superficial tale of two functional groups: On the surface propensity of aqueous carboxylic acids, alkyl amines, and amino acids, *Acc. Chem. Res.* 55 (2022) 3285–3293.
20. J.H. Ryu, P.B. Messersmith, H. Lee, Polydopamine surface chemistry: A decade of discovery, *ACS Appl. Mater. Interfaces.* 10 (2018) 7523–7540.
21. A.J. Ragauskas, C.K. Williams, B.H. Davison, G. Britovsek, J. Cairney, C.A. Eckert, W.J. Frederick, J.P. Hallett, D.J. Leak, C.L. Liotta, J.R. Mielenz, R. Murphy, R. Templer, T. Tschaplinski, The path forward for biofuels and biomaterials, *Science* (80-.). 311 (2006) 484–489.
22. S. Bähn, S. Imm, L. Neubert, M. Zhang, H. Neumann, M. Beller, The catalytic amination of alcohols, *ChemCatChem.* 3 (2011) 1853–1864.
23. S. Hameury, H. Bensalem, K. De Oliveira Vigier, Sustainable amination of bio-based alcohols by hydrogen borrowing catalysis, *Catalysts.* 12 (2022) 1306.
24. T.E. Müller, K.C. Hultzsch, M. Yus, F. Foubelo, M. Tada, Hydroamination: Direct addition of amines to alkenes and alkynes, *Chem. Rev.* 108 (2008) 3795–3892.
25. K. Zeng, The role of N-containing carbohydrates in organic catalysis: A review, *Green Chem.* 27 (2025) 12888–12908.
26. N.A.S. Ramli, N.A.S. Amin, Catalytic conversion of carbohydrate biomass in ionic liquids to 5-hydroxymethyl furfural and levulinic acid: A review, *BioEnergy Res.* 13 (2020) 693–736.
27. D. Verboekend, Y. Liao, W. Schutyser, B.F. Sels, Alkylphenols to phenol and olefins by zeolite catalysis: a pathway to valorize raw and fossilized lignocellulose, *Green Chem.* 18 (2016) 297–306.
28. F. Ma, M.A. Hanna, Biodiesel production: A review, *Bioresour. Technol.* 70 (1999) 1–15.
29. E.F. Aransiola, T.V. Ojumu, O.O. Oyekola, T.F. Madzimbamuto, D.I.O. Ikhu-Omoregbe, A review of current technology for biodiesel production: State of the art, *Biomass Bioenergy.* 61 (2014) 276–297.
30. M. Rinaudo, Chitin and chitosan: Properties and applications, *Prog. Polym. Sci.* 31 (2006) 603–632.
31. M.N.V. Ravi Kumar, A review of chitin and chitosan applications, *React. Funct. Polym.* 46 (2000) 1–27.
32. N. Yan, X. Chen, Sustainability: Don't waste seafood waste, *Nature.* 524 (2015) 155–157.
33. T. Liu, J. Qian, Y. Shao, G. Du, J. Li, L. Yang, X. Ran, Enhancing performance of sustainable biomass-based adhesive through supramolecular crosslinked networks, *Colloids Surf. Physicochem. Eng. Asp.* 721 (2025) 137233.

34. A. Patel, S. Mekala, O.G. Kravchenko, T. Yilmaz, D. Yuan, L. Yue, R.A. Gross, I. Manas-Zloczower, Design and formulation of a completely biobased epoxy structural adhesive, *ACS Sustain. Chem. Eng.* 7 (2019) 16382–16391.
35. C. Gui, G. Wang, D. Wu, J. Zhu, X. Liu, Synthesis of a bio-based polyamidoamine-epichlorohydrin resin and its application for soy-based adhesives, *Int. J. Adhes. Adhes.* 44 (2013) 237–242.
36. A.-S. Mora, R. Tayouo, B. Boutevin, G. David, S. Caillol, Synthesis of biobased reactive hydroxyl amines by amination reaction of cardanol-based epoxy monomers, *Eur. Polym. J.* 118 (2019) 429–436.
37. P. Samyn, J. Bosmans, P. Cosemans, Benchmark study of epoxy coatings with selection of bio-based phenalkamine versus fossil-based amine crosslinkers, *Molecules.* 28 (2023) 4259.
38. İ. Güneş, B. Oktay, Fabrication of eco-friendly sustainable superhydrophobic and anticorrosive coatings from acrylated soybean oil and biobased amine by the Aza-Michael reaction, *ACS Sustain. Chem. Eng.* 11 (2023) 15253–15261.
39. L. Xia, T. Gui, J. Wang, H. Tian, Y. Wang, L. Ning, L. Wu, Bio-based coatings: Progress, challenges and future perspectives, *Polymers (Basel).* 17 (2025) 3266.

10 Bio-Based Cationic Polymers for Water Purification Derived from Bio-Amines

10.1 INTRODUCTION

Nitrogen-containing polymers have long occupied a central role in water treatment technologies due to their ability to generate positively charged functional groups capable of interacting with negatively charged contaminants. Industrial water purification systems rely heavily on cationic polymers such as polyethyleneimine (PEI), poly(diallyldimethylammonium chloride) (polyDADMAC), and cationic cationic polyacrylamides (CPAMs), which operate through charge neutralization, electrostatic attraction, and bridging flocculation mechanisms. These synthetic polyelectrolytes demonstrate high charge density, predictable molecular weight distribution, and scalable production, which have positioned them as industrial benchmarks in municipal and industrial wastewater treatment [1]. However, increasing awareness of environmental persistence, residual monomer toxicity, and oxidative transformation products has raised important concerns regarding their long-term ecological impact [2]. Simultaneously, global sustainability goals and carbon accounting frameworks have shifted scientific focus toward renewable feedstocks and low-emission material platforms. Life cycle assessment (LCA) studies comparing fossil-based and bio-based polymers reveal that petroleum-derived systems often carry higher global warming potential (GWP) and cumulative energy demand at the cradle-to-gate stage [3]. More recent sustainability analyses further emphasize that feedstock origin, processing route, and energy intensity strongly influence environmental impact categories such as greenhouse gas emissions, eutrophication potential, and acidification burden [4]. These findings have catalyzed interest in reengineering cationic polymer platforms using renewable nitrogen-containing building blocks while maintaining high purification efficiency. Bio-amines derived from lignocellulosic biomass, amino acids, vegetable oils, and fermentation pathways represent structurally diverse alternatives to petrochemical amines. Lignin, which constitutes nearly one-third of terrestrial organic carbon, provides an abundant aromatic framework that can be transformed into functional amines via catalytic reductive amination, hydrogenation–dehydrogenation sequences, and cross-coupling strategies. These routes avoid traditional nitration-reduction processes and enable selective synthesis of aromatic amines under milder conditions [5–7]. Similarly, amino acid upgrading through enzymatic or catalytic decarboxylation yields aliphatic diamines such

DOI: 10.1201/9781003731122-10

as cadaverine and putrescine, which serve as renewable monomers for polyamides and cationic polymer backbones [8]. Vegetable oil and fatty acid derivatives offer additional routes to long-chain aliphatic amines, providing tunable flexibility and hydrophobic balance in polymer networks. The structural diversity of bio-based amines enables precise control over polymer architecture. Primary, secondary, and tertiary amines exhibit distinct reactivity toward epoxides, acrylates, aldehydes, and carbonates, allowing formation of linear, branched, or crosslinked cationic networks. Aromatic amines derived from lignin typically introduce rigidity and thermal stability, whereas long-chain aliphatic amines impart flexibility and toughness. These structure–property relationships directly influence charge density, swelling behavior, crosslink density, and adsorption kinetics in water treatment applications. Through quaternization or protonation, these polymers acquire permanent or pH-responsive cationic functionality, enabling interaction with dyes, heavy metals, oil droplets, and emerging contaminants. Applications of bio-based cationic polymers in water purification extend beyond conventional flocculation. Adsorptive removal of anionic dyes, chelation of heavy metal ions, oil–water separation, antimicrobial activity, and PFAS binding have all been explored using renewable amine-derived platforms. Compared to petrochemical counterparts, these systems can incorporate hydrolysable ester or amide linkages, which enhance biodegradability and reduce long-term environmental persistence. While performance parity with industrial CPAM and polyDADMAC systems remains application-dependent, advances in molecular design have significantly narrowed the gap. Beyond bulk adsorption and flocculation, emerging material formats such as membranes, hydrogels, electrospun nanofibers, and composite systems further expand functionality. Cationic hydrogels enable high surface area adsorption and controlled swelling, while electrospun nanofibrous mats enhance mass transfer kinetics. Hybrid materials integrating bio-based polymers with carbon frameworks or metal–organic structures demonstrate synergistic adsorption and mechanical reinforcement. Importantly, recyclability and regeneration have become central design criteria. Thermoregulated poly(ionic liquid) systems demonstrate catalyst and material recovery efficiencies exceeding 95% with minimal residual contamination, illustrating how advanced cationic architectures can incorporate circular chemistry principles [8]. Such strategies highlight the broader shift from single-use flocculants toward regenerable functional materials. Nevertheless, sustainability claims must be evaluated quantitatively. LCA frameworks reveal that while bio-based polymers may reduce fossil carbon dependency, agricultural inputs, fertilizer use, land occupation, and downstream processing energy can significantly influence environmental metrics [3]. Recent analyses emphasize that renewable feedstock alone does not guarantee lower environmental impact; rather, process optimization, supply chain management, and energy source selection determine overall performance [9,10]. Therefore, direct comparison with petrochemical benchmarks must consider performance efficiency, toxicity, carbon footprint, resource intensity, and end-of-life behavior simultaneously. In this context, this chapter systematically examines bio-based cationic polymers derived from renewable amines, beginning with feedstock origins and catalytic upgrading strategies, followed by structure–reactivity relationships and polymerization mechanisms. Subsequent sections explore water purification applications, including adsorption, membrane technologies, hydrogel

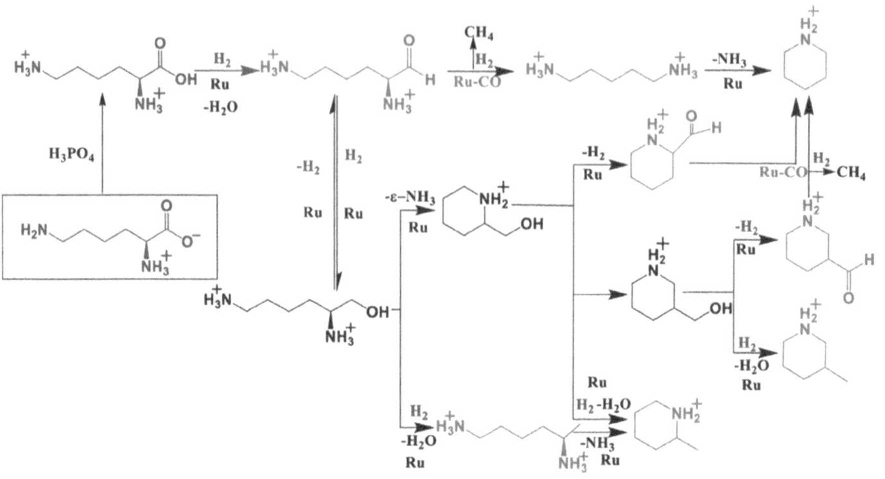

FIGURE 10.2 Overall reaction pathway of the deoxygenation of l-lysine to diamines and other N-containing chemicals. (Adapted with permission [8]. Copyright 2020 American Chemical Society.)

10.2.3 Polysaccharide-Derived Amines (Chitosan, Cellulose)

Polysaccharides provide abundant renewable backbones for amine functionalization. Chitin, the second most abundant natural polymer, contains acetamide groups that can be valorized through catalytic deoxygenation. A Ru/C–H_3PO_4 system demonstrated selective conversion of N-acetyl-D-glucosamine into cyclic amines with approximately 50% nitrogen yield [21]. Protonation of the amino group protects nitrogen functionality during high-temperature deoxygenation, preserving amine selectivity. Chitosan, obtained by deacetylation of chitin, inherently contains primary amines and is widely recognized as a renewable cationic polymer platform [15]. Its amine-rich backbone enables protonation, metal coordination, and chemical grafting for enhanced performance. Aminolysis of bio-based glycidyl ethers derived from glycerol, trimethylolpropane, and limonene has been shown to produce multifunctional β-hydroxylamines in high yield under mild conditions [22]. These materials combine hydroxyl and amine functionalities, enhancing hydrogen bonding and reactivity in subsequent polymerization. Collectively, polysaccharide-derived amines offer structural diversity, biodegradability, and intrinsic nitrogen functionality, making them valuable for sustainable polymer systems.

10.2.4 Vegetable Oil and Fatty Acid-Derived Amines

Vegetable oils such as soybean oil and cardanol represent triglyceride-based renewable platforms for synthesizing long-chain aliphatic amines. Epoxidized cardanol monomers can undergo direct amination with aqueous ammonia to yield reactive hydroxyl amines [23]. Reactivity differences between terminal and internal epoxies were confirmed by nuclear magnetic resonance analysis, illustrating steric and inductive effects. Green aminolysis of bio-based glycidyl ethers further demonstrates efficient ammonia ring-opening strategies without hazardous reagents [22]. These long-chain aliphatic amines introduce flexibility and amphiphilicity into polymer systems. Comprehensive reviews on bio-based curing agents and amine preparation emphasize the industrial relevance of fatty amines and cardanol-derived systems [16,24]. The hydrophobic alkyl chains enhance compatibility with nonpolar matrices while maintaining cationic functionality after quaternization.

10.2.5 Fermentation-Derived Diamines (Cadaverine, Putrescine)

Fermentation technologies enable large-scale production of diamines directly from renewable sugars. Reviews on sugar-to-amine conversion outline catalytic and biotechnological strategies for producing linear aliphatic amines [20]. These processes reduce dependence on fossil feedstocks and integrate into biorefinery frameworks. Lysine-derived cadaverine production [8] and protein-waste valorization pathways [18] demonstrate how nitrogen-rich biomass can be directly upgraded into industrially relevant diamines. Catalytic amination of biomass oxygenates further broadens the feedstock scope. Life cycle assessment comparisons between amine-based systems and alternative processes highlight environmental considerations associated with solvent regeneration and degradation [25], reinforcing the need for stable and renewable amine platforms.

10.3 SYNTHETIC STRATEGIES

The transformation of renewable feedstocks into functional bio-based amines relies on a combination of catalytic, enzymatic, and polymerization strategies that control nitrogen incorporation, reactivity, and molecular architecture. Among these, reductive amination represents one of the most fundamental and versatile approaches. In biomass valorization, aldehydes and ketones derived from lignin, sugars, or platform oxygenates are first converted into imine intermediates, followed by hydrogenation to yield primary or secondary amines. Catalytic reductive amination of lignin-derived phenolic aldehydes has been demonstrated using heterogeneous metal systems, enabling selective formation of aromatic amines while avoiding traditional nitration–reduction sequences [11]. Broader catalytic frameworks for biomass-derived oxygenates describe hydrogen-borrowing mechanisms, in situ imine formation, and metal–hydride transfer steps that govern selectivity and suppress over-reduction. These strategies extend to sugar-derived substrates and highlight the importance of catalyst surface properties and reaction conditions in controlling amine yield [20]. Enzymatic reductive amination further complements heterogeneous catalysis. Engineered transaminases enable selective amination of vanillin and related aldehydes under mild conditions, with docking and kinetic studies confirming substrate–enzyme hydrogen bonding interactions that stabilize transition states (Figure 10.3) [14]. Collectively, reductive amination provides a scalable and mechanistically tunable route for introducing nitrogen into renewable aromatic and aliphatic frameworks. Beyond direct amination, Mannich functionalization offers a powerful method to graft amine functionalities onto phenolic and aromatic substrates. In this three-component reaction, a phenol, aldehyde, and amine react to form β-aminomethylated products.

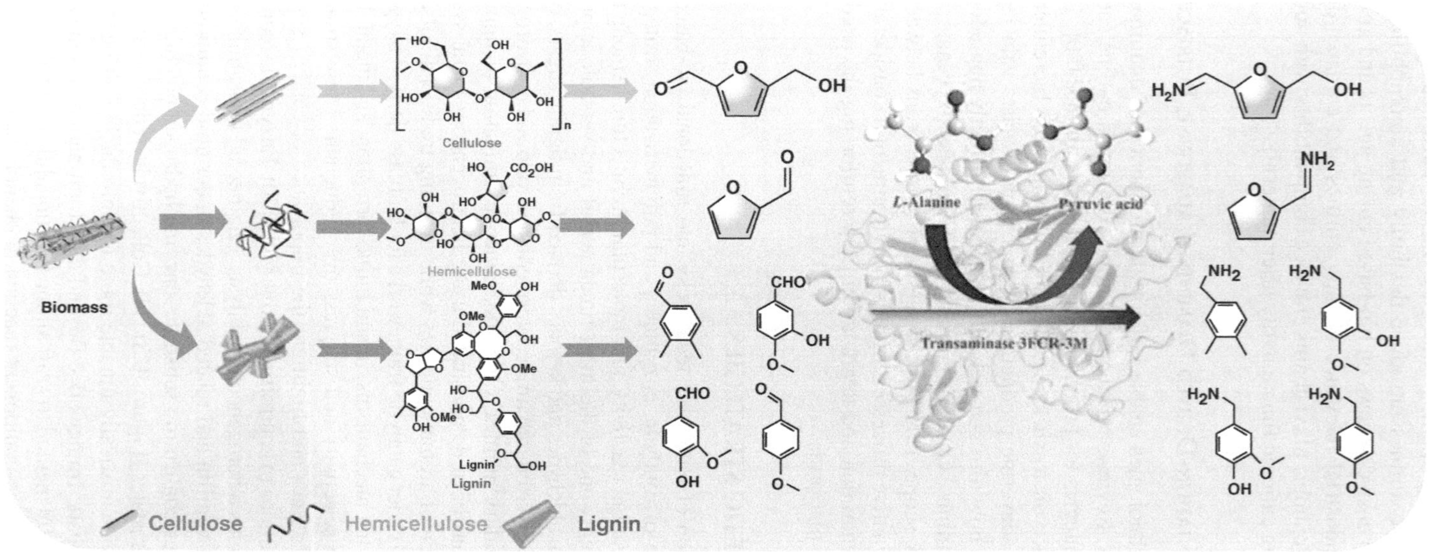

FIGURE 10.3 Schematic illustration of the biomanufacturing of bio-based organic amines from biomass-derived aldehydes via transaminase 3FCR-3M-catalyzed amination. (Adapted with permission [14]. Copyright 2026 American Chemical Society.)

Lignin-derived phenolics and cardanol derivatives have been successfully modified through Mannich reactions to generate multifunctional curing agents and reactive amine monomers, expanding nitrogen content without preactivation of the aromatic ring [16]. The reaction proceeds via iminium ion formation followed by electrophilic aromatic substitution, and its efficiency depends on phenolic activation and amine nucleophilicity. Mannich functionalization is particularly attractive for renewable aromatics because it introduces cationic character and crosslinking potential while preserving aromatic rigidity, a key factor in designing high-performance networks. Epoxide–amine polymerization constitutes another central synthetic strategy, especially in the preparation of bio-based thermosets and cationic networks. Ring-opening of epoxides by primary or secondary amines yields β-hydroxyl amines, which can further react to form crosslinked structures. Bio-based glycidyl ethers derived from glycerol, trimethylolpropane, limonene, or cardanol have been aminolyzed efficiently under mild conditions, producing multifunctional hydroxyl amines with high epoxy conversion [22,23]. The presence of newly formed hydroxyl groups accelerates subsequent reactions through autocatalytic hydrogen bonding, influencing gel time and network homogeneity. In fully bio-based epoxy–imine systems derived from vanillin, epoxy–amine curing is integrated with dynamic imine exchange, enabling vitrimeric behavior and reshaping capacity [13]. Mechanistically, epoxide–amine reactions proceed through nucleophilic attack on the less hindered carbon of the oxirane ring, with reactivity influenced by steric hindrance, amine basicity, and electronic effects of adjacent substituents. These parameters allow fine control over crosslink density and mechanical properties in polymer systems. Quaternization reactions represent a key post-functionalization strategy to enhance cationic charge density and water solubility of bio-based amines. Alkylation of tertiary amines through SN2 reactions generates quaternary ammonium salts with permanent positive charges, independent of pH. While often applied to petrochemical amines, similar transformations have been extended to renewable amine frameworks derived from lignin, fatty acids, and amino acids [16]. Quaternization increases hydrophilicity and antimicrobial activity, which is particularly relevant for water purification and membrane applications. The efficiency of alkylation depends on steric accessibility and the electronic environment of the tertiary nitrogen, and long-chain bio-based alkyl groups can impart amphiphilic behavior. Grafting and crosslinking approaches further expand the structural versatility of bio-amines. Surface grafting through epoxy ring-opening or carbodiimide mediated coupling enables attachment of amine groups to polysaccharides or aromatic backbones. Schiff base formation between aldehydes and amines produces imine crosslinks that can introduce dynamic covalent behavior, as demonstrated in bio-based vitrimer networks [13]. Catalytic decarboxylation of amino acids, such as proline to pyrrolidine, provides cyclic amines that can serve as multifunctional crosslinkers (Figure 10.4) [17]. These crosslinking reactions are influenced by amine functionality, molecular weight, and steric architecture, enabling rational design of network topology. The formation of polyamidoamines and hyperbranched systems represents an advanced synthetic pathway exploiting multifunctional amine reactivity. Step-growth reactions between diamines and diacids or Michael addition between amines and acrylates generate branched architectures with high amine density. Reviews of bio-based amines highlight the use of amino acid-derived diamines and vegetable oil-based multifunctional

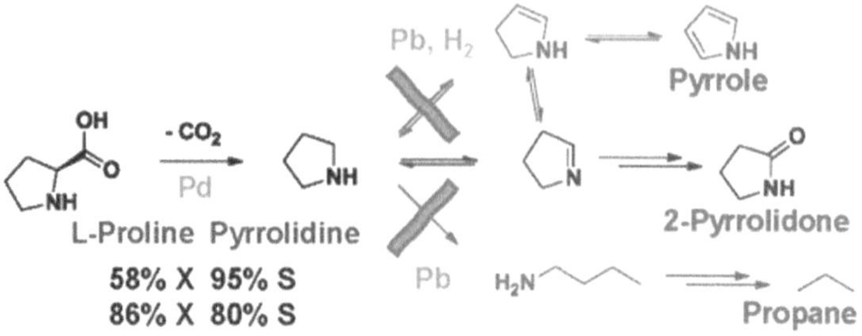

FIGURE 10.4 PdPb-catalyzed decarboxylation of proline to pyrrolidine. (Adapted with permission [17]. Copyright 2016 American Chemical Society.)

amines in producing hyperbranched and dendritic structures [15,19]. Control of stoichiometry and monomer functionality governs branching degree and gelation threshold. Hyperbranched systems exhibit high solubility, multiple reactive sites, and tunable viscosity, making them attractive for cationic polymer platforms. Overall, the synthetic strategies for bio-based amines integrate catalytic nitrogen insertion, electrophilic aromatic substitution, nucleophilic ring opening, alkylation, and controlled polycondensation. Each method offers distinct advantages in terms of selectivity, scalability, and structural tunability. By selecting appropriate feedstocks and reaction pathways, researchers can tailor amine functionality, molecular weight, branching, and charge density, enabling the design of renewable cationic polymers with controlled performance and environmental compatibility.

10.4 APPLICATIONS IN WATER TREATMENT

Bio-based amine-functional materials have gained increasing attention in water treatment because their nitrogen-containing functional groups provide multiple interaction mechanisms with dissolved and dispersed pollutants. Primary, secondary, tertiary, and quaternary amines can participate in electrostatic attraction, metal coordination, hydrogen bonding, hydrophobic association, ion exchange, and even catalytic degradation processes. The uploaded literature collectively demonstrates that renewable amine platforms derived from lignin, cellulose, chitosan, rosin, terpenes, algae, and synthetic bio-based polymers can address a broad spectrum of contaminants, including dyes, heavy metals, PFAS, oils, microorganisms, and complex industrial waste streams. Importantly, these systems combine adsorption efficiency with structural tunability, regeneration potential, and, in some cases, magnetic recoverability, making them promising candidates for sustainable water remediation. Dye removal is one of the most established applications of amine-rich bio-materials. Many textile dyes are sulfonated azo compounds that exist as anions in aqueous solution. Under mildly acidic conditions, amine groups are protonated to $-NH_3^+$, creating strong electrostatic attraction toward negatively charged dye molecules. Electrospun chitosan-based nanofibrous membranes have demonstrated extremely high adsorption capacities, exceeding 1300 mg/g for Acid Blue-113 when optimized for small fiber diameter and high surface area. Enhanced performance arises from increased accessibility of protonated amine sites and reduced mass transfer limitations. Composite systems incorporating silica nanoparticles further improve porosity and mechanical stability, achieving nearly complete removal of dyes such as Direct Red 23. These results illustrate that adsorption capacity depends not only on amine density but also on hierarchical morphology and pore connectivity. However, dye removal cannot be considered complete without understanding dye transformation products. Under anaerobic conditions, azo bonds are reductively cleaved, generating aromatic amines that may be persistent and toxic. The environmental behavior and carcinogenic risks of these aromatic amines have been extensively analyzed, emphasizing that reductive cleavage products can accumulate in wastewater systems [26]. Advanced oxidation processes such as ozonation and ultraviolet (UV) irradiation have been investigated to mineralize these amines. For example, ozonation removed approximately 85% of 4-amino-1-naphthalenesulfonic acid within minutes, while UV treatment required longer irradiation times; nevertheless, complete chemical oxygen demand (COD) removal required subsequent biological polishing [27]. These studies highlight that amine-based adsorbents should be integrated with oxidative or biological post-treatment to prevent accumulation of aromatic amine byproducts. Heavy metal adsorption represents another major domain where amine functionality provides distinct advantages. Metal ions such as Pb^{2+}, Cd^{2+}, Cu^{2+}, Co^{2+}, and Cr(VI) can coordinate with lone pairs on nitrogen atoms, forming stable inner-sphere complexes. A rosin-derived aminated resin demonstrated adsorption capacities of approximately 1.8 mmol/g for Pb(II), 1.32 mmol/g for Cd(II), and 1.12 mmol/g for Cu(II), with regeneration efficiencies exceeding 85% after multiple

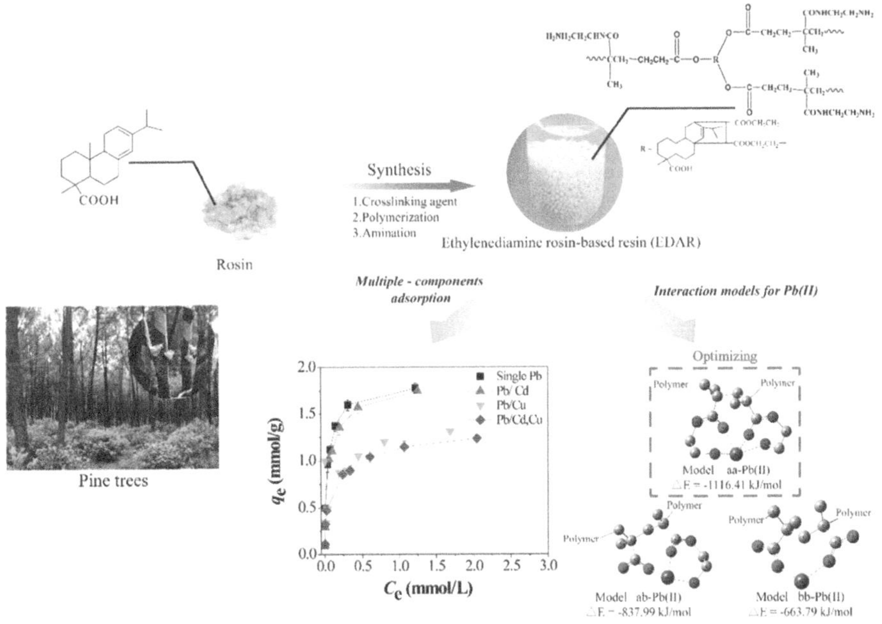

FIGURE 10.5 Amino bio-based resin derived from rosin. (Adapted with permission [28]. Copyright 2019 MDPI. Licensed under CC BY 4.0. https://creativecommons.org/licenses/by/4.0/.)

cycles (Figure 10.5) [28]. Spectroscopic analysis confirmed that adsorption involved both electrostatic attraction and coordination bonding. The pH dependence observed in these systems reflects competition between protonation of amine groups and metal complexation; optimal uptake typically occurs near neutral pH where amines are partially protonated yet still available for coordination.

Magnetic nanocomposites further enhance the practicality of heavy metal removal. Amine-functionalized magnetite nanoparticles exhibited rapid adsorption of natural organic matter, achieving substantial dissolved organic carbon and UV254 reduction within minutes, while allowing magnetic recovery of the adsorbent [29]. More recently, a magnetic nanoparticle–algal biomass hybrid achieved 96%–99% cobalt removal from battery wastewater and enabled magnetic harvesting within 60 s [30]. The synergy between biosorption and magnetic separation reduces sludge production and improves operational efficiency, illustrating how amine chemistry can be integrated into advanced recovery systems. Anion adsorption also benefits from amine functionalization. Lignin-derived adsorbents modified through Mannich-type amination exhibited nitrate adsorption capacities approaching 65.79 mg/g, with removal governed by electrostatic interaction and hydrogen bonding [31]. Protonated amines create positively charged sites that attract nitrate and other oxyanions, demonstrating that bio-based amine materials can function as effective anion exchangers. PFAS capture has become a critical research priority due to the extreme persistence of perfluorinated compounds. PFAS molecules contain hydrophobic fluorocarbon tails and negatively charged head groups (carboxylate or sulfonate). Amine-functional materials, particularly those containing quaternary ammonium groups, bind PFAS through a combination of electrostatic attraction and hydrophobic interactions. Long-chain bio-derived amines synthesized from renewable terpenes also display amphiphilic properties advantageous for interfacial PFAS capture [32]. These materials demonstrate how renewable amine chemistry can be adapted for emerging contaminants. Oil–water separation exploits the amphiphilic nature of certain bio-based amines. Renewable long-chain amines derived from β-myrcene form surfactant-like molecules capable of stabilizing or destabilizing emulsions depending on formulation and pH [32]. Porous chitosan and cellulose aerogels also exhibit high oil sorption capacity due to interconnected macropores and tunable surface wettability. By adjusting the protonation state of amine groups, materials can switch between hydrophilic and hydrophobic behavior, enabling selective oil adsorption while repelling water. This adaptability is particularly valuable in treating oily industrial effluents and marine spill scenarios. Antimicrobial water disinfection is closely linked to the cationic character of amine-containing polymers. Protonated and quaternized amines interact with negatively charged bacterial membranes, disrupting membrane integrity and leading to cell lysis. Chitosan-based nanofibers and amine-rich aerogels exhibit intrinsic antibacterial activity while simultaneously adsorbing contaminants. However, secondary and tertiary amines can participate in undesirable reactions during chlorination. Nitrosamine formation, particularly N-nitrosodimethylamine, has been linked to dimethylamine-containing precursors in wastewater treatment plants [33]. This finding underscores the importance of balancing antimicrobial effectiveness with chemical stability and minimizing formation of hazardous byproducts. Flocculation in industrial wastewater is another major application. Cationic amine polymers neutralize negatively charged colloids and promote particle aggregation through charge neutralization and bridging. Lignin-derived amine adsorbents demonstrate strong electrostatic interactions with anionic species, facilitating aggregation and removal [31]. Magnetic nanoparticle systems further enhance floc recovery and shorten settling times [29]. In amine-rich effluents from carbon capture processes, biological treatment strategies have been investigated to degrade monoethanolamine and related solvents. Adapted microbial systems achieved degradation rates up to 1.64 g/L/h and over 80% COD removal under optimized aerobic–anaerobic conditions [34]. These results demonstrate that bio-based amine materials can be integrated with biological processes to create circular treatment pathways. From a sustainability perspective, regeneration and lifecycle performance are crucial. Reviews of bio-based polymeric adsorbents emphasize that crosslinked amine networks can maintain structural stability and adsorption efficiency over repeated cycles [35]. Broader sustainability analyses highlight the importance of integrating adsorption with low-energy regeneration strategies to reduce environmental footprint [36]. Magnetic separation, pH-triggered desorption, and solvent regeneration are among the promising approaches. In summary, bio-based amine-functional materials provide a multifunctional platform for water treatment applications. Through electrostatic attraction, metal coordination, hydrophobic interaction, and antimicrobial action, these materials effectively remove dyes, heavy metals, PFAS, oils, and microbial contaminants. At the same time, the literature emphasizes the need to consider transformation products such as aromatic amines and nitrosamines, as well as regeneration and lifecycle impacts. By combining renewable feedstocks with advanced composite design and integrated treatment strategies, amine-based materials offer a pathway toward efficient and sustainable water purification technologies.

10.5 MEMBRANE AND HYDROGEL SYSTEMS

Membrane and hydrogel platforms derived from bio-amines represent an advanced class of bio-based cationic polymer systems for water purification. Unlike simple adsorbent powders, these materials are engineered into structured architectures thin films, porous membranes, electrospun mats, crosslinked hydrogels, and hybrid composites that combine adsorption, diffusion, and filtration mechanisms. The integration of amine functionalities into renewable polymer backbones such as cellulose, chitosan, lignin, and other bio-derived matrices enables the formation of cationic networks capable of selective pollutant removal. Recent reviews and experimental studies demonstrate that tailoring crosslink density, surface charge, porosity, and mechanical stability allows these systems to function across dye removal, heavy metal adsorption, oil–water separation, and emerging contaminant remediation [37,38].

Cationic surface coatings represent one of the most straightforward strategies to impart bio-amine functionality to membrane systems. In this approach, amine-rich polymers such as chitosan, PEI, or aminated lignin derivatives are deposited onto existing polymeric membranes to enhance hydrophilicity, surface charge density, and antifouling performance. The incorporation of bio-based additives into membrane matrices has been shown to reduce contact angle, improve water permeability, and increase rejection of anionic species through electrostatic repulsion [37]. Surface modification also plays a crucial role in oil–water separation. Cationic coatings increase surface hydrophilicity and reduce oil adhesion, improving separation efficiency. Reviews of biodegradable membrane systems highlight that blending or coating with bio-derived amine polymers enhances flux recovery ratios and decreases irreversible fouling [38]. Thus, cationic coatings serve as a bridge between traditional filtration and charge-assisted separation technologies.

Electrospinning provides a versatile route to fabricate nanofibrous membranes with high surface area, interconnected porosity, and tunable fiber diameter. When bio-amine polymers such as chitosan or aminated cellulose are electrospun, the resulting mats exhibit enhanced adsorption capacity due to increased exposure of functional groups. The small fiber diameter significantly reduces diffusion limitations and promotes rapid pollutant uptake. Hydrogel-based electrospun nanofibers combine structural filtration with adsorption-controlled removal. Reviews of hydrogel membranes emphasize that electrospun systems allow simultaneous size exclusion and electrostatic interaction, particularly effective for anionic dyes and heavy metals [39].

Cationic hydrogels derived from bio-amines represent three-dimensional crosslinked networks capable of high water uptake and diffusion-controlled adsorption. These systems are typically fabricated through Schiff base crosslinking, epoxide–amine reactions, ionic interactions, or interpenetrating polymer network formation [40]. Amine-functional lignin-based hydrogels demonstrate the importance of balancing charge density and mechanical integrity. A branched PEI–lignin hydrogel crosslinked via epoxide chemistry achieved adsorption capacities approaching 900 mg/g for anionic dyes, while maintaining compressive strength above 1 MPa. The high performance results from synergistic electrostatic attraction, hydrogen bonding, and π–π interactions within a stable porous network (Figure 10.6) [36].

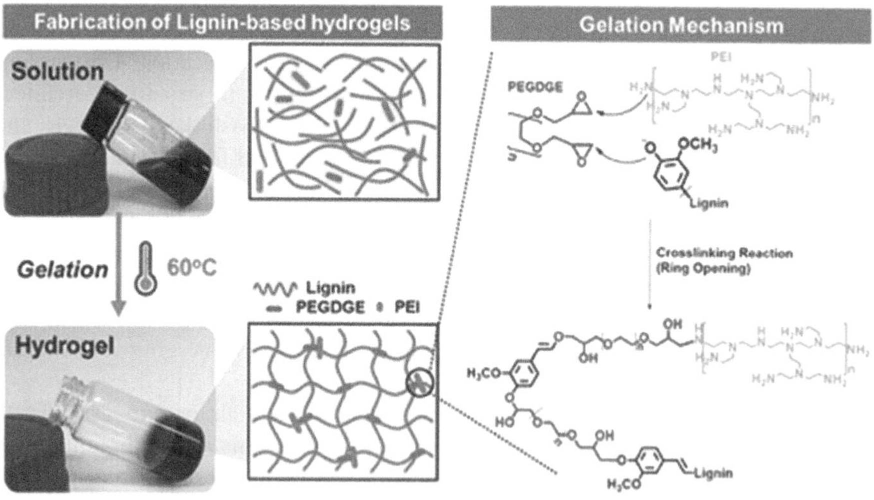

FIGURE 10.6 Synthesis of lignin-based hydrogel. (Adapted with permission [41]. Copyright 2024 Elsevier. Licensed under CC BY 4.0. https://creativecommons.org/licenses/by/4.0/.)

Cellulose-based hydrogels modified with amines or PEI exhibit heavy metal adsorption capacities exceeding 300 mg/g for Pb^{2+} and similar ions, owing to chelation and ion exchange mechanisms [42]. Swelling behavior, controlled by crosslink density, governs diffusion rates and mechanical robustness. Excessive swelling may compromise structural integrity, whereas insufficient swelling limits accessibility of active sites. Therefore, optimizing crosslink density is essential for performance. Hydrogel membranes represent an intermediate system in which a swollen polymer network is supported within a membrane configuration. These materials combine adsorption and filtration, enabling selective permeation based on both size and charge [39]. The protonation state of amine groups can be tuned by pH, providing responsiveness and selectivity in variable water conditions.

Composite systems further enhance the performance of bio-amine membranes and hydrogels. Incorporating inorganic or carbon-based fillers increases surface area, mechanical stability, and adsorption capacity. Reviews highlight the integration of functional nanoparticles, carbon materials, and porous additives within bio-based polymer matrices to produce hybrid membranes with improved rejection and flux properties [37]. In hydrogel systems, combining aminated polymers with porous structures creates interconnected pathways for rapid diffusion. Although many current studies focus on dye removal, the same design principles apply to PFAS and heavy metal remediation. Composite strategies allow tuning of hydrophilicity, charge density, and structural rigidity. Carbon-based hybrids enhance π–π interactions and structural strength, while metal organic framework–polymer hybrids introduce additional porosity and selective adsorption sites. The polymer matrix provides flexibility and processability, whereas the inorganic component enhances selectivity and adsorption kinetics. Such hybrid systems represent the next generation of cationic bio-amine membranes.

Regeneration and long-term stability are essential for practical implementation. Cationic hydrogels and membranes can often be regenerated by pH adjustment, salt treatment, or solvent washing. For example, lignin–PEI hydrogels retained over 70% removal efficiency after ten adsorption–desorption cycles, demonstrating structural resilience [41]. Similarly, cellulose-based hydrogels show stable heavy metal adsorption across multiple cycles when regenerated using mild acidic solutions [42]. Membrane systems benefit from improved fouling resistance due to hydrophilic amine groups. Reviews emphasize that bio-based additives reduce irreversible fouling and enhance flux recovery ratios [37]. Biodegradable membrane systems further offer reduced environmental impact at end-of-life [38]. However, regeneration efficiency depends on the nature of pollutant binding. Strong chelation may require harsher conditions for desorption, potentially affecting polymer stability. Balancing binding strength and recyclability remains a design challenge.

Membrane and hydrogel systems derived from bio-amines represent a structurally diverse and mechanistically versatile class of cationic materials for water purification. Cationic coatings improve antifouling and selective rejection; electrospun nanofibers enhance surface area and flux; crosslinked hydrogels enable high-capacity adsorption through swelling-controlled diffusion; and composite hybrids integrate adsorption with structural reinforcement. Regeneration studies demonstrate promising recyclability, though optimization of binding strength and structural durability

polymers across key criteria such as performance, industrial maturity, toxicity, environmental transformation, carbon footprint, energy demand, land use, recyclability, and regulatory pressure, showing that while petrochemical systems remain industrially established, bio-based alternatives offer promising sustainability and circularity advantages.

10.7 CONCLUSION

Nitrogen functionality remains one of the most powerful design elements in water purification materials. Across conventional and renewable platforms, the presence of amine groups enables electrostatic attraction, metal coordination, hydrogen bonding, ion exchange, and antimicrobial interactions. This chapter demonstrates that bio-based amines are not merely substitutes for petrochemical amines but versatile molecular building blocks capable of engineering structurally diverse and multifunctional cationic polymers. Lignin-derived aromatic amines introduce rigidity and π-interactions that enhance adsorption of aromatic pollutants, while amino acid- and fermentation-derived diamines provide scalable routes to linear and branched aliphatic amines suitable for polymer backbones. Polysaccharide-based systems such as chitosan and aminated cellulose offer inherent biodegradability and abundant reactive sites. Vegetable oil-derived long-chain amines contribute amphiphilicity and flexibility, enabling oil–water separation and interfacial adsorption. Through reductive amination, enzymatic transamination, Mannich reactions, epoxide–amine polymerization, and quaternization, renewable feedstocks can be transformed into charge-dense polymer architectures with controlled crosslink density and mechanical stability. Applications discussed throughout this chapter illustrate the broad utility of bio-based cationic systems. High-capacity dye adsorption, selective heavy metal chelation, nitrate and PFAS binding, antimicrobial activity, and flocculation performance demonstrate that renewable amine platforms can approach or match conventional materials under optimized conditions. Membrane coatings, electrospun nanofibers, crosslinked hydrogels, and composite hybrids further extend functionality by integrating adsorption with filtration, diffusion control, and mechanical reinforcement. Regeneration studies indicate that many bio-based systems retain performance across multiple cycles, particularly when designed with reversible interactions or magnetic recovery pathways. The comparative evaluation with petrochemical polymers such as PEI, polyDADMAC, and CPAM reveals a nuanced landscape. Petrochemical systems maintain advantages in industrial maturity, cost predictability, and large-scale infrastructure. However, concerns regarding residual monomers, environmental persistence, oxidative degradation products, and carbon footprint continue to drive interest in renewable alternatives. LCA frameworks show that bio-based polymers can reduce fossil carbon dependency and greenhouse gas emissions, although agricultural inputs, land use, and downstream processing energy must be carefully managed to ensure genuine environmental benefit. Future progress in bio-based cationic polymers will depend on several factors. First, structure–property optimization must continue to enhance charge density without compromising mechanical stability or biodegradability. Second, techno-economic analysis and pilot-scale validation are required to bridge laboratory demonstrations and industrial implementation. Third, comprehensive environmental impact assessments should accompany material development to prevent unintended by-product formation, such as aromatic amines or nitrosamines. Finally, integration with circular design principles, including regeneration, recyclability, and low-energy processing, will be essential for long-term sustainability. Overall, bio-amine-derived cationic polymers represent a scientifically mature and strategically important platform for next-generation water purification. By combining renewable feedstocks, controlled amine chemistry, advanced material architectures, and rigorous sustainability assessment, these systems offer a balanced approach that aligns performance with environmental responsibility.

REFERENCES

1. B.K. Amos, E.J. Suchomel, K.D. Pennell, F.E. Löffler, Microbial activity and distribution during enhanced contaminant dissolution from a NAPL source zone, *Water Res.* 42 (2008) 2963–2974.
2. M. Blanco, L. Jiménez, I. Valverde, Degradation of a stilbene-type fluorescent whitening agent with hypochlorite: Identification of the degradation products by capillary electrophoresis, *Environ. Toxicol. Chem.* 20 (2001) 2193–2197.
3. R. Long, J. Liu, C. Lu, J. Shi, J. Zhang, Coordinated optimal operation method of the regional energy internet, *Sustainability.* 9 (2017) 848.
4. S.K. Samal, M. Dash, S. Van Vlierberghe, D.L. Kaplan, E. Chiellini, C. van Blitterswijk, L. Moroni, P. Dubruel, Cationic polymers and their therapeutic potential, *Chem. Soc. Rev.* 41 (2012) 7147–7194.
5. R.S. Downing, P.J. Kunkeler, H. van Bekkum, Catalytic syntheses of aromatic amines, *Catal. Today.* 37 (1997) 121–136.
6. A.Y. Ardashnikov, I.Y. Kardash, A.N. Pravednikov, The nature of the equilibrium in the reaction of aromatic anhydrides with aromatic amines and its role in synthesis of polyimides, *Polym. Sci. U.S.S.R.* 13 (1971) 2092–2100.
7. J.G. Westra, A rapid and simple synthesis of reactive metabolites of carcinogenic aromatic amines in high yield, *Carcinogenesis.* 2 (1981) 355–357.
8. S. Xie, C. Jia, Z. Wang, S.S.G. Ong, M. Zhu, H. Lin, Mechanistic insight into selective deoxygenation of l-lysine to produce biobased amines, *ACS Sustain. Chem. Eng.* 8 (2020) 11805–11817.
9. A.T. Espinoza Pérez, M. Camargo, P.C. Narváez Rincón, M. Alfaro Marchant, Key challenges and requirements for sustainable and industrialized biorefinery supply chain design and management: A bibliographic analysis, *Renew. Sustain. Energy Rev.* 69 (2017) 350–359.
10. T.M. Mata, A.A. Martins, S.K. Sikdar, C.A.V. Costa, Sustainability considerations of biodiesel based on supply chain analysis, *Clean Technol. Environ. Policy.* 13 (2011) 655–671.
11. M. Pelckmans, T. Renders, S. Van de Vyver, B.F. Sels, Bio-based amines through sustainable heterogeneous catalysis, *Green Chem.* 19 (2017) 5303–5331.
12. E. Blondiaux, J. Bomon, M. Smoleń, N. Kaval, F. Lemière, S. Sergeyev, L. Diels, B. Sels, B.U.W. Maes, Bio-based aromatic amines from lignin-derived monomers, *ACS Sustain. Chem. Eng.* 7 (2019) 6906–6916.
13. P. Verdugo, N. Montesó, D. Guzman, D. Santiago, S. De la Flor, À. Serra, Fully bio-based epoxy–imine monomer: Synthesis and curing with amines toward vitrimeric materials, *ACS Sustain. Chem. Eng.* 14 (2026) 3222–3234.
14. J. Di, Y. Hu, Y. He, C. Ma, Biomanufacturing biobased organic amines from biomass-derived aldehydes through the amination with transaminase 3FCR-3M, *J. Agric. Food Chem.* 74 (2026) 5431–5441.

15. V. Froidevaux, C. Negrell, S. Caillol, J.P. Pascault, B. Boutevin, Biobased amines: From synthesis to polymers; present and future, *Chem. Rev.* 116 (2016) 14181–14224.
16. D.M. Correia, L.C. Fernandes, B.D.D. Cruz, G. Botelho, V. de Zea Bermudez, S. Lanceros-Méndez, Development of poly(l-lactic acid)-based bending actuators, *Polymers (Basel)*. 12 (2020) 1187.
17. J. Verduyckt, M. Van Hoof, F. De Schouwer, M. Wolberg, M. Kurttepeli, P. Eloy, E.M. Gaigneaux, S. Bals, C.E.A. Kirschhock, D.E. De Vos, PdPb-catalyzed decarboxylation of proline to pyrrolidine: Highly selective formation of a biobased amine in water, *ACS Catal.* 6 (2016) 7303–7310.
18. E. Scott, F. Peter, J. Sanders, Biomass in the manufacture of industrial products—the use of proteins and amino acids, *Appl. Microbiol. Biotechnol.* 75 (2007) 751–762.
19. A.-S. Mora, R. Tayouo, B. Boutevin, G. David, S. Caillol, Synthesis of pluri-functional amine hardeners from bio-based aromatic aldehydes for epoxy amine thermosets, *Molecules.* 24 (2019) 3285.
20. B. Vermeeren, S. Van Praet, W. Arts, T. Narmon, Y. Zhang, C. Zhou, H.P. Steenackers, B.F. Sels, From sugars to aliphatic amines: as sweet as it sounds? Production and applications of bio-based aliphatic amines, *Chem. Soc. Rev.* 53 (2024) 11804–11849.
21. S. Xie, C. Jia, S.G. Ong, Q. Wang, Y. Yang, H. Lin, S. Xie, C. Jia, S. Sergio, G. Ong, Z. Wang, M. Zhu, Q. Wang, Y. Yang, A shortcut route to close nitrogen cycle : Bio-based amines production via selective deoxygenation of chitin monomers over Ru/C in acidic solutions, *iScience.* 23 (2020) 101096.
22. H. Blattmann, R. Mülhaupt, Multifunctional β-amino alcohols as bio-based amine curing agents for the isocyanate- and phosgene-free synthesis of 100% bio-based polyhydroxyurethane thermosets, *Green Chem.* 18 (2016) 2406–2415.
23. A.-S. Mora, R. Tayouo, B. Boutevin, G. David, S. Caillol, Synthesis of biobased reactive hydroxyl amines by amination reaction of cardanol-based epoxy monomers, *Eur. Polym. J.* 118 (2019) 429–436.
24. Q. Huang, Y. Shen, Y. Wang, J. Xiao, H. Yuan, Z. Lou, N. Zhu, Synergy between denitrification and calcium bridging improves dewaterability of waste activated sludge, *J. Clean. Prod.* 242 (2020) 118438.
25. F. Saunier, S. Fradette, F. Clerveaux, S. Lefebvre, É. Madore, G. Veilleux, C. Bulle, R. Surprenant, Comparison of life-cycle assessment between bio-catalyzed and promoted potassium carbonate processes and amine-based carbon capture technologies, *Int. J. Greenh. Gas Control.* 88 (2019) 134–155.
26. H.M. Pinheiro, E. Touraud, O. Thomas, Aromatic amines from azo dye reduction: status review with emphasis on direct UV spectrophotometric detection in textile industry wastewaters, *Dye. Pigment.* 61 (2004) 121–139.
27. A.M.T. Mata, A. Ligneul, N.D. Lourenço, H.M. Pinheiro, Advanced oxidation for aromatic amine mineralization after aerobic granular sludge treatment of an azo dye containing wastewater, *Desalin. Water Treat.* 91 (2017) 168–174.
28. W. Huang, K. Diao, X. Tan, F. Lei, J. Jiang, B.A. Goodman, Y. Ma, S. Liu, Mechanisms of adsorption of heavy metal cations from waters by an amino bio-based resin derived from rosin, *Polymers (Basel).* 11 (2019) 969.
29. C.C.P. Chan, H. Gallard, P. Majewski, Fabrication of amine-functionalized magnetite nanoparticles for water treatment processes, in: M.S. Diallo, N.A. Fromer, M.S. Jhon (Eds.), *Nanotechnology for Sustainable Development*, Springer International Publishing, Cham, 2014: pp. 137–147.
30. T. Fóris, P. Koska, Á.M. Ilosvai, F. Kristály, L. Daróczi, L. Vanyorek, B. Viskolcz, An innovative bioremediation approach to heavy metal removal: Combined application of Chlorella vulgaris and amine-functionalized MgFe2O4 nanoparticles in industrial wastewater treatment, *Int. J. Mol. Sci.* 26 (2025) 5467.

31. T. Li, L. Liu, M. Li, Y. Li, Enhanced nitrate removal from aqueous solutions using amine-functionalized biowaste-derived adsorbent, *Sci. Rep.* 15 (2025) 36534.
32. A. Kampwerth, M. Terhorst, N. Kampling, D. Vogt, T. Seidensticker, Synthesis of bio-based amines via Pd-catalysed telomerisation of the renewable β-myrcene in a water/ethanol multiphase system: catalyst recycling enabled by a self-separating product phase, *Green Chem.* 25 (2023) 6345–6354.
33. W.A. Mitch, D.L. Sedlak, Characterization and fate of N-nitrosodimethylamine precursors in municipal wastewater treatment plants, *Environ. Sci. Technol.* 38 (2004) 1445–1454.
34. D. Botheju, Y. Li, J. Hovland, H.A. Haugen, R. Bakke, Biological treatment of amine wastes generated in post combustion CO2 capture, *Energy Procedia.* 4 (2011) 496–503.
35. N. Bouazizi, J. Vieillard, B. Samir, F. Le Derf, Advances in amine-surface functionalization of inorganic adsorbents for water treatment and antimicrobial activities: A review, *Polymers (Basel).* 14 (2022) 378.
36. V.M. Fulfer, C. Lo, S. Wang, M.M. Arienzo, D. Lin, L. Erdle, Airborne microfiber capture: Secondary filtration a solution to filter microfiber emissions from clothing dryers, *Environ. Toxicol. Chem.* (2025) vgaf260.
37. D. Maureira, S.R. Rodriguez, O. Romero, M. Guillén, G. Álvaro, L. Wilson, C. Ottone, Immobilization of FDH on carbon felt by affinity binding strategy for CO2 conversion, *Results Eng.* 25 (2025) 104442.
38. R. Sen Gupta, P.K. Samantaray, S. Bose, Going beyond cellulose and chitosan: Synthetic biodegradable membranes for drinking water, wastewater, and oil–water remediation, *ACS Omega.* 8 (2023) 24695–24717.
39. L. Alsaka, L. Alsaka, A. Altaee, S.J. Zaidi, J. Zhou, T. Kazwini, A review of hydrogel application in wastewater purification, *Separations.* 12 (2025) 51.
40. G. Rando, E. Scalone, S. Sfameni, M.R. Plutino, Functional bio-based polymeric hydrogels for astewater treatment: From remediation to sensing applications, *Gels.* 10 (2024) 498.
41. J.W. Heo, M.S. Kim, D.H. Oh, Y.S. Kim, Innovative construction of amine-functionalized lignin-based hydrogel for ultrafast and selective dye remediation, *Chem. Eng. J.* 498 (2024) 155506.
42. F. Persano, C. Malitesta, E. Mazzotta, Cellulose-based hydrogels for wastewater treatment: A focus on metal ions removal, *Polymers (Basel).* 16 (2024) 1292.
43. M.F. Pitzalis, J.C. Sadler, Chemical bio-manufacture from diverse C-rich waste polymeric feedstocks using engineered microorganisms, *RSC Sustain.* 3 (2025) 1672–1684.
44. B. Zhang, L. Yao, X. Liu, L. Zhang, Z. Cheng, X. Zhu, Facilely recyclable Cu(II) macrocomplex with thermoregulated poly(ionic liquid) macroligand: Serving as a highly efficient atom transfer radical polymerization catalyst, *ACS Sustain. Chem. Eng.* 4 (2016) 7066–7073.

modern catalytic polymerization techniques allow the preparation of structurally well-defined conjugated polymers with tailored electronic and optical properties. These versatile synthesis strategies provide researchers with significant flexibility in designing conducting polymer materials for specific applications. Due to their unique combination of electrical conductivity, mechanical flexibility, and chemical tunability, conducting polymers have been explored in a wide range of technological applications. These include polymer light-emitting diodes, organic photovoltaic devices, field-effect transistors, electromagnetic interference shielding materials, electrochromic displays, and energy storage systems such as rechargeable batteries and supercapacitors. Conducting polymers are also widely used in sensors, smart windows, glare-reduction systems, and printed electronic circuits, highlighting their versatility as functional materials in modern electronic technologies. In recent years, increasing attention has been directed toward the development of sustainable and bio-derived components for conducting polymer systems. Traditional conducting polymers are often synthesized using petrochemical precursors and synthetic dopants, which may raise environmental and sustainability concerns. Consequently, researchers are exploring renewable alternatives that can provide similar functionality while reducing environmental impact. In this context, bio-based amines have emerged as promising functional molecules for designing sustainable conducting polymer materials. These compounds can be derived from renewable resources such as lignocellulosic biomass, amino acids, natural polyamines, and other biological feedstocks. Bio-based amines contain nitrogen-based functional groups capable of participating in proton transfer, hydrogen bonding, and coordination interactions with polymer chains or analytes. These properties enable them to influence the structural organization, electronic properties, and interfacial behavior of conducting polymers. Depending on their molecular architecture, bio-derived amines can function as monomers, dopants, crosslinkers, or surface modifiers within conducting polymer systems. Such multifunctional roles make bio-based amines valuable components in the design of environmentally friendly electronic materials with improved sensing performance and electrochemical functionality. Furthermore, the incorporation of bio-based amines into conducting polymer systems has opened new opportunities in sensor technologies and bioelectronic applications. Amine functional groups provide active chemical sites capable of interacting with various analytes through acid–base interactions, hydrogen bonding, and charge-transfer processes. These interactions enable the development of highly sensitive sensors for detecting gases, biomolecules, and environmental pollutants. As a result, conducting polymer materials incorporating bio-derived amines are increasingly being explored for applications in food safety monitoring, environmental sensing, healthcare diagnostics, and wearable electronic devices. This chapter focuses on the emerging role of bio-based amines in conducting polymer systems and their applications in sensing and electronic technologies. It discusses the sources, structures, and synthesis of bio-based amines, their roles in conducting polymer materials, major bio-amine platforms such as chitosan-based systems, and their applications in sensors and electronic devices. In addition, this chapter highlights current challenges, research gaps, and future perspectives in the development of sustainable conducting polymer materials based on bio-derived amines.

11.2 SOURCES, STRUCTURES, AND SYNTHESIS OF BIO-BASED AMINES

Bio-based amines represent an important class of nitrogen-containing organic compounds derived from renewable resources. In contrast to conventional petroleum-derived amines, these compounds originate from biomass feedstocks such as carbohydrates, amino acids, lignocellulosic materials, vegetable oils, and naturally occurring polyamines. The increasing demand for sustainable chemicals and environmentally benign polymer precursors has significantly accelerated research on the production and utilization of bio-based amines. Because amines are widely used as intermediates in pharmaceuticals, agrochemicals, coatings, polymers, and specialty materials, the development of renewable synthetic routes is essential for reducing dependence on fossil-based resources and promoting greener chemical processes. Several natural sources serve as precursors to produce bio-based amines. Microbial metabolism and biological pathways play a crucial role in generating amine compounds from amino acids and other nitrogen-containing substrates. For example, microbial decarboxylation of amino acids can produce various biogenic amines that occur naturally in biological systems. Advances in metabolic engineering and biotechnology have enabled the microbial production of valuable amines and N-functionalized amino acid derivatives using engineered microorganisms. These systems allow the conversion of renewable carbon sources such as glucose and other sugars into amine-containing compounds through enzyme-catalyzed biochemical pathways, offering highly selective and environmentally friendly production strategies [8–10]. Another important route for producing bio-based amines involves the use of amino acid-derived intermediates and microbial fermentation processes. Microbial cell factories have been developed to produce amine molecules through engineered biosynthetic pathways, enabling the transformation of renewable substrates into industrially valuable amines. These systems can produce compounds such as diamines, polyamines, and other nitrogen-containing molecules that serve as key building blocks in polymer chemistry and pharmaceutical synthesis. The development of microbial platforms for amine production represents an important step toward sustainable nitrogen chemistry and renewable chemical manufacturing [10]. Naturally occurring polyamines represent another significant class of bio-based amines. These molecules, including compounds such as putrescine, spermidine, and spermine, are widely distributed in living organisms and play essential roles in biological functions such as cell growth, gene regulation, and metabolic processes. Structurally, polyamines contain multiple amine groups attached to aliphatic carbon chains, allowing them to participate in strong electrostatic interactions with negatively charged biomolecules such as deoxyrebonucleic acid acid (DNA) and proteins. Because of their multifunctional amine groups, polyamines are also valuable precursors for the synthesis of advanced polymer materials and functional chemical intermediates [11,12]. In addition to biological production routes, thermochemical catalytic methods have been widely explored for synthesizing bio-based amines from biomass-derived oxygenates. Biomass components such as lignocellulosic sugars, alcohols, aldehydes, and ketones can be converted into amines through catalytic amination reactions using ammonia or other nitrogen donors. These reactions often proceed through reductive amination mechanisms in which carbonyl compounds react with ammonia in the presence of hydrogen and metal catalysts to form primary amines. Various heterogeneous catalysts based on transition metals such as ruthenium, nickel, cobalt, or copper have been used to facilitate these transformations efficiently. Such catalytic processes enable the conversion of renewable platform molecules into valuable amine intermediates used in the production of polymers, pharmaceuticals, and specialty chemicals [13]. Structurally, bio-based amines encompass a wide range of molecular architectures depending on their origin and synthetic route. These compounds may include primary, secondary, or tertiary amines, as well as multifunctional amines containing multiple nitrogen groups. The presence of amino functional groups ($-NH_2$, $-NHR$, or $-NR_2$) imparts strong nucleophilicity and chemical reactivity, enabling these molecules to participate in various chemical transformations such as polymerization, crosslinking, and curing reactions. In polymer chemistry, bio-based diamines and polyamines are particularly valuable because they can serve as monomers or curing agents in the synthesis of polyamides, polyurethanes, epoxy resins, and other high-performance materials [14]. Overall, the development of bio-based amines integrates biological, catalytic, and chemical approaches for converting renewable resources into valuable nitrogen-containing molecules. Advances in metabolic engineering, catalytic chemistry, and biomass conversion technologies continue to expand the range of available bio-derived amines and improve production efficiency. These developments are expected to play a significant role in enabling sustainable chemical manufacturing and supporting the transition toward a bio-based nitrogen economy [10].

11.3 CONDUCTING POLYMERS

Conducting polymers represent a unique class of materials that combine the structural and mechanical advantages of polymers with the electronic properties typically associated with semiconductors and metals. Traditionally, polymers were considered electrically insulating materials due to the absence of free charge carriers and the presence of localized σ-bonds in their molecular structures. However, the discovery of electrically conductive organic polymers fundamentally changed this perception and opened a new area of materials science focused on organic electronic materials. Conducting polymers exhibit a diverse range of functional properties, including electrical conductivity, optical activity, mechanical flexibility, and surface wettability, which make them suitable for various electronic and sensing applications. These materials also demonstrate unique magnetic behavior and microwave absorption capabilities, further expanding their technological potential (Figure 11.2).

The development of conducting polymers dates back to the late 1970s when Hideki Shirakawa, Alan MacDiarmid, and Alan Heeger discovered that polyacetylene could exhibit dramatically enhanced electrical conductivity when exposed to oxidizing agents such as halogen vapors [15]. Their pioneering work demonstrated that the electrical conductivity of conjugated polymers could increase by several orders of magnitude through a process known as doping, transforming these materials from insulating polymers into semiconducting or even metallic conductors. This groundbreaking discovery significantly expanded the understanding of electronic transport

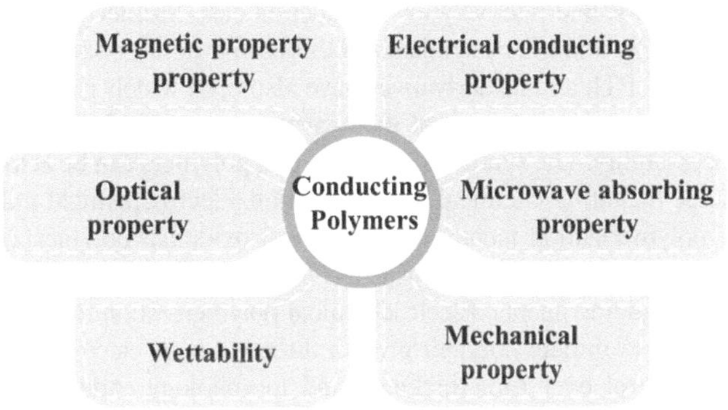

FIGURE 11.2 Properties of conducting polymers.

in organic materials and led to the awarding of the Nobel Prize in Chemistry in 2000 to the three scientists for their contributions to the discovery and development of conducting polymers. Conducting polymers evolved from early inorganic conductive materials such as (SN)x , which demonstrated that conjugated structures can support electronic conduction. The discovery of polyacetylene and subsequent development of polymers such as PANI and PPy expanded the field significantly, enabling tunable electronic properties through doping and various synthetic strategies [16].

The key structural feature responsible for electrical conductivity in these polymers is the presence of a conjugated backbone, consisting of alternating single and double bonds along the polymer chain. This arrangement allows the overlap of p-orbitals along the polymer backbone, enabling the delocalization of π-electrons across the molecular structure. In such systems, electrons are not confined to individual bonds but can move more freely along the chain, creating pathways for electrical conduction. Despite this conjugation, many conducting polymers in their pristine state exhibit relatively low conductivity because the number of charge carriers is limited. As a result, external doping is often required to significantly enhance electrical transport properties. Doping in conducting polymers involves the oxidation or reduction of the polymer backbone, which introduces charge carriers into the system. These charge carriers may exist as polarons or bipolarons that can migrate along the conjugated chain, thereby facilitating charge transport. Chemical doping typically involves the use of electron-accepting or electron-donating species, whereas electrochemical doping involves applying an external potential to control the oxidation state of the polymer. Through this mechanism, the electrical conductivity of polymers can increase from values typical of insulators to levels approaching those of metals. In addition to modifying conductivity, doping can also influence optical properties, mechanical characteristics, and electrochemical behavior of the material.

Among the various conducting polymers that have been developed, several materials have received significant attention due to their stability, conductivity, and ease of synthesis. The most widely studied conducting polymers include PANI, PPy, PTh, and PEDOT. These materials possess conjugated structures that support efficient charge transport and can be synthesized using relatively simple chemical or electrochemical techniques. PANI, for example, is particularly attractive due to its environmental stability, tunable electrical conductivity, and relatively low production cost. Similarly, PPy exhibits high electrical conductivity and can be synthesized through straightforward oxidative polymerization processes. PTh and its derivatives have also been widely explored for applications in organic electronics because of their favorable semiconducting properties and improved processability. The synthesis of conducting polymers can be achieved through several methods, including chemical polymerization, electrochemical polymerization, and catalytic polymerization techniques. Chemical oxidative polymerization is commonly used to produce materials such as PANI and PPy by initiating polymerization in the presence of oxidizing agents. Electrochemical polymerization involves applying an electrical potential to initiate polymer growth directly on an electrode surface, which allows precise control over film thickness and morphology. Additionally, transition metal-catalyzed coupling reactions such as Suzuki or Stille coupling are used to synthesize more structurally defined conjugated polymers with tailored electronic properties. These different synthetic approaches provide researchers with significant flexibility in designing conducting polymers with desired structural and functional characteristics. Another important feature of conducting polymers is that their electronic properties are strongly influenced by molecular organization and intermolecular interactions. Unlike inorganic semiconductors, which possess highly ordered crystalline lattices, conducting polymers consist of long molecular chains that interact through relatively weak forces such as van der Waals interactions and π–π stacking. The arrangement of these chains within the material can significantly influence charge transport, since electron mobility often depends on the degree of conjugation and chain packing. As a result, factors such as polymer morphology, crystallinity, and chain alignment play a critical role in determining the electrical performance of these materials [17–19].

Advances in molecular design have also enabled the development of conducting polymers with improved processability and environmental stability. The introduction of side chains, heteroatoms, and functional groups into the polymer backbone allows tuning of solubility, mechanical flexibility, and electronic characteristics. These modifications have enabled conducting polymers to be processed using solution-based methods such as spin coating, printing, and inkjet deposition. Such processing capabilities make conducting polymers particularly attractive for large-area and flexible electronic devices. Conducting polymers have found applications in a wide range of technologies due to their unique combination of electronic conductivity, lightweight structure, and mechanical flexibility. Early applications focused on antistatic coatings, electromagnetic shielding materials, and corrosion protection systems. However, the field has rapidly expanded to include advanced technologies such as organic light-emitting diodes, organic photovoltaic cells, organic field-effect transistors (OFETs), and energy storage devices. In addition, conducting polymers have been increasingly explored in the development of electrochemical sensors, biosensors, and flexible electronic devices. A particularly important advantage of conducting polymers is their ability to support mixed ionic and electronic conductivity, which makes them highly suitable for electrochemical applications. In many devices, the transport of ions and electrons occurs simultaneously within the polymer matrix, enabling efficient charge transfer processes. This property is especially useful in sensing platforms and bioelectronic interfaces, where interactions between the polymer and chemical or biological species are required to generate measurable electrical signals. The versatility of conducting polymers has also led to their integration with other functional materials, including nanoparticles, carbon nanomaterials, and biomolecules. These hybrid systems can exhibit enhanced conductivity, improved stability, and increased sensitivity in sensing applications. As research continues to advance, conducting polymers are increasingly being considered key components in emerging technologies such as wearable electronics, smart sensors, and environmentally sustainable electronic devices [16,20].

Overall, conducting polymers have evolved from a scientific curiosity into an important class of functional materials with broad technological relevance. Their tunable electronic properties, structural versatility, and compatibility with solution processing make them highly attractive for applications in electronics, sensing, and energy systems. Continued developments in polymer chemistry and materials engineering are expected to further expand the capabilities of conducting polymers, enabling new opportunities for the design of advanced electronic materials.

11.4 ROLES OF BIO-BASED AMINES IN CONDUCTING POLYMER SYSTEMS

Bio-based amines have recently attracted increasing attention as sustainable functional components in conducting polymer systems. These nitrogen-containing molecules derived from renewable resources such as lignocellulosic biomass, amino acids, chitosan, dopamine, and other bio-derived feedstocks possess versatile chemical functionalities that enable multiple roles in polymer chemistry and electronic materials. The presence of primary, secondary, or tertiary amine groups allows strong intermolecular interactions, proton transfer capability, hydrogen bonding, and coordination with polymer chains or analytes. These features make bio-based amines particularly valuable in the design of next-generation organic electronic materials, biosensors, and sustainable conducting polymer systems. In conducting polymers, the incorporation of bio-based amines can significantly influence polymer structure, electrical conductivity, charge transport, interfacial properties, and sensing performance. Depending on their molecular architecture and chemical functionality, bio-based amines may function as monomers forming the polymer backbone, dopants facilitating charge transport, crosslinkers stabilizing polymer networks, or surface modifiers providing selective analyte recognition sites. Such multifunctional roles enable the development of environmentally friendly materials with improved performance for electronic and sensing applications. Moreover, replacing petroleum-derived amines with bio-based alternatives contributes to the development of sustainable organic electronics [21].

11.4.1 As Monomers or Polymer Backbones

One of the most fundamental roles of bio-based amines in conducting polymer systems is their function as monomeric building blocks or polymer backbone components. Bio-derived amines containing conjugated aromatic structures can participate in polymerization reactions that generate electrically active conjugated polymers. Among these materials, polyazomethines and Schiff-base polymers have emerged as promising examples of conducting or semiconducting polymers synthesized from amine-containing monomers. Polyazomethines are typically synthesized through the condensation reaction between diamines and dialdehydes, producing imine (–C=N–) linkages along the polymer backbone. This reaction proceeds without the need for catalysts and generates water as the only by-product, making it an environmentally friendly polymerization strategy. Such condensation-based synthesis enables the preparation of conjugated polymers with tunable optoelectronic properties by modifying the molecular structure of the amine and aldehyde precursors. These polymers are particularly attractive for organic electronic applications because the imine linkage contributes to π-conjugation along the polymer chain, facilitating charge transport and optical activity (Figure 11.3) [22].

Bio-derived aromatic amines obtained from lignin or other biomass sources can also serve as precursors for conjugated polymer backbones. Biomass provides a diverse set of molecular building blocks that can be transformed into functional monomers for electronic materials. Compared with petroleum-based monomers, these bio-derived molecules offer advantages such as reduced environmental impact, structural diversity, and potential biodegradability. The incorporation of such monomers into π-conjugated polymer systems enables the development of sustainable organic electronic materials without compromising device performance. Recent studies have demonstrated that biomass-derived poly(azomethine) polymers can be used in electronic devices such as photonic transistor memories and optoelectronic systems. These materials exhibit reversible optical switching, fluorescence, and charge transport behavior due to the conjugated imine structures within the polymer backbone. Furthermore, the presence of nitrogen atoms in the imine linkages enhances electronic interactions within the polymer chain and facilitates charge transport processes. In addition to polyazomethines, other bio-based amine-containing polymers such as poly(imine-amide) networks have been developed as conductive materials.

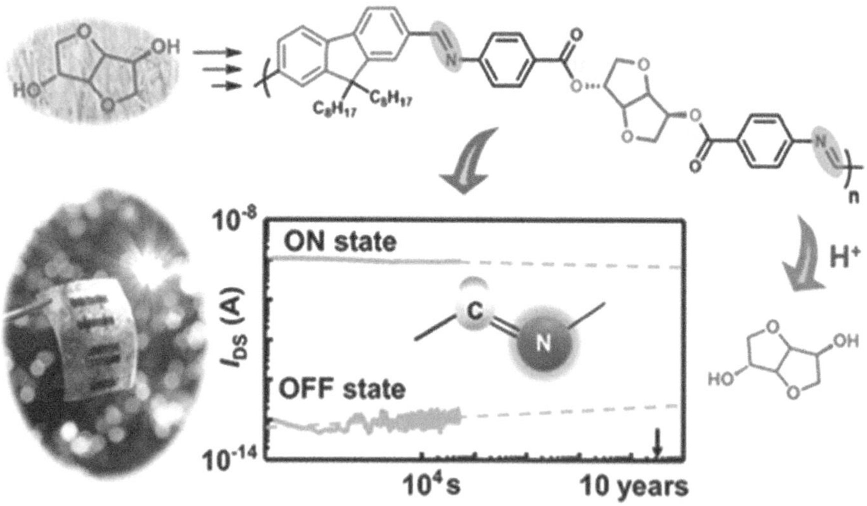

FIGURE 11.3 Flexible bistable photonic memory from degradable biomass-derived poly(azomethine)s. (Adapted with permission [22]. Copyright 2022 American Chemical Society.)

These polymers incorporate dynamic covalent imine bonds that enable structural adaptability and self-healing properties. Such dynamic networks are particularly valuable for flexible electronics and wearable sensors, where mechanical durability and long-term stability are required. Overall, the use of bio-based amines as monomers provides an effective pathway toward the development of sustainable conjugated polymers with tunable electrical and optical properties. By carefully selecting bio-derived amine structures, researchers can design polymer backbones that combine environmental sustainability with high electronic performance [23,24].

11.4.2 As Dopants or Proton/Ion Transport Mediators

In addition to serving as monomeric components, bio-based amines can also function as dopants or charge transport mediators in conducting polymer systems. Doping is a critical process that enhances the electrical conductivity of conjugated polymers by introducing charge carriers such as polarons and bipolarons into the polymer backbone. Traditional doping agents often involve strong acids or inorganic species, which may pose environmental or stability concerns. Bio-based amines offer a more sustainable alternative for tuning the electronic properties of conducting polymers. The amine functional group possesses a lone pair of electrons that can participate in protonation or electron-donation processes, enabling effective interaction with conjugated polymer chains. When incorporated into conducting polymer matrices, protonated amines can facilitate charge transfer by stabilizing charge carriers along the polymer backbone. This interaction improves the electrical conductivity and electrochemical performance of the resulting materials. For example, amine-containing biomolecules such as chitosan and dopamine derivatives have been widely explored in conducting polymer systems. These materials contain multiple amine groups capable of forming hydrogen bonds and electrostatic interactions with polymer chains. Such interactions promote the formation of proton conduction pathways, which are beneficial for electrochemical sensors, fuel cell membranes, and flexible electronic devices [25].

Another important advantage of bio-based amine dopants is their ability to enhance ionic conductivity and proton transport. The presence of protonated amine groups creates ion-transport channels within the polymer matrix, facilitating the movement of protons or other ionic species. This property is particularly valuable in electrochemical devices where efficient ion transport is essential for device operation. Furthermore, bio-based amines can influence the morphology and microstructure of conducting polymer films. Through hydrogen bonding and electrostatic interactions, these molecules can improve polymer chain ordering and promote the formation of conductive pathways. As a result, the electrical and electrochemical performance of the polymer system can be significantly improved. Therefore, the use of bio-based amines as dopants or proton transport mediators provides a sustainable approach for enhancing the conductivity and functionality of conducting polymers while maintaining environmental compatibility.

11.4.3 As Crosslinkers, Binders, and Interface Builders

Bio-based amines also play a crucial role as crosslinkers and structural stabilizers in conducting polymer systems. Crosslinking is often required to improve the mechanical strength, thermal stability, and durability of polymer materials used in electronic devices. Amine groups can react with aldehydes, epoxides, or other functional groups to form covalent crosslinked networks that enhance the structural integrity of conducting polymer composites. One promising approach involves the formation of dynamic imine networks, where amine groups react with aldehydes to form reversible imine bonds. These dynamic covalent interactions allow the polymer network to rearrange under mechanical stress, providing self-healing capabilities and improved mechanical resilience. Such materials are particularly attractive for flexible electronics and wearable devices, where mechanical deformation is unavoidable. Bio-based poly(imine-amide) materials have demonstrated the ability to form adaptable networks capable of maintaining electrical conductivity while providing mechanical robustness. In conductive composites, these materials can act as binders that hold conductive fillers such as carbon nanotubes, graphene, or conductive polymers together within a stable matrix. This improves electrical connectivity and prevents phase separation in composite materials (Figure 11.4) [24].

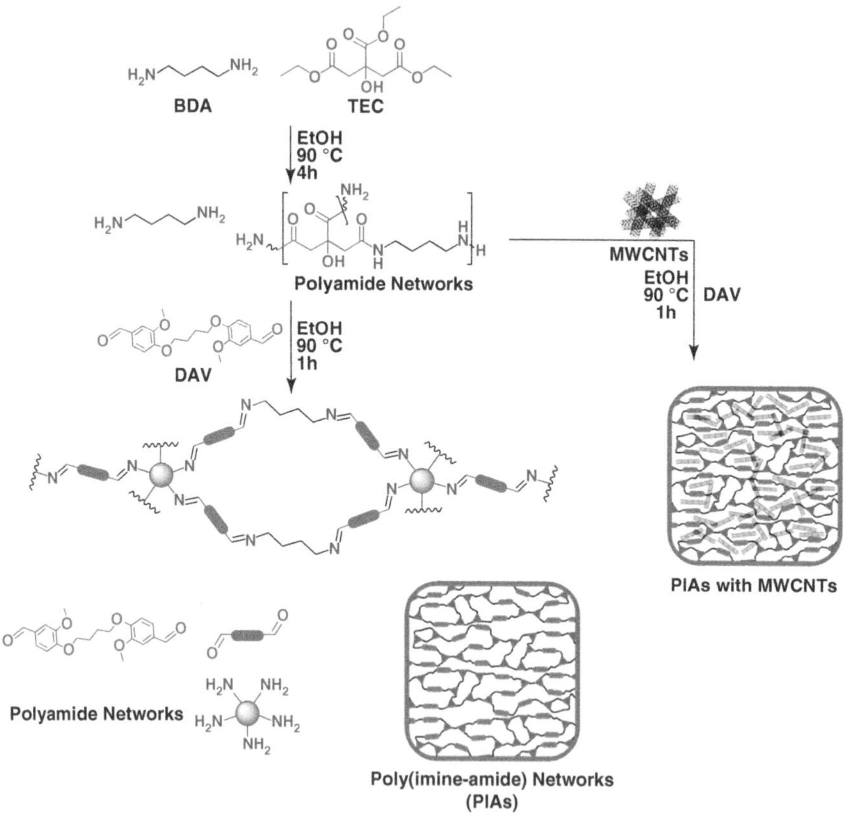

FIGURE 11.4 Synthesis of poly(imine-amide) networks with/without multiwalled carbon nanotubes (MWCNTs). (Adapted with permission [24]. Copyright 2022 American Chemical Society.)

In addition, amine-containing biomolecules can serve as interfacial coupling agents that enhance adhesion between different components of electronic devices. For example, amine groups can interact with metal surfaces, oxide layers, or polymer substrates through coordination bonds or hydrogen bonding. These interactions improve interfacial stability and facilitate efficient charge transfer between layers in electronic devices. Another important advantage of amine-based crosslinking systems is their environmental compatibility. Many traditional crosslinking agents involve toxic chemicals or harsh reaction conditions, whereas bio-derived amines can form crosslinked networks under mild conditions. This characteristic aligns well with the principles of green chemistry and sustainable materials design. Consequently, the incorporation of bio-based amines as crosslinkers and interface builders enables the development of mechanically robust and environmentally friendly conducting polymer systems suitable for advanced electronic application [23,26].

11.4.4 As Surface Modifiers and Analyte-Recognition Elements

Beyond structural and electronic roles, bio-based amines are widely used as surface modifiers and analyte-recognition elements in conducting polymer-based sensors. The presence of amine functional groups on polymer surfaces provides active sites capable of interacting with target molecules through hydrogen bonding, electrostatic interactions, or covalent bonding. These interactions enable selective detection of chemical and biological species. One of the most widely studied examples is polydopamine, a bio-inspired polymer derived from the oxidation of dopamine. Polydopamine contains both catechol and amine functional groups, allowing strong adhesion to a wide variety of surfaces. This universal adhesion property makes polydopamine an effective surface coating for modifying electrodes and sensor interfaces. Once deposited, the amine groups within polydopamine can bind analytes or immobilize biomolecules such as enzymes, antibodies, or DNA probes. Similarly, amino acid-derived polymers and chitosan-based materials have been widely used in biosensor platforms. The amine groups present in these materials provide reactive sites for the immobilization of biological recognition elements, enabling the detection of biomolecules such as glucose, proteins, and nucleic acids. These materials also improve sensor sensitivity by enhancing electron transfer between the analyte and the electrode surface [27].

Bio-based amines also contribute to the selective detection of environmental pollutants and toxic chemicals. For instance, amine-functionalized conducting polymers can interact strongly with acidic gases, heavy metal ions, or organic contaminants through acid–base interactions or coordination chemistry. This property enables the development of highly sensitive chemical sensors for environmental monitoring and industrial safety applications. Furthermore, the use of bio-derived amine modifiers can improve the biocompatibility of conducting polymer surfaces. This is particularly important for biomedical devices and implantable sensors, where materials must interact safely with biological systems. The incorporation of naturally derived amine groups helps create surfaces that support cell adhesion and minimize toxicity. Overall, bio-based amines provide versatile chemical functionality for designing highly selective and sensitive conducting polymer sensors. Their ability to interact

ion transport and structural stability. Nanofillers influence the conductivity of polymer electrolytes by increasing the amorphous phase of the polymer matrix, improving salt dissociation, and immobilizing anions to promote proton or cation transport. For example, chitosan electrolytes doped with ammonium thiocyanate salts and Al_2O_3 nanoparticles have shown significant improvements in ionic conductivity due to enhanced proton transport pathways and increased polymer chain mobility. Optimal filler concentrations can produce conductivities approaching 10^{-4}S/cm, while excessive filler loading may increase crystallinity and reduce ionic mobility. These nanocomposite electrolytes demonstrate the importance of structural tuning in optimizing the electrical properties of bio-based polymer systems [32]. Blending chitosan with synthetic polymers is another widely explored method for designing functional conducting materials. Polymer blending allows the combination of desirable properties from different polymers, such as mechanical strength, flexibility, and electrical performance. For instance, blends of chitosan with poly(vinyl alcohol) (PVA) have been investigated for their electrical and dielectric behavior. Strong hydrogen bonding interactions occur between the hydroxyl groups of PVA and the amine groups of chitosan, forming homogeneous polymer networks with improved thermal and structural stability. Electrical conductivity in these blends is generally governed by thermally activated charge carrier hopping mechanisms, where charge carriers move through localized states within the polymer matrix. The dielectric properties of such systems often exhibit frequency-dependent behavior, with high dielectric constants at low frequencies due to space-charge polarization effects. These polymer blends highlight the potential of chitosan as a functional component in electrically responsive polymer materials [33]. Beyond these systems, chitosan has also been incorporated with carbon-based nanomaterials, metal nanoparticles, and conductive fillers to fabricate advanced conductive composites. The abundant functional groups present on chitosan chains facilitate strong interactions with nanomaterials through hydrogen bonding, electrostatic attraction, or coordination interactions. Such interactions help disperse conductive fillers uniformly within the polymer matrix and prevent aggregation, thereby maintaining continuous electron or ion transport pathways. These hybrid materials have been explored for a wide range of applications, including biosensors, wearable electronics, energy storage devices, and electrochemical sensors. Despite these advantages, certain challenges remain in the development of chitosan-based conducting systems. Pure chitosan exhibits relatively low intrinsic electrical conductivity, which necessitates the incorporation of conductive polymers, salts, or nanomaterials. In addition, the hydrophilic nature of chitosan can lead to excessive swelling in humid environments, potentially affecting the long-term stability of conductive devices. To address these limitations, current research focuses on chemical functionalization, crosslinking strategies, nanocomposite engineering, and molecular design approaches to tailor the electrical and mechanical properties of chitosan-based materials. Overall, chitosan provides a versatile and sustainable platform for the design of conducting polymer systems. Its unique combination of bio-derived origin, chemical functionality, and compatibility with various conductive components enables the fabrication of hybrid materials with tunable electrical properties. Through strategies such as conducting polymer integration, polymer electrolyte formation, nanofiller incorporation, and polymer blending, chitosan-based materials have demonstrated significant potential in applications ranging from biomedical electronics to energy storage and fuel cell technologies. As interest in green and bio-based electronic materials continues to grow, chitosan-based conducting systems are expected to play an increasingly important role in the development of sustainable electrochemical and electronic devices.

11.6 SENSOR APPLICATIONS

Bio-based amines, commonly referred to as biogenic amines, are nitrogen-containing organic compounds formed mainly through the microbial decarboxylation of amino acids during the degradation of protein-rich materials. These compounds, including histamine, putrescine, cadaverine, and tyramine, are commonly produced during the spoilage of foods such as fish, meat, and dairy products and therefore serve as important chemical indicators of biological decomposition and quality deterioration. Monitoring these compounds is essential for ensuring food safety, reducing food waste, and preventing health risks associated with the consumption of spoiled food products. Consequently, the development of sensitive and rapid detection technologies for biogenic amines has received considerable attention in recent years. Organic π-conjugated materials have emerged as promising sensing platforms because their electronic structures respond sensitively to interactions with amine molecules, producing measurable optical or electrical signals that enable detection. These sensing systems enable real-time monitoring of food freshness and may eventually replace traditional fixed "best-before" labeling approaches with intelligent sensing technologies that more accurately reflect the actual condition of food products [25]. Fluorescence-based sensors represent another important class of amine detection systems. In these sensors, the interaction between amine molecules and conjugated sensing materials modifies the photophysical properties of the material, resulting in fluorescence quenching, enhancement, or emission shifts. These optical responses arise from electron-transfer processes, proton transfer reactions, or structural changes in the sensing material induced by amine adsorption. Because of their high sensitivity and rapid response, fluorescent sensors are widely explored for detecting volatile amines released during food spoilage and other biological processes. Conjugated polymers, covalent organic frameworks (COFs), and covalent organic polymers (COPs) have been widely studied as fluorescent sensing materials due to their extended π-electron networks, tunable structures, and high chemical stability [25]. COF- and COP-based sensing systems provide additional advantages for amine detection because of their highly porous structures and large surface areas. These features allow efficient diffusion and adsorption of amine molecules within the framework pores, enhancing interactions between the analyte and the sensing material. When amine molecules enter the porous network, they interact with electron-deficient sites in the framework through charge-transfer interactions, hydrogen bonding, or acid–base reactions, leading to changes in fluorescence intensity or emission wavelength. As a result, COF-based sensors have shown strong potential for detecting volatile biogenic amines associated with the decomposition of fish and meat products, making them promising materials for intelligent food packaging and real-time freshness monitoring systems [34]. In addition to optical sensors, electrochemical sensing systems have also been developed for the detection of biogenic amines. Electrochemical sensors operate by converting chemical interactions between the analyte and the electrode surface into measurable electrical signals such as current, voltage, or impedance changes. When amine molecules interact with conductive sensing materials at the electrode interface, they modify the charge transfer properties of the material and produce detectable electrical responses. Conductive polymers, carbon nanomaterials, and porous organic frameworks are commonly used as electrode materials because of their high surface area and excellent electron-transfer capabilities. These sensors offer rapid response times, high sensitivity, and compatibility with portable electronic devices, making them suitable for practical sensing applications [25]. Advanced sensing platforms such as chemiresistive sensors and OFET devices have also been explored for detecting bio-based amines. In chemiresistive sensors, adsorption of amine molecules on the sensing material alters its electrical resistance due to changes in charge carrier density or electron transport pathways. In OFET-based sensors, interactions between amine molecules and the semiconducting polymer channel influence charge carrier transport and modulate the device current. These sensing platforms provide highly sensitive detection and can be integrated with wireless technologies, including radio-frequency identification systems, enabling the development of intelligent packaging systems capable of monitoring food freshness in real time [25]. Polymer-based sensing materials have also been extensively investigated for amine detection due to their tunable electronic properties and strong interactions with amine molecules. Conjugated polymers containing aromatic backbones can interact with biogenic amines through charge-transfer interactions because amines act as strong electron donors due to the presence of lone-pair electrons on nitrogen atoms. These interactions modify the electronic structure of the polymer and lead to detectable optical responses such as fluorescence quenching, emission shifts, or color changes. Because these interactions occur rapidly at the material surface, polymer-based fluorescent sensors can provide fast and sensitive detection of volatile amines associated with food spoilage and biological degradation processes [35]. More recently, porous organic polymers and conjugated frameworks have been developed as advanced sensing materials for amine detection. These materials possess extended π-conjugated structures and abundant functional groups that promote strong interactions with volatile amine molecules through charge-transfer interactions, hydrogen bonding, and acid–base reactions. The porous structure of these materials facilitates efficient adsorption and diffusion of amine molecules, increasing the interaction between the analyte and the sensing surface. As a result, even low concentrations of volatile amines released from decomposing food products can be detected. These materials therefore show strong potential for applications in food freshness monitoring, environmental sensing, and real-time chemical detection systems where rapid and selective detection of biogenic amines is essential [36]. Overall, the strong electron-donating nature and high chemical reactivity of bio-based amines make them ideal targets for a wide range of sensing technologies. Optical, electrochemical, and transistor-based sensors developed using conjugated polymers, porous frameworks, and nanomaterials offer rapid and sensitive detection of volatile amines associated with food spoilage and biological degradation processes. The integration of these sensing materials into smart packaging systems, portable devices, and wireless monitoring technologies represents an important advancement in intelligent food quality monitoring and highlights the growing importance of bio-based amine detection technologies in modern analytical science.

11.7 CHALLENGES, LIMITATIONS, AND RESEARCH GAPS

Despite the rapid progress in the development of bio-based amines for conducting polymers, sensing systems, and electronic materials, several scientific and technological challenges remain that limit their widespread industrial implementation. One of the primary challenges lies in the availability and consistency of biomass-derived feedstocks. Unlike petrochemical resources, biomass sources often exhibit variability in chemical composition depending on plant species, geographical origin, and processing methods. This variability can influence the purity, yield, and functional group distribution of the resulting bio-based amines, which may subsequently affect polymerization efficiency, electrical conductivity, and sensor performance. Achieving consistent quality of bio-derived amines therefore remains a significant challenge for large-scale manufacturing. Another important limitation is related to the complexity of synthesis and purification processes. Many bio-based amine production routes rely on multistep catalytic or microbial processes that require precise reaction conditions, specialized catalysts, or engineered microorganisms. While these approaches offer environmentally friendly alternatives to conventional synthetic routes, they can also increase production costs and reduce overall scalability. Efficient catalytic systems, improved reaction selectivity, and simplified purification strategies are still required to make bio-based amine synthesis economically competitive with petrochemical counterparts. From a materials perspective, integration of bio-based amines into conducting polymer systems presents additional challenges. Conducting polymers rely heavily on controlled electronic structures and charge transport pathways. Variations in amine structure, functionality, or molecular weight can influence polymer chain alignment, doping efficiency, and charge carrier mobility. As a result, bio-based amines may sometimes produce conducting polymers with lower electrical conductivity or reduced stability compared with traditional synthetic amines. Understanding the structure–property relationships between bio-derived amines and conjugated polymer systems remains an important research gap. Another key limitation is the long-term stability of bio-based materials in electronic and sensing applications. Conducting polymer sensors and electronic devices must operate under varying environmental conditions such as humidity, temperature fluctuations, and chemical exposure. Bio-derived amines may introduce additional functional groups that increase sensitivity but also make the materials more susceptible to degradation or oxidation over time. Improving environmental stability while maintaining high sensitivity and conductivity is therefore a major challenge. In sensor technologies, issues related to selectivity and interference also remain unresolved. Many amine-responsive sensing materials can interact with multiple volatile compounds simultaneously, which may lead to false signals or reduced detection accuracy. Designing sensing systems that selectively respond to specific bio-based amines while avoiding interference from other gases or volatile organic compounds is an important research direction. Finally, limited industrial translation represents a broader research gap in this field. While numerous laboratory-scale studies have demonstrated the potential of bio-based amines in sensors, conductive materials, and electronic devices, only a few systems have progressed toward commercialization. Bridging the gap between academic research and practical applications will require interdisciplinary collaboration between chemists, materials scientists, engineers, and industry partners. Addressing these challenges will be essential for realizing the full potential of bio-based amines in sustainable electronic materials and smart sensing technologies.

11.8 FUTURE OUTLOOK AND CONCLUSION

The development of bio-based amines represents a promising direction in the transition toward sustainable materials and environmentally responsible chemical technologies. As global efforts continue to reduce dependence on fossil resources and minimize environmental impact, renewable nitrogen-containing compounds derived from biomass are expected to play an increasingly important role in advanced materials research. Bio-based amines offer unique opportunities for designing next-generation conducting polymers, smart sensors, and functional electronic materials with improved sustainability and tunable chemical properties. Future research is likely to focus on advanced biomass conversion technologies that enable efficient production of high-purity bio-based amines. Emerging catalytic systems, including heterogeneous metal catalysts, enzyme-based processes, and engineered microbial pathways, are expected to significantly improve the efficiency and scalability of amine production from renewable feedstocks. The integration of metabolic engineering and synthetic biology approaches may further enable the development of microbial cell factories capable of producing tailored amine structures for specific material applications. Another important direction involves the rational molecular design of bio-based amines to enhance their compatibility with conducting polymer systems. By controlling the molecular architecture, functional group distribution, and electronic properties of bio-derived amines, researchers can tailor polymer conductivity, stability, and sensing performance. Computational chemistry, machine learning, and data-driven materials design are expected to accelerate the discovery of optimized amine structures for electronic and sensing applications. In sensor technologies, bio-based amines will likely contribute to the development of next-generation smart sensing platforms, particularly for food safety monitoring, environmental detection, and healthcare diagnostics. Conducting polymers incorporating bio-derived amines may enable highly sensitive detection systems capable of responding to chemical signals through changes in electrical conductivity, fluorescence, or optical absorption. These materials could be integrated into flexible electronics, wearable devices, and intelligent packaging systems that provide real-time monitoring of environmental or biological conditions. Another promising area is the development of multifunctional hybrid materials that combine bio-based amines with nanomaterials such as graphene, carbon nanotubes, metal–organic frameworks, and COFs. Such hybrid systems may exhibit enhanced charge transport properties, improved mechanical stability, and increased sensing sensitivity. These materials could play a key role in the design of flexible electronics, energy storage devices, and advanced chemical sensors. From a broader perspective, the continued advancement of bio-based amine research aligns closely with global sustainability goals and the development of a bio-based circular economy. The utilization of renewable feedstocks, reduction of hazardous chemicals, and development of environmentally friendly synthesis routes are expected to significantly reduce the environmental footprint of chemical manufacturing. As research in this field progresses, bio-based amines may emerge as critical building blocks for sustainable polymers, green electronics, and intelligent sensing technologies. In conclusion, bio-based amines offer significant potential for advancing the field of conducting polymers, chemical sensing, and electronic materials. Although several challenges remain in terms of synthesis, stability, and industrial scalability, ongoing developments in catalytic chemistry, biotechnology, and materials science are expected to overcome these limitations. Continued interdisciplinary research will be essential to fully realize the potential of bio-derived amines and to translate laboratory discoveries into practical technologies that support sustainable materials innovation.

REFERENCES

1. K. Namsheer, C.S. Rout, Conducting polymers: A comprehensive review on recent advances in synthesis, properties and applications, *RSC Adv.* 11 (2021) 5659–5697.
2. A.K. Mishra, Conducting polymers : Concepts and applications, *J. At. Mol. Condens. Nano Phys.* 5 (2018) 159–193.
3. A.G. MacDiarmid, "Synthetic metals": A novel role for organic polymers (nobel lecture), *Angew. Chemie Int. Ed.* 40 (2001) 2581–2590.
4. H. Shirakawa, The discovery of polyacetylene film: The dawning of an era of conducting polymers (nobel lecture), *Angew. Chemie Int. Ed.* 40 (2001) 2574–2580.
5. S. Kirchmeyer, K. Reuter, Scientific importance, properties and growing applications of poly(3,4-ethylenedioxythiophene), *J. Mater. Chem.* 15 (2005) 2077–2088.
6. T.-H. Le, Y. Kim, H. Yoon, Electrical and electrochemical properties of conducting polymers, *Polymers (Basel).* 9 (2017) 150.
7. I. Salzmann, G. Heimel, M. Oehzelt, S. Winkler, N. Koch, Molecular electrical doping of organic semiconductors: Fundamental mechanisms and emerging dopant design rules, *Acc. Chem. Res.* 49 (2016) 370–378.
8. A.R. Alcántara, G. de Gonzalo, Green pathways: Enhancing amine synthesis using deep eutectic solvents, *Catalysts.* 15 (2025) 586.
9. A. Todea, D.M. Dreavă, I.C. Benea, I. Bîtcan, F. Peter, C.G. Boeriu, Achievements and trends in biocatalytic synthesis of specialty polymers from biomass-derived monomers using lipases, *Processes.* 9 (2021) 646.
10. H. Gröger, Biocatalytic concepts for synthesizing amine bulk chemicals: Recent approaches towards linear and cyclic aliphatic primary amines and ω-substituted derivatives thereof, *Appl. Microbiol. Biotechnol.* 103 (2019) 83–95.
11. N. Zavradashvili, C. Sarisozen, G. Titvinidze, G. Otinashvili, T. Kantaria, D. Tugushi, J. Puiggali, V.P. Torchilin, R. Katsarava, Library of cationic polymers composed of polyamines and arginine as gene transfection gents, *ACS Omega.* 4 (2019) 2090–2101.
12. R. Katsarava, N. Zavradashvili, T. Kotorashvili, T. Kantaria, N. Kupatadze, N. Ochkhikidze, N. Medzmariashvili, N. Kutsiava, M. Gurielidze, Polyamines and arginine-based cationic polymers as antimicrobial agents, in: A.K. H aghi, T. Tatrishvili (Eds.), *Advances in Engineering Materials*, 1st Ed., Taylor & Francis, Waretown, New Jersey, 2025: p. 12.

13. M. Pelckmans, T. Renders, S. Van de Vyver, B.F. Sels, Bio-based amines through sustainable heterogeneous catalysis, *Green Chem.* 19 (2017) 5303–5331.
14. V. Froidevaux, C. Negrell, S. Caillol, J.P. Pascault, B. Boutevin, Biobased amines: From synthesis to polymers; present and future, *Chem. Rev.* 116 (2016) 14181–14224.
15. H. Shirakawa, A. MacDiarmid, A. Heeger, Twenty-five years of conducting polymers, *Chem. Commun.* 39 (2003) 1–4.
16. X. Guo, A. Facchetti, The journey of conducting polymers from discovery to application, *Nat. Mater.* 19 (2020) 922–928.
17. Y. Long, Z. Chen, N. Wang, Z. Zhang, M. Wan, Resistivity study of polyaniline doped with protonic acids, *Phys. B Condens. Matter.* 325 (2003) 208–213.
18. Y. Long, Z. Chen, J.L. Duvail, Z. Zhang, M. Wan, Electrical and magnetic properties of polyaniline/Fe3O4 nanostructures, *Phys. B Condens. Matter.* 370 (2005) 121–130.
19. Z. Zhang, J. Deng, J. Shen, M. Wan, Z. Chen, Chemical one step method to prepare polyaniline nanofibers with electromagnetic function, *Macromol. Rapid Commun.* 28 (2007) 585–590.
20. T.K. Das, S. Prusty, Review on conducting polymers and their applications, *Polym. Plast. Technol. Eng.* 51 (2012) 1487–1500.
21. A.A. Nada, A.E. Andicsová, J. Mosnáček, Irreversible and self-healing electrically conductive hydrogels made of bio-based polymers, *Int. J. Mol. Sci.* 23 (2022) 842.
22. C.-K. Chen, Y.-C. Lin, J.-C. Ho, W.-C. Yang, W.-C. Chen, Biomass-derived degradable poly(azomethine)s for flexible bistable photonic transistor memories, *ACS Sustain. Chem. Eng.* 10 (2022) 5268–5277.
23. L. Giraud, S. Grelier, E. Grau, G. Hadziioannou, C. Brochon, H. Cramail, E. Cloutet, Upgrading the chemistry of π-conjugated polymers toward more sustainable materials, *J. Mater. Chem.* C. 8 (2020) 9792–9810.
24. Y.-H. Chen, Y.-C. Cheng, S.-P. Rwei, Bio-based poly(imine-amide) materials with dynamic covalent adaptable networks: Toward conductive composites and thermally moldable microcellular foams, *ACS Sustain. Chem. Eng.* 10 (2022) 14794–14805.
25. M.J. Grant, K.M. Wolfe, C.R. Harding, G.C. Welch, Biogenic amine sensors using organic π-conjugated materials as active sensing components and their commercialization potential, *J. Mater. Chem. C.* 11 (2023) 9749–9767.
26. H. Azak, E. Guler, U. Can, D.O. Demirkol, H.B. Yildiz, O. Talaz, S. Timur, Synthesis of an amine-functionalized naphthalene-containing conducting polymer as a matrix for biomolecule immobilization, *RSC Adv.* 3 (2013) 19582–19590.
27. T. Kotnik, S. Kovačič, Modern methods for conjugated polymers preparation: A concept of transition metal-free synthesis, *European J. Org. Chem.* 28 (2025) e202400874.
28. B.K. Roy, I. Tahmid, T.U. Rashid, Chitosan-based materials for supercapacitor applications: A review, *J. Mater. Chem. A.* 9 (2021) 17592–17642.
29. C. Ulutürk, N. Alemdar, Electroconductive 3D polymeric network production by using polyaniline/chitosan-based hydrogel, *Carbohydr. Polym.* 193 (2018) 307–315.
30. S.B. Aziz, A.S.F.M. Asnawi, P.A. Mohammed, R.T. Abdulwahid, Y.M. Yusof, R.M. Abdullah, M.F.Z. Kadir, Impedance, circuit simulation, transport properties and energy storage behavior of plasticized lithium ion conducting chitosan based polymer electrolytes, *Polym. Test.* 101 (2021) 107286.
31. N.A. Rosli, K.S. Loh, W.Y. Wong, R.M. Yunus, T.K. Lee, A. Ahmad, S.T. Chong, Review of chitosan-based polymers as proton exchange membranes and roles of chitosan-supported ionic liquids, *Int. J. Mol. Sci.* 21 (2020) 632.
32. N.A. Aziz, S.R. Majid, R. Yahya, A.K. Arof, Conductivity, structure, and thermal properties of chitosan-based polymer electrolytes with nanofillers, *Polym. Adv. Technol.* 22 (2011) 1345–1348.
33. N. Gupta, R. Nagorny, Tensile properties of glass microballoon-epoxy resin syntactic foams, *J. Appl. Polym. Sci.* 102 (2006) 1254–1261.

34. Z. Zhou, C. Li, J. He, X. Hou, A dual-color ratiometric fluorescence sensor of biogenic amines with dye encapsulated covalent organic framework for meat freshness, *LWT.* 209 (2024) 116788.
35. D.M.G. Preethichandra, M.D. Gholami, E.L. Izake, A.P. O'Mullane, P. Sonar, Conducting polymer based ammonia and hydrogen sulfide chemical sensors and their suitability for detecting food spoilage, *Adv. Mater. Technol.* 8 (2023) 2200841.
36. S. Wang, H. Li, H. Huang, X. Cao, X. Chen, D. Cao, Porous organic polymers as a platform for sensing applications, *Chem. Soc. Rev.* 51 (2022) 2031–2080.

12 Life Cycle Assessment and Environmental Impact of Bio-Based Amines

12.1 INTRODUCTION

The global transition toward a low-carbon chemical industry has intensified interest in bio-based alternatives to conventional petrochemical intermediates. Amines constitute a particularly important class of chemicals, serving as precursors for polyamides, polyurethanes, pharmaceuticals, agrochemicals, surfactants, corrosion inhibitors, and performance additives. Conventional amine production relies heavily on fossil feedstocks such as benzene, ethylene, and ammonia derived from steam methane reforming, processes that are energy intensive and associated with significant greenhouse gas (GHG) emissions. As renewable carbon platforms mature, bio-based amines have emerged as promising candidates to reduce fossil carbon dependency while enabling integration within biorefinery systems. However, the environmental performance of bio-based amines cannot be assumed a priori; rigorous life cycle assessment (LCA) is required to quantify trade-offs and identify true sustainability benefits. Recent catalytic advances demonstrate that renewable alcohols, carbonyl compounds, and biomass-derived platform molecules can be converted into primary and secondary amines via hydrogen borrowing or reductive amination strategies [1]. These approaches offer improved atom economy and reduced salt waste relative to classical petrochemical routes. Likewise, green-chemistry evaluations of amine synthesis highlight progress in minimizing by-products and improving reaction efficiency [2]. Nevertheless, reaction-level green metrics alone do not capture upstream feedstock cultivation, hydrogen production, energy demand, or end-of-life impacts [3]. A molecule synthesized from renewable carbon is not automatically environmentally superior; its full cradle-to-grave performance must be quantified. LCA provides a structured framework to evaluate environmental burdens from raw material extraction through manufacturing, use, and disposal [4]. Meta-analyses of bio-based chemicals indicate that while most biomass-derived pathways reduce fossil carbon use, reported GHG emissions vary widely depending on feedstock type, system boundaries, and allocation methods [5]. After harmonization to cradle-to-gate boundaries, median global warming potential (GWP) values for bio-based chemicals typically fall between 1 and 5 kg CO_2-eq per kilogram of product, yet variability remains high due to agricultural inputs and process energy demand. Feedstock selection has

 DOI: 10.1201/9781003731122-12

been identified as the most statistically significant driver of environmental performance, particularly when nitrogen fertilizer application and indirect land-use change (iLUC) are considered. Land occupation and nitrogen emissions represent recurring hotspots in biomass-based systems. A detailed LCA comparison of dedicated versus waste biomass routes showed that agricultural feedstocks can introduce substantial burdens in eutrophication, acidification, and land use categories, whereas waste-derived feedstocks often perform more favorably [6]. Similar patterns were observed in the cradle-to-gate assessment of bio-based 1,4-butanediol from wheat straw, where feedstock cultivation contributed up to 70% of certain impact categories and ammonia volatilization dominated acidification potential [7]. These findings are directly relevant to fermentation-derived amines, particularly those produced from sugar crops or amino acids, where nitrogen fertilizer intensity can offset climate benefits. A prominent example of amine-specific LCA is the prospective industrial-scale assessment of bio-based aniline production from biomass [8]. Compared with fossil-derived aniline (6.0 kg CO_2-eq/kg), bio-based production reduced cradle-to-grave GWP to 3.9 kg CO_2-eq/kg using corn feedstock and as low as 1.9 kg CO_2-eq/kg using corn stover. However, acidification and eutrophication potentials increased substantially for first-generation feedstocks due to agricultural nitrogen emissions. Even after inclusion of iLUC, bio-based aniline maintained lower climate impacts, yet multi-impact trade-offs remained evident. This case illustrates a central theme in bio-based amine assessment: climate mitigation potential is strongly pathway-dependent and must be evaluated alongside non-climate categories. Manufacturing energy intensity further influences environmental performance. Studies of fermentation-derived platform chemicals, such as 1,3-propanediol, consistently identify downstream separation and steam demand as major contributors to cradle-to-gate emissions. Similarly, catalytic amination routes often require elevated temperatures and hydrogen pressures, making hydrogen source a decisive variable [9]. If hydrogen is produced via fossil routes, GHG advantages may diminish; conversely, integration of renewable electricity and green hydrogen can substantially improve performance. Thus, process configuration and energy sourcing are as important as renewable carbon origin. Beyond climate change, broader sustainability metrics must be considered. Multi-impact LCA studies of bio-based polymer systems reveal that increasing bio-content does not necessarily correlate with lower total environmental burden [10]. While GWP and fossil resource depletion often decrease, categories such as eutrophication, acidification, and water use may increase. These trade-offs underscore the importance of harmonized functional units, transparent allocation methods, and sensitivity analyses. In this chapter, we critically examine the life cycle environmental performance of bio-based amines across feedstock platforms, manufacturing pathways, and end-of-life scenarios. Emphasis is placed on quantitative benchmarking against petrochemical analogues, identification of environmental hotspots, and interpretation of trade-offs across multiple impact categories. By integrating recent meta-analyses, process-level simulations, and amine-specific case studies, this chapter aims to provide a balanced and evidence-based evaluation of whether, when, and under what conditions bio-based amines contribute to meaningful environmental improvement within the chemical sector.

12.2 BIO-BASED AMINES: SOURCES AND PRODUCTION PATHWAYS

The environmental performance of bio-based amines is intrinsically linked to their feedstock origin and conversion pathway. Unlike fossil-derived amines, which are typically synthesized from petrochemical intermediates such as benzene, ethylene, or propylene followed by nitration, hydrogenation, or ammonolysis, bio-based amines emerge from diverse biomass platforms. These platforms differ in chemical functionality, nitrogen incorporation routes, hydrogen demand, and processing intensity. Consequently, understanding the upstream source and production chemistry is essential before interpreting life cycle impacts.

12.2.1 Biomass Platform Families Relevant to Amine Production

Bio-based amines can be derived from four primary biomass platform families: (i) lipid-based systems, (ii) carbohydrate-derived systems, (iii) polyol platform molecules such as glycerol, and (iv) lignocellulosic or residue-based feedstocks.

Lipid-derived platforms: Vegetable oils, such as soybean or castor oil, are primarily used to generate epoxidized intermediates and polyols that can subsequently be aminated or transformed into nitrogen-containing derivatives. Although more commonly associated with polyurethanes and surfactants, lipid-based feedstocks may serve as precursors for fatty amines through catalytic amination of fatty alcohols. These systems benefit from high functionality and tunable chain length but may involve energy-intensive epoxidation and hydrogenation steps. The catalytic amination of bio-derived alcohols has been widely reviewed, particularly using heterogeneous catalysts under hydrogen-borrowing or reductive amination conditions. Reaction temperatures frequently range between 120°C and 250°C, and hydrogen pressures can significantly influence process energy intensity.

Carbohydrate-derived platforms: Sugars and sugar alcohols constitute the dominant feedstock for fermentation-based amine production. Amino acids such as lysine and glutamic acid can be produced from glucose fermentation and subsequently decarboxylated to yield diamines such as cadaverine and putrescine. This pathway integrates biological nitrogen fixation via ammonia or ammonium salts during fermentation. The environmental implications of this route depend heavily on agricultural nitrogen inputs and fermentation energy demand, as demonstrated in meta-analyses of bio-based chemicals [5]. Carbohydrate platforms offer scalability and established fermentation infrastructure but introduce upstream land use and eutrophication burdens.

Polyol platform molecules (glycerol): Glycerol, primarily generated as a by-product of biodiesel production, represents a low-cost renewable polyol that can be upgraded into nitrogen-containing compounds. Amination of glycerol-derived intermediates, including glycerol carbonate or glycerol-derived alcohols, provides access to various amine structures. The green synthesis of amines from renewable polyols has been critically assessed using green chemistry metrics, emphasizing solvent choice, catalyst selection, and reaction temperature [2]. While reaction-level metrics may be favorable, full life cycle performance depends on glycerol purification, carbonate synthesis routes, and downstream separation.

Lignocellulosic and residue-based platforms: Second-generation feedstocks such as corn stover, wheat straw, and forestry residues reduce direct competition with food crops and may lower land-use impacts. These feedstocks can be converted into platform molecules such as furfural, levulinic acid, or aromatic intermediates, which are subsequently functionalized into amines. A notable example is the prospective LCA of bio-based aniline derived from biomass, which demonstrated that lignocellulosic feedstocks such as corn stover can reduce GWP to 1.9 kg CO_2-eq/kg compared to 6.0 kg CO_2-eq/kg for fossil aniline [8]. This pathway illustrates the potential of second-generation feedstocks to mitigate both climate and fossil depletion impacts.

12.2.2 Fermentation-Based Amine Production

Fermentation is one of the most mature and industrially established routes for producing bio-based amines, particularly aliphatic diamines used in polyamide manufacturing. In this approach, sugars derived from biomass are first fermented into amino acids such as lysine or glutamic acid, which are then converted into diamines through catalytic or enzymatic decarboxylation. The resulting amines are subsequently purified, typically by crystallization and/or distillation, to achieve polymer-grade quality. From a life cycle perspective, studies of fermentation-derived chemicals consistently show that aeration energy, nutrient inputs, and downstream separation steps are major contributors to environmental impact. For instance, LCAs of fermentation-based platform molecules such as 1,3-propanediol highlight the substantial energy demand associated with evaporation and distillation, particularly during broth concentration. Comparable trends are expected for amino acid-derived amines, where removal of water and salts from dilute fermentation broths requires energy-intensive processing. In addition, agricultural nitrogen fertilizer used during sugar cultivation can significantly increase acidification and eutrophication impacts, as reported in bio-based chemical assessments [7]. Nevertheless, fermentation offers important advantages, including operation under relatively mild conditions, high selectivity, and the potential to integrate renewable electricity for reduced carbon intensity. When powered by low-carbon energy sources, fermentation-based pathways can achieve meaningful reductions in fossil carbon emissions compared with petrochemical routes [11].

12.2.3 Catalytic Amination of Biomass-Derived Intermediates

An alternative approach involves direct catalytic conversion of biomass-derived alcohols, aldehydes, or nitriles into amines. Hydrogen borrowing and reductive amination reactions are particularly prominent. In hydrogen borrowing, alcohols are temporarily dehydrogenated to carbonyl intermediates, reacted with ammonia or amines, and re-hydrogenated to form the target amine. This route often produces water as the sole stoichiometric by-product, enhancing atom economy [12]. However, catalytic systems frequently operate at elevated temperatures and under hydrogen pressure, making hydrogen source a decisive environmental parameter [13]. If hydrogen is produced via steam methane reforming, cradle-to-gate GWP may remain substantial. Integration of green hydrogen derived from electrolysis can significantly improve environmental performance [14,15]. The aniline case study provides an example of catalytic conversion following fermentation-derived intermediates. Process simulation demonstrated meaningful GWP reductions compared to fossil-based production, while also highlighting trade-offs in non-climate impact categories [8].

12.2.4 Nitrogen Incorporation Pathways

A critical distinction between bio-based amines and other bio-based chemicals lies in nitrogen incorporation. Nitrogen is typically introduced via ammonia or ammonium salts, the production of which is energy intensive in the Haber–Bosch process. Unless green ammonia is employed, upstream ammonia production can offset part of the renewable carbon advantage [16]. Emerging approaches aim to preserve biogenic nitrogen in protein-rich biomass or integrate biological nitrogen fixation pathways. Such strategies could reduce dependence on fossil-derived ammonia but remain at early development stages.

12.2.5 Implications for Life Cycle Assessment

The diversity of feedstocks and production pathways implies that bio-based amines cannot be evaluated as a single category. Fermentation-derived diamines, catalytic amination routes, and aromatic amines from lignocellulosic biomass each exhibit distinct environmental profiles. Feedstock origin, hydrogen source, process heat integration, and separation intensity collectively determine cradle-to-gate performance [17]. Therefore, subsequent sections of this chapter will evaluate these pathways through a structured LCA framework, identifying hotspots and benchmarking performance against petrochemical analogues.

12.3 LIFE CYCLE ASSESSMENT METHODOLOGY FOR BIO-BASED AMINES

LCA provides the most systematic and widely accepted framework for evaluating the environmental performance of bio-based amines. Because these systems combine agricultural production, biological conversion, catalytic processing, and energy-intensive purification, their environmental profile cannot be inferred from feedstock origin alone. Instead, a structured methodology is required to capture material and energy flows from biomass cultivation through chemical synthesis and, where relevant, end-of-life treatment [18]. Methodological decisions particularly system boundaries, functional units, allocation procedures, and carbon accounting assumptions can significantly alter the conclusions drawn regarding environmental superiority [19]. The definition of system boundaries is particularly critical. Many studies of bio-based chemicals apply a cradle-to-gate boundary, covering biomass cultivation, transportation, chemical conversion, and purification up to the factory gate. However, for nitrogen-containing compounds such as amines, cradle-to-grave boundaries are often necessary for climate impact assessment. The prospective LCA of bio-based aniline by Winter et al. adopted cradle-to-grave accounting for GWP while applying cradle-to-gate boundaries for other categories [8]. This distinction reflects the importance of accounting for biogenic carbon release at end-of-life. Under cradle-to-gate analysis, biomass-derived carbon may be credited as carbon-neutral due to uptake during plant growth, occasionally yielding negative GWP values. However, unless carbon is permanently sequestered, subsequent release during combustion or degradation neutralizes this benefit. Therefore, cradle-to-grave modeling provides a more balanced climate evaluation for bio-based amines used in polymers, coatings, or other applications with defined lifetimes. The functional unit must also be harmonized to ensure comparability. Most LCAs of bio-based chemicals employ 1 kg of product at the factory gate as the reference unit. A large-scale meta-regression of 65 LCA studies on priority biochemicals demonstrated that inconsistent functional units and system boundaries contribute substantially to variability in reported GWP values [5]. For bio-based amines, functional equivalence should reflect comparable purity and technical performance relative to petrochemical analogues, particularly when used in polymer synthesis. The life cycle inventory (LCI) stage compiles quantitative data on material and energy flows. For bio-based amines, the inventory typically includes agricultural inputs such as fertilizers, pesticides, irrigation water, and diesel; biomass transport; fermentation or catalytic conversion energy; hydrogen production (where applicable); process heat and electricity; solvent consumption and recovery; catalyst production; and wastewater treatment. Agricultural nitrogen inputs are especially significant. In the cradle-to-gate LCA of bio-based 1,4-butanediol from wheat straw, ammonia volatilization and fertilizer application were major contributors to acidification and eutrophication impacts [7]. Comparable burdens can arise in sugar-based fermentation routes for diamines derived from lysine or glutamic acid. For catalytic amination pathways, hydrogen production becomes a dominant parameter. Many heterogeneous amination reactions operate at elevated temperatures and hydrogen pressures; if hydrogen is sourced from steam methane reforming, associated emissions must be fully integrated into the LCI. Impact assessment translates inventory flows into environmental indicators. While climate change remains the primary focus in many studies, comprehensive assessment requires multiple impact categories. Methods such as ReCiPe 2016, EF 3.0, CML, and IMPACT World+ are commonly applied to bio-based systems [6]. For bio-based amines, relevant categories include GWP, fossil resource depletion, acidification, eutrophication (freshwater and marine), photochemical ozone formation, land occupation, water use, and toxicity indicators. The aniline case study illustrates the importance of multi-impact evaluation: although bio-based production reduced GWP by 34%–69% depending on feedstock, acidification and eutrophication increased substantially for first-generation crops [8]. Similarly, LCA of partially bio-based polyester resins demonstrated that increased bio-content does not guarantee improvement across all categories [10]. These findings underscore the necessity of assessing trade-offs rather than focusing solely on carbon metrics. Biogenic carbon accounting remains one of the most debated methodological elements in bio-based chemical LCA. Carbon sequestration during biomass growth can significantly reduce reported cradle-to-gate GWP values. Meta-regression analysis indicates that carbon uptake credits reduce GWP by approximately 0.5 kg CO_2-eq per kilogram of product on average [5,20]. However, this benefit depends on end-of-life treatment assumptions and temporal boundaries.

For durable applications in polymer matrices, delayed carbon release may provide temporary climate mitigation, whereas short-lived uses offer limited long-term benefit. Allocation procedures introduce additional variability. Bio-based amines are often produced in multifunctional biorefineries that generate fuels, coproduct chemicals, and energy streams. Allocation may be based on mass, energy content, economic value, or system expansion via substitution. The meta-analysis of priority biochemicals showed that system expansion approaches can increase reported GWP relative to allocation methods [5]. Transparent reporting of allocation assumptions is therefore essential when comparing bio-based amines with petrochemical benchmarks. Uncertainty and sensitivity analysis are equally important due to variability in agricultural yields, fertilizer emission factors, electricity carbon intensity, hydrogen sourcing, and process efficiency. Monte Carlo simulations applied in bio-based chemical LCAs demonstrate that heat source and agricultural emissions frequently dominate variability ranges [7]. Sensitivity analyses of bio-based aniline showed that even with iLUC inclusion, bio-based pathways maintained lower GWP than fossil routes, although non-climate impacts remained sensitive to feedstock choice [8]. For bio-based amines, critical sensitivity parameters include fertilizer application rates, N_2O and NH_3 emission factors, electricity mix, hydrogen production pathway, and process yields. Prospective LCA combined with process simulation has become increasingly important for emerging amine pathways. Industrial-scale modeling using Aspen Plus, as applied in the bio-based aniline study, allows estimation of optimized process performance rather than relying on pilot-scale inefficiencies [8]. Nevertheless, early-stage data uncertainty must be acknowledged, and scenario analysis is recommended. Overall, the environmental evaluation of bio-based amines demands methodological rigor. Harmonized functional units, transparent allocation rules, explicit carbon accounting, multi-impact assessment, and comprehensive sensitivity analysis are necessary to avoid misleading conclusions. The following sections apply this framework to analyze feedstock-related impacts, manufacturing energy intensity, and comparative performance relative to fossil-derived amines. Table 12.1 compares the cradle-to-gate climate impacts, fossil reference reductions, and key environmental trade-offs of selected bio-based amines and related platform chemicals relative to their conventional counterparts.

12.4 FEEDSTOCK-RELATED ENVIRONMENTAL IMPACTS

Feedstock selection is one of the most influential determinants of environmental performance in bio-based amine production. Unlike fossil-derived amines, which originate from relatively standardized hydrocarbon streams, bio-based amines are derived from diverse biomass resources that differ in land requirements, fertilizer intensity, regional agricultural practices, and carbon sequestration potential. Meta-regression analysis of 65 LCA studies covering priority biochemicals identified feedstock type as the most statistically significant driver of GWP variability [5]. This conclusion is directly applicable to bio-based amines, particularly those produced via sugar fermentation or lignocellulosic conversion routes. Table 12.2 summarizes the key feedstock choices and methodological assumptions that influence the environmental performance of bio-based amines, highlighting their relative effects on climate impact, acidification, eutrophication, and overall sensitivity.

TABLE 12.1
Comparative Climate and Resource Impacts of Bio-Based Amines and Related Platform Chemicals

Product/ Chemical	Feedstock	System Boundary	GWP (kg CO_2-eq/ kg)	Fossil Reference (kg CO_2-eq/ kg)	% Reduction Vs Fossil	Key Environmental Trade-Offs	Ref.
Aniline	Corn	Cradle-to-grave	3.9	6.0	34%	↑ Acidification, ↑ eutrophication	[8]
Aniline	Corn stover	Cradle-to-grave	1.9	6.0	69%	Slight ↑ nutrient impacts	[8]
1,4-Butanediol	Wheat straw	Cradle-to-gate	~65% lower than fossil	—	~65%	↑ Acidification, ↑ marine eutrophication	[7]
Partially bio-based polyester resin	Mixed bio-feedstocks	Cradle-to-gate	60%–77% lower	Oil-based reference	60%–77%	↑ Acidification, ↑ renewable energy demand	[10]
Priority biochemicals (median range)	Various	Cradle-to-gate	1–5	Varies	~80% cases lower	High variability depending on feedstock	[5]

TABLE 12.2
Feedstock and Methodological Drivers Influencing Environmental Performance of Bio-Based Amines

Key Variable	Observed Effect on GWP	Effect on Acidification/ Eutrophication	Sensitivity Level	Ref.
First-generation crops (corn, beet)	Moderate reduction	Significant increase due to fertilizer	High	[8]
Residue feedstocks (stover, straw)	Strong reduction	Lower nutrient burden	Very high positive effect	[8]
Agricultural NH_3 volatilization	Minor climate effect	Major driver of acidification and marine eutrophication	Very high	[7]
Biogenic carbon credit (cradle-to-gate)	~0.5 kg CO_2-eq reduction (avg.)	No direct nutrient impact	Moderate	[5,21]
Allocation method (system expansion vs mass)	Can shift GWP several kg CO_2-eq	Indirect effect	Moderate–high	[5]

contributes to cumulative carbon intensity reduction across value chains. A second dimension concerns durability and lifetime extension. Amines are frequently used as curing agents, crosslinkers, and chain extenders in thermoset and elastomer systems. Improved crosslink density, chemical resistance, or adhesion performance can extend product lifetime and reduce replacement frequency. Life cycle studies of bio-based polymer systems demonstrate that functional performance may offset higher upstream burdens if durability increases sufficiently [10]. Although that study focused on polyester resins rather than amines directly, the principle remains applicable: enhanced performance during use can shift overall environmental balance. For example, bio-based amine curing agents that improve corrosion resistance in coatings could reduce maintenance cycles and material consumption over time. Third, reduced toxicity and safer chemical profiles may provide indirect environmental and occupational benefits. Traditional petrochemical amine production often involves hazardous intermediates and energy-intensive chlorination or nitration routes. Reviews of green amination strategies highlight the potential to reduce salt waste, corrosive by-products, and hazardous reagents through hydrogen borrowing and catalytic methodologies [2]. While toxicity is not always directly quantified in LCA impact categories, lower hazardous emissions during manufacture and application may reduce human toxicity and ecotoxicity indicators under comprehensive LCIA methods. The energy performance of downstream products is another consideration. Amines are widely used in insulation foams, adhesives, and high-performance composites. If bio-based amines contribute to improved foam structure, thermal insulation efficiency, or mechanical performance, operational energy savings during building use may outweigh incremental differences in production emissions. Such use-phase benefits must be evaluated through system expansion or avoided burden approaches within cradle-to-grave modeling. Although detailed amine-specific case studies remain limited, the methodological precedent is established in polymer and materials LCAs where use-phase energy dominates total life cycle impacts. Carbon accounting during use also warrants careful interpretation. For bio-based amines incorporated into durable materials, biogenic carbon may remain stored for years or decades. This temporary storage can reduce atmospheric carbon concentrations over the product lifetime. However, as emphasized in meta-analyses of bio-based chemicals, biogenic carbon neutrality assumptions must consider eventual end-of-life release [5]. Temporary carbon storage provides time-shifted mitigation but not permanent sequestration unless materials are landfilled under conditions that prevent degradation or are recycled into long-lived applications. Another performance-driven advantage relates to process compatibility with renewable energy systems. Fermentation-based amine production can be integrated with low-carbon electricity, thereby reducing fossil energy demand relative to petrochemical routes. When downstream users adopt renewable electricity for curing or processing, the cumulative life cycle carbon intensity decreases further. In contrast, fossil-based amines are intrinsically tied to hydrocarbon feedstocks, limiting decarbonization flexibility. Nevertheless, use-phase benefits are not universal and must be demonstrated quantitatively. If bio-based amines exhibit equivalent performance but higher eutrophication or land-use burdens during production, net environmental advantage may be limited. Therefore, performance-driven environmental benefits should be evaluated through scenario analysis that includes service life, substitution ratios, and functional equivalence. In summary, the use phase can significantly influence the environmental profile of bio-based amines. Fossil carbon displacement, enhanced durability, potential toxicity reduction, and integration with renewable energy systems represent key pathways through which environmental benefits may extend beyond production-stage improvements. However, these benefits are context-dependent and require cradle-to-grave modeling to avoid overestimation. The following section addresses end-of-life considerations, completing the life cycle perspective for bio-based amines.

12.7 COMPARATIVE LCA CASE INSIGHTS AND DISCUSSION

Comparative LCA provides the most direct means of evaluating whether bio-based amines offer meaningful environmental advantages over their petrochemical counterparts. While individual case studies provide valuable insights, broader interpretation requires harmonization of system boundaries, functional units, and impact categories. Recent meta-regression analysis of 65 LCA studies on priority biochemicals showed that approximately 80% of bio-based pathways demonstrated lower cradle-to-gate GWP than fossil-based equivalents, but variability remained high due to feedstock, allocation, and methodological differences [5]. These findings frame the comparative discussion for bio-based amines. A prominent amine-specific example is the prospective industrial-scale LCA of bio-based aniline [8]. Fossil-based aniline exhibited cradle-to-grave GWP of approximately 6.0 kg CO_2-eq/kg, whereas bio-based production reduced emissions to 3.9 kg CO_2-eq/kg when corn was used as feedstock and to as low as 1.9 kg CO_2-eq/kg when corn stover was employed. This corresponds to a 34%–69% reduction in climate impact depending on feedstock. Even after incorporating iLUC, bio-based routes maintained lower GWP than fossil-based production. However, significant trade-offs emerged in non-climate categories: acidification increased by more than 200%, and eutrophication increased up to sevenfold when first-generation crops were used. These results demonstrate that climate mitigation potential does not necessarily translate into overall environmental superiority. Feedstock sensitivity emerges as a consistent theme across case studies. The meta-regression by Tang et al. confirmed that feedstock type is the most statistically significant determinant of GWP variability [5]. Residue-based and lignocellulosic feedstocks typically exhibit stronger climate performance and reduced land competition compared to first-generation crops. This trend aligns with the aniline case, where corn stover significantly outperformed corn grain. For aliphatic diamines derived from sugar fermentation, similar patterns can be expected: first-generation sugar crops introduce fertilizer-driven acidification and eutrophication burdens, whereas second-generation feedstocks may reduce these impacts but potentially increase preprocessing energy requirements. Comparative evidence from non-amine bio-based chemicals further reinforces these conclusions. In the cradle-to-gate LCA of bio-based 1,4-butanediol from wheat straw, climate change impact was reduced by approximately 65% relative to fossil BDO, while fossil depletion decreased by roughly 72% [7]. Nevertheless, terrestrial acidification and marine eutrophication were strongly influenced by agricultural nitrogen emissions. These findings mirror the trade-offs observed in the aniline system and suggest that nutrient-related categories often become limiting factors in biomass-based nitrogen chemistry. Energy intensity during manufacturing also influences comparative outcomes. Fermentation-based chemicals frequently exhibit significant steam and electricity demand during downstream separation, while catalytic amination routes require elevated temperatures and hydrogen input. If hydrogen is sourced from steam methane reforming, climate advantages may diminish. Conversely, integration of renewable electricity and green hydrogen substantially improves environmental performance. Sensitivity analysis in the aniline study confirmed that electricity mix and process heat source materially affect GWP outcomes [8]. Thus, decarbonization of energy inputs represents a key lever for enhancing comparative advantage. Another important insight concerns carbon accounting and system boundaries. Cradle-to-gate assessments often report negative or near-zero GWP values due to biogenic carbon uptake. However, cradle-to-grave accounting typically neutralizes these apparent benefits unless long-term carbon storage occurs. The meta-analysis of priority biochemicals estimated that carbon sequestration credits reduce reported GWP by approximately 0.5 kg CO_2-eq/kg on average [5]. For bio-based amines incorporated into durable polymer systems, temporary carbon storage may provide time-shifted mitigation, but permanent sequestration is rarely achieved. Allocation methods further complicate comparative interpretation. Bio-based amines are frequently produced in multifunctional biorefineries generating fuels, heat, and coproduct streams. System expansion approaches can alter GWP results relative to mass or economic allocation [5]. Transparent reporting of allocation assumptions is therefore essential when benchmarking against fossil-derived amines. Across comparative studies, several consistent patterns emerge. First, bio-based amines generally demonstrate reduced fossil resource depletion and lower climate impact under optimized conditions. Second, first-generation feedstocks often increase acidification and eutrophication due to fertilizer and nitrogen emissions. Third, second-generation feedstocks improve climate robustness and reduce land competition but may require additional preprocessing energy. Fourth, energy source particularly hydrogen production can shift results substantially. Finally, methodological transparency and harmonization are critical to avoid misleading conclusions. In summary, comparative LCA evidence indicates that bio-based amines can deliver meaningful climate benefits relative to petrochemical routes, especially when residue-based feedstocks and low-carbon energy inputs are employed. However, trade-offs in nutrient-related and land-use categories remain significant for first-generation systems. Robust comparative evaluation therefore requires multi-impact assessment, sensitivity analysis, and cradle-to-grave carbon accounting. These insights provide the foundation for informed decision-making regarding the sustainable deployment of bio-based amines within the broader chemical sector.

12.8 CONCLUSIONS

The environmental assessment of bio-based amines demonstrates that renewable origin alone does not guarantee sustainability. Instead, life cycle performance is governed by a combination of feedstock selection, nitrogen management, energy sourcing, conversion efficiency, allocation methodology, and end-of-life treatment. Across the available evidence, bio-based amines show clear potential to reduce fossil carbon dependency and lower GHG emissions compared to petrochemical routes, but these benefits are highly pathway-specific and often accompanied by trade-offs in other environmental categories. Comparative analyses consistently show that second-generation feedstocks such as agricultural residues or lignocellulosic biomass provide stronger and more robust climate benefits than first-generation sugar or starch crops. While biomass-derived pathways frequently reduce GWP and fossil resource depletion, they may increase acidification and eutrophication due to fertilizer use and nitrogen emissions during cultivation. For fermentation-based amines, environmental burdens are amplified by both agricultural nitrogen inputs and nutrient requirements during biochemical processing. These findings underscore the importance of evaluating multiple impact categories rather than relying solely on carbon metrics. Manufacturing-stage energy demand also plays a decisive role. Fermentation pathways are often dominated by aeration and downstream separation energy, whereas catalytic amination routes depend heavily on heat and hydrogen input. The carbon intensity of electricity and hydrogen production can significantly alter comparative outcomes. Integration of renewable electricity and green hydrogen infrastructure represents one of the most effective levers for improving environmental performance across both fermentation and catalytic systems. Methodological transparency remains essential. Reported environmental results are sensitive to system boundaries, allocation rules, and biogenic carbon accounting. Cradle-to-gate assessments may overstate climate benefits if end-of-life carbon release is excluded, while cradle-to-grave modeling provides a more balanced perspective. Sensitivity and uncertainty analyses are necessary to account for variability in agricultural practices, emission factors, process yields, and energy mixes. Use-phase performance can further influence overall sustainability. When bio-based amines enhance durability, improve material efficiency, or displace fossil-derived intermediates in downstream applications, indirect environmental benefits may accrue. However, such advantages must be demonstrated quantitatively through scenario-based life cycle modeling rather than assumed. Overall, bio-based amines represent a promising pathway toward reducing the carbon intensity of nitrogen-containing chemicals, particularly when derived from residue-based feedstocks and produced with low-carbon energy inputs. Nonetheless, nutrient-related impacts and land-use considerations remain critical challenges for first-generation systems. Future progress will depend on improved nitrogen management, expanded renewable energy integration, optimized biorefinery design, and harmonized life cycle methodologies. In summary, the environmental superiority of bio-based amines is conditional rather than inherent. A comprehensive life cycle perspective encompassing climate change, resource depletion, land use, and nutrient impacts is essential to ensure that renewable nitrogen chemistry contributes meaningfully to sustainable chemical production.

REFERENCES

1. N.K. Gupta, P. Reif, P. Palenicek, M. Rose, Toward renewable amines: Recent advances in the catalytic amination of biomass-derived oxygenates, *ACS Catal.* 12 (2022) 10400–10440.
2. A.M. Afanasenko, N. Deak, J. October, R. Sole, K. Barta, 'Green' synthesis of amines from renewable resources? A detailed analysis of case studies using the CHEM21 green metrics toolkit, *Green Chem.* 27 (2025) 5947–5981.

3. T. Terlouw, L. Rosa, C. Bauer, R. McKenna, Future hydrogen economies imply environmental trade-offs and a supply-demand mismatch, *Nat. Commun.* 15 (2024) 7043.
4. B. He, H. Duan, W. Yang, H. Yabar, J. Song, Decarbonizing polyamide textile production in China: Footprints and mitigation pathways from life cycle perspective, *Resour. Conserv. Recycl.* 208 (2024) 107705.
5. Z. Tang, W. Zhang, Q. Tu, Life cycle assessment of priority biochemicals: A review and meta-regression analysis, *Resour. Conserv. Recycl.* 215 (2025) 108087.
6. D. Cespi, Bio-based chemicals from dedicated or waste biomasses: Life cycle assessment for evaluating the impacts on land, *Sustain. Chem.* 4 (2023) 184–196.
7. A. Forte, A. Zucaro, R. Basosi, A. Fierro, LCA of 1,4-butanediol produced via direct fermentation of sugars from wheat straw feedstock within a territorial biorefinery, *Materials (Basel).* 9 (2016) 563.
8. B. Winter, R. Meys, A. Bardow, Towards aromatics from biomass: Prospective life cycle assessment of bio-based aniline, *J. Clean. Prod.* 290 (2021) 125818.
9. M. Pelckmans, T. Renders, S. Van de Vyver, B.F. Sels, Bio-based amines through sustainable heterogeneous catalysis, *Green Chem.* 19 (2017) 5303–5331.
10. A.T. Shahid, M.A. Hofmann, J.D. Silvestre, M. Garrido, J.R. Correia, Life cycle assessment of novel partially bio-based unsaturated polyester resins, *J. Polym. Environ.* 33 (2025) 3329–3347.
11. C. Minke, M. Suermann, B. Bensmann, R. Hanke-Rauschenbach, Is iridium demand a potential bottleneck in the realization of large-scale PEM water electrolysis?, *Int. J. Hydrogen Energy.* 46 (2021) 23581–23590.
12. S. Hameury, H. Bensalem, K. De Oliveira Vigier, Sustainable amination of bio-based alcohols by hydrogen borrowing catalysis, *Catal.* 12 (2022) 1306.
13. A. Odenweller, F. Ueckerdt, The green hydrogen ambition and implementation gap, *Nat. Energy.* 10 (2025) 110–123.
14. A.O. Oni, K. Anaya, T. Giwa, G. Di Lullo, A. Kumar, Comparative assessment of blue hydrogen from steam methane reforming, autothermal reforming, and natural gas decomposition technologies for natural gas-producing regions, *Energy Convers. Manage.* 254 (2022) 115245.
15. X. Wei, S. Sharma, A. Waeber, D. Wen, S.N. Sampathkumar, M. Margni, F. Maréchal, J. Van herle, Comparative life cycle analysis of electrolyzer technologies for hydrogen production: Manufacturing and operations, *Joule.* 8 (2024) 3347–3372.
16. S. Vinardell, P. Nicolas, A.M. Sastre, J.L. Cortina, C. Valderrama, Sustainability assessment of green ammonia production to promote industrial decarbonization in Spain, *ACS Sustain. Chem. Eng.* 11 (2023) 15975–15983.
17. J. Boyce, R. Sacchi, E. Goetheer, B. Steubing, A prospective life cycle assessment of global ammonia decarbonisation scenarios, *Heliyon.* 10 (2024) e27547.
18. M. Erakca, M. Baumann, C. Helbig, M. Weil, Systematic review of scale-up methods for prospective life cycle assessment of emerging technologies, *J. Clean. Prod.* 451 (2024) 142161.
19. T. Nemecek, D. Bretscher, F. Cirone, Biogenic carbon accounting in LCA (response to Silva et al. 2024), *Int. J. Life Cycle Assess.* 30 (2025) 2569–2572.
20. A. Levasseur, P. Lesage, M. Margni, R. Samson, Biogenic carbon and temporary storage addressed with dynamic life cycle assessment, *J. Ind. Ecol.* 17 (2013) 117–128.
21. J. Leifeld, S.G. Keel, Quantifying negative radiative forcing of non-permanent and permanent soil carbon sinks, *Geoderma.* 423 (2022) 115971.
22. J. Blanco-Cejas, S. Martín, M. Linares, J. Iglesias, J. Moreno, Life cycle assessment applied to bio-based platform molecules: Critical review of methodological practices, *J. Clean. Prod.* 414 (2023) 137513.

13 Industrial Landscape, Regulatory Considerations, and Future Outlook of Bio-Based Amines

13.1 INTRODUCTION

Amines constitute one of the most versatile and industrially indispensable classes of nitrogen-containing organic compounds [1]. Their high reactivity originates from the presence of a lone electron pair on nitrogen, which governs both their basicity and nucleophilicity. These two fundamental properties dictate their participation in acid–base equilibria and bond-forming reactions, respectively. While basicity is a thermodynamic parameter describing proton affinity, nucleophilicity represents a kinetic concept related to reaction rate. Although these properties often correlate, they are strongly influenced by substituent effects, steric hindrance, electronic delocalization, and solvent environment. Aliphatic amines typically exhibit higher basicity and nucleophilicity than ammonia due to electron-donating substituents, whereas aromatic amines show reduced basicity because of resonance delocalization of the nitrogen lone pair into the aromatic ring. Solvent polarity, hydrogen bonding, charge state, and α-substituent effects further modulate reactivity. These fundamental chemical characteristics ultimately determine how amines behave in polymerization, curing, and functionalization reactions. From an industrial perspective, amines are critical intermediates in agrochemicals, pharmaceuticals, detergents, surfactants, lubricants, and especially polymer systems [2,3].

In polymer chemistry, amines function as monomers, chain extenders, crosslinkers, and curing agents in polyamides, polyureas, epoxy resins, polyurethanes, and emerging non-isocyanate polyhydroxyurethanes. Their nucleophilic reactivity enables amide formation, epoxide ring opening, Michael addition, and cyclic carbonate aminolysis. Primary amines are particularly important in polymer design because they can participate in hydrogen bonding and bifunctional crosslinking, while secondary amines offer controlled reactivity and reduced crosslink density. Tertiary amines, in contrast, are often used as catalysts or bases rather than structural monomers. The global production of polyamides and epoxy resins, each in the multi-million-ton-per-year range, reflects the substantial and growing demand for amine-based building blocks across automotive, aerospace, construction, electronics,

DOI: 10.1201/9781003731122-13

and healthcare sectors. Traditionally, commercial amines are synthesized from fossil-derived feedstocks via ammonia alkylation, nitrile hydrogenation, nitro compound reduction, and hydroamination processes. Ammonia serves as the foundational nitrogen source for most industrial amine production [4,5].

However, increasing environmental concerns, carbon reduction targets, and volatility in fossil resource supply have stimulated interest in renewable alternatives [6–8]. In contrast to oxygenated biomass derivatives, naturally occurring amines suitable for direct polymer-scale applications are rare. Notable examples include poly(ε-L-lysine) produced by microbial fermentation and chitosan derived from chitin deacetylation. These materials demonstrate the feasibility of biologically sourced amines but remain limited in structural diversity and scalability for broad industrial deployment. The transition toward bio-based amines therefore relies primarily on transforming renewable carbon feedstocks such as carbohydrates, lignocellulosic biomass, terpenes, vegetable oils, and fermentation intermediates into structurally diverse primary and secondary amines. Advances in catalytic reductive amination, hydrogen-borrowing strategies, enzymatic transamination, and hybrid chemo-biological processes have expanded the accessible chemical space of renewable amines. These developments align with the principles of green chemistry, particularly the use of renewable resources, atom economy, and reduced hazardous by-products. Market projections for green chemistry and sustainable materials indicate strong economic incentives for integrating renewable feedstocks into existing chemical value chains [9–11].

Beyond market drivers, regulatory pressures are increasingly shaping the evolution of amine chemistry. Frameworks such as Registration, Evaluation, Authorisation and Restriction of Chemicals (REACH) in the European Union, Toxic Substances Control Act (TSCA) in the United States, and analogous systems across Asia regulate chemical substances based on hazard profile, tonnage, and exposure risk rather than feedstock origin. Consequently, bio-based amines must meet the same toxicological, occupational, and environmental standards as their petrochemical counterparts. At the same time, sustainability certification systems, life cycle assessment (LCA) methodologies, and carbon accounting standards are becoming central to commercialization strategies. The distinction between molecule identity and carbon origin introduces both opportunities and compliance challenges for renewable amine producers. Although previous reviews have addressed specific families of bio-based amines particularly those derived from vegetable oils there remains a need for a comprehensive perspective that integrates chemical reactivity, renewable synthesis routes, industrial scale-up, regulatory governance, market dynamics, and emerging innovation pathways. This chapter addresses that gap by examining bio-based primary and secondary amines from a systems-level standpoint. Tertiary and quaternary amines are excluded due to their limited relevance in structural polymer design [12].

This chapter begins with an overview of industrial production trends and supply chain architecture, followed by techno-economic considerations and scale-up challenges. Subsequent sections examine regulatory frameworks, sustainability certification standards, market dynamics, and commercialization models. Emerging technologies including advanced catalysis, electrochemical amination, synthetic biology, artificial intelligence (AI)-assisted process design, and circular economy integration are then discussed as critical innovation drivers. Finally, this chapter evaluates future opportunities and challenges in the transition toward a bio-based nitrogen economy aligned with global decarbonization and net-zero strategies.

13.2 INDUSTRIAL LANDSCAPE OF BIO-BASED AMINES

13.2.1 Global Market Evolution, Industrial Players, and Supply Chain Architecture

The global amines industry represents a multibillion-dollar sector supplying essential intermediates to polymers, pharmaceuticals, agrochemicals, coatings, and specialty chemicals. Traditionally dominated by petrochemical feedstocks such as ethylene, propylene, and benzene derivatives, the industry is increasingly witnessing strategic integration of renewable carbon pathways. Bio-based amines are emerging not merely as substitutes but as structurally tunable molecules derived from sugars, lignocellulosic biomass, vegetable oils, and fermentation intermediates [13,14]. Although precise global production capacities for "bio-based amines" are not separately reported in most commodity statistics, several indicators demonstrate rapid industrial acceleration. Renewable feedstock availability has expanded substantially over the last decade. Global bioethanol production alone exceeds tens of billions of gallons annually, providing scalable C2 building blocks that can be catalytically aminated to ethanolamines, ethylenediamines, and higher amines [15]. Carbohydrate-derived intermediates such as glycolaldehyde and HMF have also been demonstrated as viable precursors for aliphatic amine synthesis through reductive amination and hydrogen-borrowing strategies [16]. Industrial players are increasingly investing in bio-derived diamines for polymer applications. Commercial developments in bio-based hexamethylenediamine (HMDA) and 1,5-diaminopentane (PDA) demonstrate that fermentation and hybrid catalytic routes can reach industrial relevance, particularly for nylon-6,6 and bio-polyamide value chains [17]. The production of PDA via engineered microbial systems further exemplifies scalable bio-amine manufacturing [18]. The supply chain architecture of bio-based amines differs from petrochemical systems primarily in feedstock sourcing and integration strategy. Biomass-derived sugars, lignin streams, and fatty acid derivatives introduce regional variability in raw material availability. However, integration into biorefineries enables co-production of fuels, chemicals, and nitrogenated intermediates, improving overall carbon efficiency [19]. Lignin valorization toward aromatic diamines and nitrogen-containing monomers further enhances circularity and waste stream utilization. Unlike linear petrochemical supply chains, bio-based systems increasingly operate under circular bioeconomy models where carbon flow, nitrogen efficiency, and co-product valorization are optimized simultaneously. This structural transformation underpins long-term scalability and resilience.

13.2.2 Techno-Economic Feasibility, Scale-Up Dynamics, and Process Intensification

The techno-economic competitiveness of bio-based amines is governed by several interconnected variables, including feedstock cost, catalytic efficiency, energy separation requirements, and capital expenditure. In contrast, petrochemical amines continue to benefit from long-established infrastructure, optimized supply chains, and large-scale economies that reduce unit production costs. Nevertheless, renewable pathways are progressively narrowing this economic gap through improvements in catalytic selectivity, advanced enzyme engineering, and hydrogen-efficient reaction strategies. Techno-economic assessments of bio-derived diamines demonstrate that overall process viability is highly sensitive to product yield, catalyst durability, and hydrogen consumption, particularly in reductive amination routes [19]. Likewise, biocatalytic transamination platforms for chiral amine production have shown competitive manufacturing economics when coupled with membrane-based extraction systems and solvent recycling schemes that minimize downstream separation costs [20]. At the scale-up stage, multiple technical challenges must be addressed to achieve industrial robustness. Catalyst deactivation in heterogeneous reductive amination systems remains a critical concern, particularly under prolonged exposure to ammonia and reactive intermediates. Efficient ammonia management and recycling are necessary to maintain atom economy and reduce operational expenses. In fermentation-based routes, downstream purification of dilute broth streams significantly influences energy demand and process complexity. Additional complications arise from variability in biomass composition, which affects reproducibility and reactor performance, as well as corrosion and materials compatibility issues in high-temperature or high-pressure amination environments. Process intensification strategies, including hydrogen-borrowing methodologies and one-pot cascade systems, reduce reaction steps and improve atom efficiency while minimizing solvent use [21]. Continuous flow processing and hybrid chemo-enzymatic platforms further enhance volumetric productivity, operational stability, and capital efficiency. Modern process optimization increasingly integrates rational catalyst design for selective amination of biomass-derived oxygenates, protein engineering to improve transaminase stability and substrate scope, and advanced modeling approaches that couple heat and mass transfer simulations with kinetic analysis. Digital process simulation tools and integrated modeling frameworks are becoming central to evaluating scalability and economic feasibility prior to commercial deployment. The convergence of catalytic upgrading technologies with industrial biomanufacturing is particularly significant for large-volume aliphatic amines, where improvements in feedstock utilization, hydrogen efficiency, and process integration are driving the transition from laboratory-scale demonstrations to pilot and early commercial operations [14,15].

13.2.3 End-Use Sectors and Strategic Market Positioning

Bio-based amines are increasingly positioned across multiple high-value application sectors in which molecular structure directly governs performance, durability, and sustainability metrics. In polymer science, bio-derived diamines play a central role in the synthesis of polyamides, polyurethanes, and non-isocyanate polyurethanes, where chain length, functionality, and rigidity determine mechanical strength, thermal stability, and network architecture. Renewable HMDA and PDA have enabled the development of partially or fully bio-based nylon systems with competitive tensile properties, crystallinity control, and heat resistance, demonstrating that renewable monomers can match or approach petrochemical analogues in demanding structural applications [17]. In polyurethane chemistry, amine structure strongly influences crosslink density, hydrogen-bonding interactions, curing kinetics, and dynamic covalent exchange behavior, particularly in advanced thermosets and reprocessable systems. The ability to tailor aliphatic versus aromatic bio-amines provides a powerful handle for balancing flexibility and stiffness in polymer networks. In coating and resin systems, renewable amine curing agents derived from lignin, vanillin, guaiacol, and cardanol offer enhanced sustainability profiles while maintaining high-performance characteristics. Aromatic bio-based amines contribute to elevated glass transition temperatures, improved modulus, and superior corrosion resistance in epoxy-amine thermosets. Epoxy systems formulated with cardanol-derived or lignin-derived amines have demonstrated promising thermal stability and chemical resistance, supporting their adoption in protective and structural coatings [22]. The incorporation of renewable nitrogen-containing hardeners not only reduces fossil carbon content but also introduces structural diversity that can enhance char formation and long-term durability. In the pharmaceutical sector, chiral amines remain indispensable intermediates in active pharmaceutical ingredient synthesis. Biocatalytic transamination has become an industrially established platform for enantioselective amine production, offering high selectivity, lower waste generation, and improved process sustainability compared with conventional stoichiometric methods [12,23]. These enzymatic processes reduce hazardous by-products and enable milder reaction conditions, reinforcing the strategic importance of renewable nitrogen chemistry in high-value pharmaceutical manufacturing. As enzyme engineering advances, broader substrate scopes and improved operational stability further enhance commercial viability. Nitrogen-containing intermediates are also central to agrochemical formulations, including herbicides, fungicides, and plant growth regulators. Renewable amines derived from biomass oxygenates provide potential decarbonization pathways for agrochemical production by integrating sustainable carbon feedstocks with established nitrogen chemistry. Although large-scale implementation remains at an early-stage relative to polymer and pharmaceutical sectors, catalytic upgrading of biomass-derived intermediates offers a technically feasible route toward lower-carbon agrochemical manufacturing. In surfactants and specialty chemical applications, fatty amines derived from vegetable oils serve as cationic surfactants, corrosion inhibitors, flotation agents, and specialty additives. Their renewable origin, tunable hydrophobic chain length, and inherent compatibility with aqueous systems make them attractive for personal care, water treatment, and industrial cleaning formulations. The structural versatility of long-chain aliphatic bio-amines allows precise control over surface activity, emulsification behavior, and antimicrobial performance, reinforcing their commercial relevance in both consumer and industrial markets. Overall, the industrial landscape of bio-based amines is evolving from isolated renewable substitutions toward a coordinated bioeconomic strategy that integrates catalytic upgrading, biomanufacturing, and advanced materials design. While petrochemical amines continue to benefit from mature infrastructure and cost advantages, renewable routes are demonstrating increasing competitiveness in polymer-grade diamines, specialty intermediates, and high-value chiral amines. The convergence of catalytic innovation, biorefinery integration, and rigorous techno-economic modeling positions bio-based amines as foundational building blocks for sustainable chemical manufacturing. Future competitiveness will depend on continued process intensification, feedstock diversification, digital process optimization, and regulatory frameworks that favor low-carbon and circular chemical systems.

13.3 REGULATORY CONSIDERATIONS

Regulatory governance for bio-based amines is determined primarily by chemical identity, tonnage, hazard profile, and use pattern rather than by feedstock origin. While renewable carbon sourcing affects sustainability claims and carbon accounting, compliance obligations under major chemical control regimes remain substance-based. This section reviews the principal regulatory frameworks, environmental and occupational requirements, certification standards, and policy drivers shaping commercialization.

13.3.1 Regulatory Frameworks for Bio-Based Chemicals

13.3.1.1 REACH (EU)

Under Regulation (EC) No 1907/2006 (REACH), any substance manufactured or imported into the EU at ≥1 ton per year per company must be registered with the European Chemicals Agency (ECHA). Registration requires technical dossiers, hazard data, exposure scenarios, and risk management measures depending on tonnage band (1–10, 10–100, 100–1000, >1000 t/y). For bio-based amines, a change in feedstock does not exempt the substance from REACH if the molecular identity is unchanged. However, route modifications may alter impurity profiles, potentially affecting classification under the Classification, Labelling and Packaging (CLP) Regulation (EC) No 1272/2008. Novel fermentation-derived amines not previously listed on the EU inventory require full registration and data generation [24,25].

13.3.1.2 TSCA (USA)

The U.S. TSCA requires submission of a Pre-Manufacture Notice (PMN) to the Environmental Protection Agency (EPA) at least 90 days before manufacture or import of a new chemical substance not listed on the TSCA Inventory. Drop-in bio-attributed amines (e.g., ethanolamines, ethyleneamines) typically remain on the Inventory because the molecule is unchanged. EPA evaluates hazard, exposure potential, and risk before allowing commercialization [26].

13.3.1.3 Asian Regulatory Landscape

Across Asia, chemical control systems are broadly comparable in structure to the European REACH framework, but they are administered at the national level and differ in procedural detail, data requirements, and timelines. In Japan, the Chemical Substances Control Law (CSCL) requires prior evaluation of new chemical substances before manufacture or import. The review process emphasizes persistence, bioaccumulation potential, and long-term environmental impact, particularly for substances that may present risks to human health or ecosystems. New chemicals must undergo screening and, depending on risk classification, may be subject to additional monitoring or restrictions [27].

In South Korea, the Act on the Registration and Evaluation of Chemicals (K-REACH) establishes a registration system similar in structure to EU REACH, with phased implementation based on annual tonnage bands. Manufacturers and importers must submit hazard data, risk assessments, and use information for new chemicals and, progressively, for existing substances above specified thresholds. The system places increasing emphasis on data transparency and downstream communication across supply chains [28]. China regulates new chemical substances under the Measures for the Environmental Management Registration of New Chemical Substances (MEE Order No. 12). Substances not listed on the Inventory of Existing Chemical Substances in China (IECSC) require registration, record filing, or simplified notification depending on volume and hazard profile. The framework includes environmental risk assessment and post-registration supervision, with increasing alignment toward global data-sharing principles [29].

In India, chemical regulation is evolving toward a REACH-like model under the proposed Chemicals (Management and Safety) Rules (CMSR), issued by the Ministry of Environment, Forest and Climate Change. The framework introduces notification and registration requirements for substances placed on the Indian market above defined tonnage thresholds, with hazard communication and data submission obligations similar in intent to EU REACH. Although full implementation timelines continue to develop, companies marketing specialty bio-based amines in India should anticipate registration, safety data generation, and inventory listing requirements [30].

For producers of bio-based amines, global commercialization therefore requires parallel compliance with multiple national inventories and registration systems. Even when the molecular structure is identical to an existing petrochemical amine, differences in impurity profiles, production routes, or regional volume thresholds can trigger additional reporting or data obligations. The regulatory burden becomes particularly significant for low-volume specialty amines, where multi-jurisdiction registration costs may represent a substantial proportion of total commercialization expenditure.

13.3.2 Environmental and Safety Compliance

13.3.2.1 Toxicological Assessment

Amines often exhibit corrosivity, irritation potential, and aquatic toxicity depending on structure. Regulatory dossiers for common amines (e.g., ethanolamine, certain diamines) demonstrate hazard classifications under EU CLP for skin corrosion and serious eye damage.

13.3.2.2 Occupational Exposure Limits

Worker exposure management is central for volatile and corrosive amines. Occupational Safety and Health Administration (OSHA) and American Conference of Governmental Industrial Hygienists (ACGIH) provide permissible exposure limits (PELs) and threshold limit values (TLVs) for many industrial amines. Where a bio-based amine is chemically identical to a petro-based counterpart, existing occupational limits apply unchanged [31,32].

13.3.3 Sustainability and Certification Standards

13.3.3.1 Life Cycle Assessment

LCA provides standardized methodology for evaluating environmental impacts across cradle-to-gate or cradle-to-grave boundaries. ISO 14044 specifies requirements and reporting guidelines for LCA studies. Recent LCA work on bio-ethylene illustrates that climate benefits depend heavily on land-use change assumptions and carbon accounting methodology. This has direct implications for bio-based amines derived from ethylene oxide or related intermediates [33,34].

13.3.3.2 Carbon Footprint and GHG Accounting

Mass-balance approaches allow renewable feedstocks to be attributed to downstream products in integrated chemical systems. International Sustainability and Carbon Certification (ISCC) PLUS provides chain-of-custody certification for renewable and circular feedstocks. Carbon footprint reporting increasingly requires product-level data consistent with International Organization for Standardization (ISO) LCA standards and audited chain-of-custody frameworks [35].

13.3.3.3 Bio-Based Certification Schemes

Verification of bio-based carbon content often uses radiocarbon testing under American Society for Testing and Materials (ASTM) D6866 [36]. ISO 16620 and EN 16640 provide comparable testing frameworks. In the United States, the United States Department of Agriculture (USDA) BioPreferred Program supports federal procurement of certified bio-based products. These schemes distinguish between physically measurable biogenic carbon and mass-balance attribution models [37].

13.3.4 Regulatory Barriers and Approval Challenges

The regulatory landscape presents several structural barriers that can slow the commercialization of bio-based amines. One of the most significant challenges is the high cost of data generation required for novel substances under frameworks such as REACH in the European Union and TSCA in the United States. Toxicological testing, ecotoxicity studies, exposure assessments, and dossier preparation can require substantial financial and technical resources, particularly for new molecular entities. In addition, companies seeking global market access often face duplicate registration obligations across major jurisdictions, including the EU, US, China, Japan, and South Korea, each with distinct procedural requirements and timelines. Further complexity arises when bio-based production routes introduce different impurity profiles compared to petrochemical counterparts, potentially affecting hazard classification and regulatory status. There is also increasing scrutiny regarding sustainability claims, requiring alignment between "bio-based" marketing language and substantiation through recognized standards such as ASTM radiocarbon testing, ISO methodologies, or ISCC certification. For low-volume specialty bio-based amines, these multi-region compliance costs can create disproportionately high fixed expenditures relative to projected sales volumes, thereby affecting overall economic feasibility.

13.3.5 Policy Incentives and Green Chemistry Promotion

Policy frameworks increasingly encourage safer and more sustainable chemical design. The European Commission's Chemicals Strategy for Sustainability promotes safe and sustainable-by-design chemicals as part of the European Green Deal. In the U.S., USDA BioPreferred establishes demand-side incentives through mandatory federal purchasing requirements. However, policy instruments based solely on bio-based content may not guarantee climate benefit unless lifecycle performance is verified, as demonstrated in recent LCA studies of bio-derived chemical platforms [38,39].

Regulatory obligations for bio-based amines are substance-driven rather than feedstock-driven. Sustainability claims increasingly depend on standardized lifecycle assessment, radiocarbon testing, and chain-of-custody certification. While green chemistry policy creates incentives for renewable feedstocks, market acceptance ultimately requires compliance with hazard classification, occupational safety standards, and robust carbon accounting methodologies.

13.4 MARKET TRENDS AND COMMERCIAL DYNAMICS

The commercial development of bio-based amines must be interpreted within the broader expansion of sustainable polymers and renewable chemical intermediates. Although bio-based amines are not typically reported as a standalone statistical market category, they serve as critical intermediates in polyamides, polyurethanes, epoxy thermosets, coatings, adhesives, surfactants, and high-performance functional materials. The global transition from fossil-derived carbon toward renewable feedstocks has significantly accelerated interest in bio-based polymer systems, particularly where amine chemistry governs curing kinetics, crosslink density, hydrogen bonding, thermal stability, and degradation behavior. Joseph et al. emphasize that the growth of bio-based polymers is strongly influenced by sustainability regulations, crude oil price volatility, environmental compliance pressures, and consumer demand for lower-carbon materials, and they argue that replacing fossil carbon with renewable carbon represents a structural transformation of the chemical industry rather than a temporary substitution trend [40].

Because "bio-based amines" are embedded within larger product families, the most appropriate way to frame their market dynamics is to anchor them in the overall amines market and in clearly bio-linked subsegments such as fatty (oleochemical) amines. Mordor Intelligence estimates the global amines market at approximately USD 16.91 billion in 2026, projected to reach USD 21.25 billion by 2031, corresponding to a compound annual growth rate (CAGR) of 4.69% [41]. Fortune Business Insights reports a similar scale, estimating USD 16.8 billion in 2025 and USD 24.7 billion by 2034 with a CAGR of 4.20% [42]. Within this broader framework, fatty amines typically derived from vegetable oils and therefore bio-based by feedstock origin represent one of the most commercially mature renewable amine segments. Grand View Research estimates the global fatty amines market at USD 4.66 billion in 2024, forecast to reach USD 7.42 billion by 2033 (CAGR 5.3%) [43]. These figures are particularly relevant because they demonstrate that bio-origin amines already operate at multi-billion-dollar industrial scale, with growth driven by surfactants, agrochemicals, personal care, and water treatment applications.

Technological advances have strengthened the feasibility of expanding renewable amine production beyond fatty amines. The review on bio-based amines from synthesis to polymer applications highlights progress in catalytic hydrogenation, lignin valorization, and fermentation-derived diamine production, which collectively improve yield, selectivity, and scalability of renewable amine platforms. However, techno-economic analyses by Mueller and Sanders show that competitiveness depends strongly on fermentation efficiency, downstream separation energy demand, biomass logistics, and capital investment intensity. Fermentation-derived diamines, for example, often face higher purification costs compared to petrochemical routes, and process integration becomes decisive for commercial viability. These economic constraints are particularly significant for specialty bio-based amines produced at lower volumes, where fixed compliance and production costs represent a larger share of total cost [44,45].

Regional dynamics further shape commercialization pathways. Asia-Pacific is consistently identified as the largest and fastest-growing region for amines overall. Mordor Intelligence reports that Asia-Pacific accounted for approximately 38.55% of global amines revenue in 2025 and is projected to show the highest growth rate through 2031 (CAGR 5.64%). For fatty amines, Grand View Research also identifies Asia-Pacific as the leading region, with strong demand growth across China and India driven by personal care, household products, and agrochemicals. Europe, in contrast, demonstrates strong demand pull for bio-attributed drop-in chemicals due to corporate decarbonization targets and product carbon footprint reporting requirements. This often manifests in certification-led commercialization. For example, Nouryon announced ISCC PLUS certification for renewable ethylene oxide, ethanolamines, and ethylene amines at its Stenungsund site in Sweden, positioning them as performance-equivalent intermediates with reduced carbon footprint [46]. Badische Anilin- und Soda-Fabrik (BASF) has similarly expanded biomass balance certified intermediates across Europe and North America, reinforcing the strategic role of certification in renewable amine market penetration [47].

Investment patterns in bio-based amines generally follow two complementary models. The first is incremental decarbonization of existing petrochemical platforms via certified renewable feedstocks and mass-balance accounting. Because the molecular structure remains unchanged, this pathway reduces customer requalification risk and accelerates commercialization. The second is biotechnology-led scale-up of structurally distinct bio-based diamines, particularly for polyamide and thermoset applications. Industry collaborations illustrate this model clearly. Covestro and Genomatica reported production of plant-based HMDA, a key nylon-6,6 intermediate, demonstrating feasibility at meaningful volumes. Similarly, Asahi Kasei and Genomatica announced collaboration to commercialize bio-based HMDA for nylon-6,6 production. These partnerships show how biotechnology developers establish route feasibility while polymer manufacturers accelerate downstream validation and scale-up [48,49].

13.5 EMERGING TECHNOLOGIES AND INNOVATION PATHWAYS

The long-term competitiveness of bio-based amines will depend on continuous innovation in catalytic design, process intensification, biological engineering, digital optimization, and circular integration. Emerging technologies are no longer limited to replacing fossil feedstocks; they aim to redesign nitrogen incorporation pathways, reduce energy intensity, utilize carbon dioxide and lignin streams, and enable molecular precision in amine synthesis. Recent advances in catalytic chemistry, electrochemical conversion, metabolic engineering, and data-driven optimization collectively define the next phase of industrial bio-based amine development.

13.5.1 Advanced Catalytic Systems

Catalytic upgrading of biomass-derived intermediates into amines has advanced significantly in the past decade. Heterogeneous hydrogenation catalysts, bifunctional metal–acid systems, and transfer hydrogenation routes now enable efficient reductive amination of platform molecules derived from sugars, lignin, and fatty acids. Pelckmans et al. provide a comprehensive overview of catalytic strategies for converting bio-based feedstocks into primary, secondary, and tertiary amines, emphasizing the importance of selectivity control, catalyst stability, and suppression of over-hydrogenation or condensation side reactions [13]. Particularly promising developments include catalytic amination of lignin-derived phenolics, which enables production of aromatic amines with rigid backbones suitable for high-performance polymer applications. Metal-supported catalysts such as Ru, Ni, and Pd systems have shown improved efficiency for reductive amination under milder conditions, reducing energy demand compared to conventional petrochemical routes. The integration of continuous flow reactors further enhances productivity, scalability, and heat management, which are critical for industrial deployment. Additionally, carbon dioxide utilization pathways are gaining relevance. Green catalytic systems that incorporate CO_2 into value-added nitrogen-containing products represent a strategic intersection between decarbonization and amine chemistry. Recent reviews on CO_2 sequestration into value-added products highlight catalytic and bio-based approaches that could complement renewable amine synthesis, particularly in urea and carbamate formation strategies [50].

13.5.2 Electrochemical and Photocatalytic Amination

Electrochemical and photocatalytic amination technologies offer low-temperature, potentially low-carbon alternatives to thermally driven catalytic routes. These methods leverage renewable electricity to activate nitrogen sources and biomass-derived carbon frameworks under milder conditions. Electrochemical reductive amination allows precise control of electron transfer steps, reducing overreaction and minimizing by-products. Emerging green technology perspectives emphasize the role of electrochemical systems in carbon dioxide conversion and nitrogen functionalization, particularly when coupled with renewable energy inputs [50]. Photocatalytic systems, using semiconductor materials under visible light irradiation, are also

than feedstock-driven. Even when bio-based amines are chemically identical to petrochemical counterparts, impurity profiles and new production routes may trigger additional registration or data submission requirements under REACH, TSCA, and parallel Asian frameworks. Multi-jurisdiction compliance, toxicological testing, and sustainability certification generate significant fixed costs, particularly for low-volume specialty amines. Moreover, sustainability claims must be substantiated through standardized LCA methodologies, carbon accounting frameworks, and certification schemes such as ISCC PLUS or ASTM radiocarbon testing. The alignment between regulatory approval and verified climate benefit remains a decisive factor for market acceptance. Economic competitiveness remains closely tied to feedstock pricing, hydrogen cost, capital expenditure, and separation energy demand. Petrochemical amines benefit from mature infrastructure and economies of scale, while bio-based routes must offset higher upstream variability with process intensification and integration strategies. Nevertheless, opportunities in green and sustainable manufacturing are expanding. Corporate decarbonization targets, product carbon footprint reporting, and demand for biomass-balanced intermediates are accelerating the adoption of drop-in renewable amines. Advances in catalyst design, enzyme engineering, AI-assisted optimization, and circular integration offer pathways to reduce production cost and improve yield. The convergence of renewable carbon sourcing, green ammonia production, and digital process simulation positions bio-based amines not merely as substitutes, but as strategic molecules within a broader sustainable manufacturing ecosystem.

13.7 FUTURE OUTLOOK AND CONCLUSION

The future trajectory of bio-based amines will depend on coordinated advances in technology, policy alignment, and market adoption. In the short term (approximately five years), industrial development is expected to focus on incremental decarbonization of existing amine value chains. Biomass-balanced and ISCC-certified intermediates will likely expand across ethanolamines, ethyleneamines, and selected diamines, allowing producers to reduce product carbon footprint without altering molecular identity. Continuous improvement in catalytic efficiency and fermentation yields will gradually lower production cost, while corporate sustainability commitments will drive demand in polymers, coatings, and specialty chemicals. Early commercial deployment of renewable HMDA and related diamines for polyamide production is expected to strengthen integration between biotechnology firms and established polymer manufacturers. Over the mid-term horizon (approximately ten years), broader market expansion scenarios become plausible. Increased investment in advanced biorefineries, green hydrogen infrastructure, and low-carbon ammonia production could significantly reduce the carbon intensity of nitrogen incorporation. Scale-up of fermentation-derived diamines, lignin-based aromatic amines, and electrochemical amination technologies may enable diversification beyond fatty amines and drop-in intermediates. Regional growth patterns are expected to reflect policy drivers: Europe emphasizing carbon accounting and certification, Asia-Pacific expanding capacity in high-growth polymer sectors, and North America integrating bio-based amines within sustainable materials strategies. Economic competitiveness will increasingly depend on integrated process optimization, digital modeling, and supply chain resilience rather than feedstock substitution alone. In the long term, the transition toward a bio-based nitrogen economy represents a structural transformation of the chemical industry. Such a system would integrate renewable carbon feedstocks, green ammonia production, carbon dioxide utilization, and circular polymer design into a unified low-carbon manufacturing model. Bio-based amines could serve as key molecular connectors linking biomass valorization, advanced materials, and sustainable agriculture. The incorporation of dynamic covalent chemistry and recyclable thermoset systems further supports circular material flows, extending the relevance of renewable amine chemistry beyond initial synthesis. From a decarbonization perspective, bio-based amines have the potential to contribute meaningfully to net-zero strategies, particularly when combined with renewable electricity, green hydrogen, and lifecycle-verified carbon accounting. However, climate benefit must be demonstrated through robust LCA and transparent certification rather than assumed from feedstock origin alone. The ultimate competitiveness of bio-based amines will depend on their ability to match or exceed petrochemical performance while delivering measurable reductions in greenhouse gas emissions and resource intensity. In conclusion, bio-based amines are transitioning from niche renewable alternatives to strategically important components of sustainable chemical manufacturing. Their future will be defined by technological convergence, regulatory clarity, and integration within circular and low-carbon industrial ecosystems. The evolution toward a bio-based nitrogen economy is not merely a substitution challenge but a systems-level redesign of how carbon and nitrogen cycles intersect in modern materials production.

REFERENCES

1. S.A. Lawrence, *Amines: Synthesis, Properties and Applications*, Cambridge University Press, Cambridge, 2004.
2. C.-T. Cao, S. Chen, C. Cao, General equation to estimate the physicochemical properties of aliphatic amines, *ACS Omega*. 8 (2023) 49088–49097.
3. J. Li, Z. Weng, Q. Cao, Y. Qi, B. Lu, S. Zhang, J. Wang, X. Jian, Synthesis of an aromatic amine derived from biomass and its use as a feedstock for versatile epoxy thermoset, *Chem. Eng. J.* 433 (2022) 134512.
4. M.G. Olayo, E.J. Alvarado, M. González-Torres, L.M. Gómez, G.J. Cruz, Quantifying amines in polymers by XPS, *Polym. Bull.* 81 (2024) 2319–2328.
5. P.R. Sruthi, S. Anas, An overview of synthetic modification of nitrile group in polymers and applications, *J. Polym. Sci.* 58 (2020) 1039–1061.
6. L. Pereira, P.K. Mondal, M. Alves, Aromatic amines sources, environmental impact and remediation, in: E. Lichtfouse, J. Schwarzbauer, D. Robert (Eds.), *Pollutants in Buildings, Water and Living Organisms*, Springer International Publishing, Cham, 2015: pp. 297–346.
7. L.I. Helgesen, E. Gjernes, A Way of qualifying amine based capture technologies with respect to health and environmental properties, *Energy Procedia*. 86 (2016) 239–251.
8. S.A. Gheni, M.M. Ali, G.C. Ta, H.J. Harbin, S.A. Awad, Toxicity, hazards, and safe handling of primary aromatic amines, *ACS Chem. Heal. Saf.* 31 (2024) 8–21.
9. J. Becker, C. Wittmann, Diamines for bio-based materials, in: C. Wittmann, J. C. Liao (Eds.), *Industrial Biotechnology: Products and Processes*, Wiley-VCH Verlag GmbH & Co. KGaA, Weinheim, 2017: pp. 391–409.
10. E. Blondiaux, J. Bomon, M. Smoleń, N. Kaval, F. Lemière, S. Sergeyev, L. Diels, B. Sels, B.U.W. Maes, Bio-based aromatic amines from lignin-derived monomers, *ACS Sustain. Chem. Eng.* 7 (2019) 6906–6916.

11. A.-S. Mora, R. Tayouo, B. Boutevin, G. David, S. Caillol, Vanillin-derived amines for bio-based thermosets, *Green Chem.* 20 (2018) 4075–4084.
12. V. Froidevaux, C. Negrell, S. Caillol, J.P. Pascault, B. Boutevin, Biobased amines: From synthesis to polymers; present and future, *Chem. Rev.* 116 (2016) 14181–14224.
13. M. Pelckmans, T. Renders, S. Van de Vyver, B.F. Sels, Bio-based amines through sustainable heterogeneous catalysis, *Green Chem.* 19 (2017) 5303–5331.
14. B. Vermeeren, S. Van Praet, W. Arts, T. Narmon, Y. Zhang, C. Zhou, H.P. Steenackers, B.F. Sels, From sugars to aliphatic amines: as sweet as it sounds? Production and applications of bio-based aliphatic amines, *Chem. Soc. Rev.* 53 (2024) 11804–11849.
15. N.K. Gupta, P. Reif, P. Palenicek, M. Rose, Toward renewable amines: Recent advances in the catalytic amination of biomass-derived oxygenates, *ACS Catal.* 12 (2022) 10400–10440.
16. W. Faveere, T. Mihaylov, M. Pelckmans, K. Moonen, F. Gillis-D'Hamers, R. Bosschaerts, K. Pierloot, B.F. Sels, Glycolaldehyde as a bio-based C2 platform chemical: Catalytic reductive amination of vicinal hydroxyl aldehydes, *ACS Catal.* 10 (2020) 391–404.
17. X. Wang, S. Gao, J. Wang, S. Xu, H. Li, K. Chen, P. Ouyang, The production of bio-based diamines from renewable carbon sources: Current advances and perspectives, *Chin. J. Chem. Eng.* 30 (2021) 4–13.
18. S. Kind, C. Wittmann, Bio-based production of the platform chemical 1,5-diaminopentane, *Appl. Microbiol. Biotechnol.* 91 (2011) 1287–1296.
19. A.B. Dros, O. Larue, A. Reimond, F. De Campo, M. Pera-Titus, Hexamethylenediamine (HMDA) from fossil- vs. bio-based routes: An economic and life cycle assessment comparative study, *Green Chem.* 17 (2015) 4760–4772.
20. J. Yang, A. Buekenhoudt, M. Van Dael, P. Luis, Y. Satyawali, R. Malina, S. Lizin, A techno-economic assessment of a biocatalytic chiral amine production process integrated with in situ membrane extraction, *Org. Process Res. Dev.* 26 (2022) 2052–2066.
21. S. Hameury, H. Bensalem, K. De Oliveira Vigier, Sustainable amination of bio-based alcohols by hydrogen borrowing catalysis, *Catal.* 12 (2022) 1306.
22. F. Jaillet, E. Darroman, A. Ratsimihety, R. Auvergne, B. Boutevin, S. Caillol, New biobased epoxy materials from cardanol, *Eur. J. Lipid Sci. Technol.* 116 (2014) 63–73.
23. I.V. Pavlidis, M.S. Weiß, M. Genz, P. Spurr, S.P. Hanlon, B. Wirz, H. Iding, U.T. Bornscheuer, Identification of (S)-selective transaminases for the asymmetric synthesis of bulky chiral amines, *Nat. Chem.* 8 (2016) 1076–1082.
24. European Commission, REACH Regulation, 2026. https://environment.ec.europa.eu/topics/chemicals/reach-regulation_en.
25. ECHA, Do I Reach the One Tonne a Year Threshold?, 2026. https://echa.europa.eu/support/registration/your-registration-obligations/do-i-reach-the-one-tonne-a-year-threshold.
26. EPA United States Environmental Protection Agency, Filing a Pre-manufacture Notice with EPA, 2025. https://www.epa.gov/reviewing-new-chemicals-under-toxic-substances-control-act-tsca/filing-pre-manufacture-notice-epa.
27. Government of Japan, Act on the Regulation of Manufacture and Evaluation of Chemical Substances, Act No. 117, Government of Japan, Tokyo, Japan, 1973.
28. S. Ha, T. Seidle, K.-M. Lim, Act on the registration and evaluation of chemicals (K-REACH) and replacement, reduction or refinement best practices, *Environ. Health Toxicol.* 31 (2016) e2016026.
29. Industrial Chemical, China REACH: New Chemical Substance Registration under MEE Order No. 12, 2026. https://en.reach24h.com/service/chemical/china-new-chemical-substance-notification.
30. M. Gowda, Environmental regulations governing bio-chemical industries in India, *Indian J. Law Legal Res.* VII (n.d.) 7995–8015. https://www.ijllr.com/post/environmental-regulations-governing-bio-chemical-industries-in-india.

31. Occupational Safety and Health Administration, 1,6–Hexanediamine, 2020. https://www.osha.gov/chemicaldata/182.
32. Occupational Safety and Health Administration, Permissible Exposure Limits – OSHA Annotated Table Z-1, n.d. https://www.osha.gov/annotated-pels/table-z-1.
33. M. Zanon-Zotin, C. Bergman-Fonte, T. Nogueira Morais, P.L. Barbosa Maia, L. Carvalho, G. Angelkorte, A.C. Oliveira Fiorini, P. Rua Rodriguez Rochedo, J. Portugal-Pereira, A. Szklo, R. Schaeffer, Unpacking bio-based alternatives to ethylene production in Brazil, Europe, and the United States: A comparative life cycle assessment, *J. Clean. Prod.* 428 (2023) 139376.
34. ISO, Environmental Management — Life Cycle Assessment — Requirements and Guidelines, 2006. https://www.iso.org/standard/38498.html.
35. International Sustainability & Carbon Certification, Sustainability Certification for the Circular Economy and Bioeconomy, n.d. https://www.iscc-system.org/certification/iscc-certification-schemes/iscc-plus/.
36. ASTM, Standard Test Methods for Determining the Biobased Content of Solid, Liquid, and Gaseous Samples Using Radiocarbon Analysis, ASTM D6866-24, 2025.
37. United States Department of Agriculture, Selling Biobased Products in the Federal Marketplace, n.d. https://www.biopreferred.gov/BioPreferred/faces/pages/SellingToFederalGovt.xhtml.
38. European Commission, Chemicals Strategy, n.d.. https://environment.ec.europa.eu/strategy/chemicals-strategy_en.
39. Biopreffered United States Department of Agriculture, Mandatory Federal Purchasing, n.d. https://www.biopreferred.gov/BioPreferred/faces/pages/MandatoryFederalPurchasing.xhtml.
40. T.M. Joseph, A.B. Unni, K.S. Joshy, D. Kar Mahapatra, J. Haponiuk, S. Thomas, Emerging bio-based polymers from lab to market: Current strategies, market dynamics and research trends, *C.* 9 (2023) 30.
41. Mordor Intelligence, Amines Market Size & Share Analysis - Growth Trends and Forecast (2026–2031), 2026. https://www.mordorintelligence.com/industry-reports/amines-market.
42. Fortune Business Insights, Amines Market Size, Share & Industry Analysis, By Product (Ethanolamines {Triethanolamine, Diethanolamine, Monoethanolamine, and Methyl Diethanolamine}, Alkylamines {Ethylamines, Methylamines, and Others}, Ethyleneamines {Ethylenediamine, Tetraethylene En, 2026. https://www.fortunebusinessinsights.com/amines-market-106578.
43. Market Analysis Report, Fatty Amines Market (2025–2033), 2026. https://www.grandviewresearch.com/industry-analysis/fatty-amines-market.
44. J. Sanders, E. Scott, R. Weusthuis, H. Mooibroek, Bio-refinery as the bio-inspired process to bulk chemicals, *Macromol. Biosci.* 7 (2007) 105–117.
45. U. Mueller, S. Huebner, Economic aspects of amino acids production, in: R. Faurie, J. Thommel, B. Bathe, V.G. Debabov, S. Huebner, M. Ikeda, E. Kimura, A. Marx, B. Möckel, U. Mueller, W. Pfefferle (Eds.), *Microbial Production of l-Amino Acids*, Springer Berlin Heidelberg, Berlin, Heidelberg, 2003: pp. 137–170.
46. Nouryon, Nouryon obtains ISCC PLUS Certification for Green Ethylene Oxide, Ethanolamines, and Ethylene Amines, enabling the Production of More Sustainable Surfactants, 2024. https://www.nouryon.com/news-and-events/news-overview/2024/certified-ISCCPLUS-for-green-ethyleneoxide-ethanolamines-and-ethyleneamines/.
47. BASF, Biomass Balanced (BMBcert®) Intermediates, n.d. https://chemicals.basf.com/global/en/Intermediates/sustainability/biomass-balance.
48. Covestro Global Corporate Website, Biomass as Resource – Sustainable by Nature, 2026. https://www.covestro.com/en/sustainability/what-drives-us/circular-economy/alternative-resources/biomass (accessed March 16, 2026).

49. G.I. Asahi Kasei Corporation, Genomatica and Asahi Kasei Partner on Renewably-Sourced Nylon 6,6, 2022. https://www.asahi-kasei.com/news/2021/e220316_2.html.
50. E.V Rebrov, P.-Z. Gao, Molecular catalysts for OER/ORR in Zn–air batteries, *Catalysts* 13 (2023) 1289.
51. S. Deng, A. Chen, W. Chen, J. Lai, Y. Pei, J. Wen, C. Yang, J. Luo, J. Zhang, C. Lei, S.N. Varma, C. Liu, Fabrication of biodegradable and biocompatible functional polymers for anti-infection and augmenting wound repair, *Polymers (Basel).* 15 (2023) 120.
52. P. Schwaller, T. Laino, T. Gaudin, P. Bolgar, C.A. Hunter, C. Bekas, A.A. Lee, Molecular transformer: A model for uncertainty-calibrated chemical reaction prediction, *ACS Cent. Sci.* 5 (2019) 1572–1583.
53. D.T. Ahneman, J.G. Estrada, S. Lin, S.D. Dreher, A.G. Doyle, Predicting reaction performance in C–N cross-coupling using machine learning, *Science* (80-.). 360 (2018) 186–190.
54. Y. Zhang, A.A. Broekhuis, F. Picchioni, Thermally self-healing polymeric materials: The next step to recycling thermoset polymers?, *Macromolecules.* 42 (2009) 1906–1912.
55. M. Aresta, A. Dibenedetto, A. Angelini, The changing paradigm in CO2 utilization, *J. CO2 Util.* 3–4 (2013) 65–73.
56. T. Keijer, V. Bakker, J.C. Slootweg, Circular chemistry to enable a circular economy, *Nat. Chem.* 11 (2019) 190–195.

Index

Note: **Bold** page numbers refer to tables; *italic* page numbers refer to figures

D

E

F

G

H

I

K

L

M

Q

R

S

T

U

V

W

X

Y

For Product Safety Concerns and Information please contact our
EU representative GPSR@taylorandfrancis.com Taylor & Francis
Verlag GmbH, Kaufingerstraße 24, 80331 München, Germany